Fragments of Science for Unscientific People
by John Tyndall

Address:
HardPress
8345 NW 66TH ST #2561
MIAMI FL 33166-2626
USA
Email: info@hardpress.net

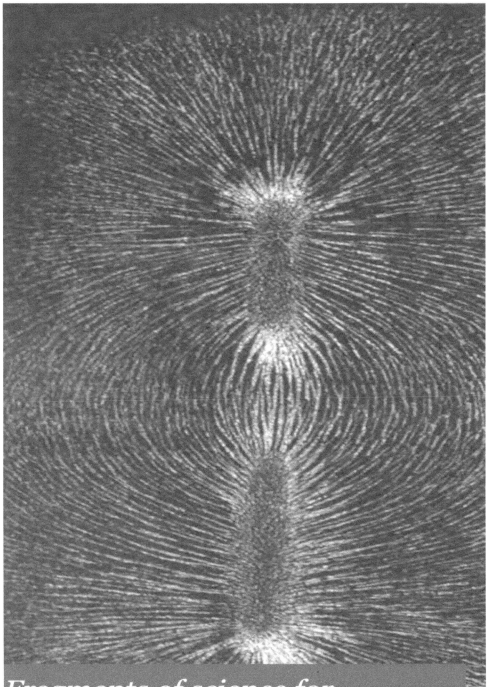

Fragments of science for unscientific people

John Tyndall

FRAGMENTS OF SCIENCE.

LONDON: PRINTED BY
SPOTTISWOODE AND CO., NEW-STREET SQUARE
AND PARLIAMENT STREET

FRAGMENTS OF SCIENCE

FOR

UNSCIENTIFIC PEOPLE;

A SERIES OF DETACHED ESSAYS, LECTURES, AND REVIEWS.

BY

JOHN TYNDALL, LL.D. F.R.S.

LONDON:
LONGMANS, GREEN, AND CO.
1871.

PREFACE.

My MOTIVE in writing these papers was mainly that which prompted the publication of my Royal Institution lectures ; a desire, namely, to extend sympathy for science beyond the limits of the scientific public.

The fulfilment of this desire has caused a temporary and sometimes reluctant deflection of thought from the line of original research. But considering the result aimed at, and in part I trust achieved, I do not regret the price paid for it.

I have carefully looked over all the articles here printed, added a little, omitted a little—in fact, tried as far as my time permitted to render the work presentable. Most of the essays are of a purely scientific character, and from those which are not, I have endeavoured, without veiling my convictions, to exclude every word that could cause needless irritation.

From America the impulse came which induced me to gather these 'Fragments' together, and to my friends in the United States I dedicate them.

JOHN TYNDALL.

ATHENÆUM CLUB: *March* 1871.

CONTENTS.

———⋅◦⋅———

Erratum.

At page 276, 8th line from bottom, *instead of* at right angles to the solar beams to scatter perfectly polarised light, *read* to scatter perfectly polarised light at right angles to the solar beams.

I.

THE CONSTITUTION OF NATURE.

AN ESSAY.

[*Fortnightly Review, vol.* iii. *p.* 129.]

'The gentle Mother of all
Showed me the lore of colours and of sounds;
The innumerable tenements of beauty;
The miracle of generative force;
Far-reaching concords of Astronomy
Felt in the plants and in the punctual birds;
Mainly, the linked purpose of the whole;
And, chiefest prize, found I true liberty
The home of homes plain-dealing Nature gave.'

RALPH WALDO EM1

I.

THE CONSTITUTION OF NATURE.

WE cannot think of space as finite, for wherever in imagination we erect a boundary we are compelled to think of space as existing beyond that boundary. Thus by the incessant dissolution of limits we arrive at a more or less adequate idea of the infinity of space. But though compelled to think of space as unbounded, there is no mental necessity to compel us to think of it either as filled or as empty ; whether it is filled or empty must be decided by experiment and observation. That it is not entirely void, the starry heavens declare ; but the question still remains, are the stars themselves hung in vacuo ? Are the vast regions which surround them, and across which their light is propagated, absolutely empty ? A century ago the answer to this question would be, 'No, for particles of light are incessantly shot through space.' The reply of modern science is also negative, but on a somewhat different ground. It has the best possible reasons for rejecting the idea of luminiferous particles ; but in support of the conclusion that the celestial spaces are occupied by matter, it is able to offer proofs almost as cogent as those which can be adduced for the existence of an atmosphere round the earth. Men's minds, indeed, rose to a conception of the celestial and universal atmosphere through the study of the terrestrial and local one. From the phenomena of sound as displayed in the air, they ascended to the phenomena of light as displayed in the *æther*; which is the name given to the interstellar medium.

The notion of this medium must not be considered as a vague or fanciful conception on the part of scientific men. Of its reality most of them are as convinced as they are of the existence of the sun and moon. The luminiferous æther has definite mechanical properties. It is almost infinitely more attenuated than any known gas, but its properties are those of a solid rather than of a gas. It resembles jelly rather than air. A body thus constituted may have its boundaries; but, although the æther may not be co-extensive with space, we at all events know that it extends as far as the most distant visible stars. In fact it is the vehicle of their light, and without it they could not be seen. This all-pervading substance takes up their molecular tremors, and conveys them with inconceivable rapidity to our organs of vision. It is the transported shiver of bodies countless millions of miles distant, which translates itself in human consciousness into the splendour of the firmament at night.

If the æther have a boundary, masses of ponderable matter might be conceived to exist beyond it, but they could emit no light. Beyond the æther dark suns might burn; there, under proper conditions, combustion might be carried on; fuel might consume unseen, and metals be heated to fusion in invisible fires. A body, moreover, once heated there, would continue for ever heated; a sun or planet once molten, would continue for ever molten. For the loss of heat being simply the abstraction of molecular motion by the æther, where this medium is absent no cooling could occur. A sentient being, on approaching a heated body in this region, would be conscious of no augmentation of temperature. The gradations of warmth dependent on the laws of radiation would not exist, and actual contact would first reveal the heat of an extra æthereal sun.

Imagine a paddle-wheel placed in water and caused to rotate. From it as a centre waves would issue in all directions, and a wader as he approached the place of disturbance would be met by stronger and stronger waves. This gradual augmentation of the impressions made upon the wader's body is exactly analogous to the augmentation of light when we approach a luminous source. In the one case, however, the coarse common nerves of the body suffice; for the other we must have the finer optic nerve. But suppose the water withdrawn; the action at a distance would then cease, and as far as the sense of touch is concerned, the wader would be first rendered conscious of the motion of the wheel by the actual blow of the paddles. The transference of motion from the paddles to the water is mechanically similar to the transference of molecular motion from the heated body to the æther; and the propagation of waves through the liquid is mechanically similar to the propagation of light and radiant heat.

As far as our knowledge of space extends, we are to conceive it as the holder of the luminiferous æther, through which is interspersed, at enormous distances apart, the ponderous nuclei of the stars. Associated with the star that most concerns us we have a group of dark planetary masses revolving at various distances round it, each again rotating on its own axis; and, finally, associated with some of these planets we have dark bodies of minor note

—the moons. Whether the other fixed stars have similar planetary companions or not is to us a matter of pure conjecture, which may or may not enter into our conception of the universe. But probably every thoughtful person believes, with regard to those distant suns, that there is in space something besides our system on which they shine.

Having thus obtained a general view of the present condition of space, and of the bodies contained in it, we may enquire whether things were so created at the beginning. Was space furnished at once, by the fiat of Omnipotence, with these burning orbs? To this question the man of science, if he confine himself within his own limits, will give no answer, though it must be remarked that in the formation of an opinion he has better materials to guide him than anybody else. He can clearly show, however, that the present state of things *may* be derivative. He can even assign reasons which render probable its derivative origin—that it was not originally what it now is. At all events, he can prove that out of common non-luminous matter this whole pomp of stars might have been evolved.

The law of gravitation enunciated by Newton is, that every particle of matter in the universe attracts every other particle with a force which diminishes as the square of the distance increases. Thus the sun and the earth mutually pull each other; thus the earth and the moon are kept in company; the force which holds every respective pair of masses together being the integrated force of their component parts. Under the operation of this force a stone falls to the ground and is warmed by the shock; under its operation meteors plunge into our atmosphere and rise to incandescence. Showers of such doubtless fall incessantly upon the sun. Acted on by this force, were it stopped in its orbit to-morrow, the earth would rush towards, and finally combine with, the sun. Heat would also be developed by this collision, and Mayer, Helmholtz, and Thomson have calculated its amount. It would equal that produced by the combustion of more than 5,000 worlds of solid coal, all this heat being generated at the instant of collision. In the attraction of gravity, therefore, acting upon non-luminous matter, we have a source of heat more powerful than could be derived from any terrestrial combustion. And were the matter of the universe cast in cold detached fragments into space, and there abandoned to the mutual gravitation of its own parts, the collision of the fragments would in the end produce the fires of the stars.

The action of gravity upon matter originally cold may, in fact, be the origin of *all* light and heat, and the proximate source of such other powers as are generated by light and heat. But we have now to enquire what is the light and what is the heat thus produced? This question has already been answered in a general way. Both light and heat are modes of motion. Two planets clash and come to rest; their motion, considered as masses, is destroyed, but it is really continued as a motion of their ultimate particles. It is this motion, taken up by the æther, and propagated through it with a velocity of 185,000 miles a second, that comes to us as the light and heat of suns and stars. The atoms of a hot body swing with inconceivable rapidity, but this power of vibration necessarily implies the operation of forces between the atoms themselves. It reveals to us that while they are held together by one force, they are kept asunder by another, their position at any moment depending on the equilibrium of attraction and repulsion. The atoms are virtually connected by elastic springs, which oppose at the same time their approach and their retreat, but which tolerate the vibration called heat. When two bodies drawn together by the force of gravity strike each other, the intensity of the ultimate vibration, or, in other words, the amount of heat generated, is proportional to the *vis viva* destroyed by the collision. The molecular motion once set up is instantly shared with the æther, and diffused by it throughout space.

We on the earth's surface live night and day in the midst of æthereal commotion. The medium is never still. The cloud canopy above us may be thick enough to shut out the light of the stars, but this canopy is itself a warm body, which radiates its motion through the æther. The earth also is warm, and sends its heat-pulses incessantly forth. It is the waste of its molecular motion in space that chills the earth upon a clear night; it is the return of its motion from the clouds which prevents the earth's temperature on a cloudy night from falling so low. To the conception of space being filled, we must therefore add the conception of its being in a state of incessant tremor. The sources of vibration are the ponderable masses of the universe. Let us take a sample of these and examine it in detail. When we look to our planet we find it to be an aggregate of solids, liquids, and gases. When we look at any one of these, we generally find it composed of still more elementary parts. We learn, for example, that the water of our rivers is formed by the union, in definite proportions, of two gases, oxygen and hydrogen. We know how to bring these constituents together, and to cause them to form water : we also know how to analyse the water, and recover from it its two constituents. So, likewise, as regards the solid proportions of the earth. Our chalk hills, for example, are formed by a combination of carbon, oxygen, and calcium. These are elements the union of which, in definite proportions, has resulted in the formation of chalk. The flints within the chalk we know to be a compound of oxygen and silicium, called silica ; and our ordinary clay is, for the most part, formed by the union of silicium, oxygen, and the well-known light metal, aluminium. By far the greater portion of the earth's crust is compounded of the elementary substances mentioned in these few lines.

The principle of gravitation has been already described as an attraction which every particle of matter, however small, has for every other particle. With gravity there is no selection ; no particular atoms choose, by preference, other particular atoms as objects of attraction ; the attraction of gravitation is proportional to the *quantity* of the attracting matter, regardless of its quality. But in the molecular world which we have now entered matters are otherwise arranged. Here we have atoms between which a strong attraction is exercised, and also atoms between which a weak attraction is exercised. One atom can jostle another out of its place in virtue of a superior force of attraction. But though the amount of force exerted varies thus from atom to atom, it is still an attraction of the same mechanical quality, if I may use the term, as that of gravity itself. Its intensity might be measured in the same way, namely, by the amount of motion which it can impart in a certain time.   Thus the attraction of gravity at the earth's surface is expressed by the number 32, because, when acting freely on a body for a second of time, it imparts to the body a velocity of thirty-two feet a second.   In like manner the mutual attraction of oxygen and hydrogen might be measured by the velocity imparted to the atoms in their rushing together.   Of course such a unit of time as a second is not here to be thought of, the whole interval required by the atoms to cross the minute spaces which separate them not amounting probably to more than an inconceivably small fraction of a second.

It has been stated that when a body falls to the earth it is warmed by the shock.   Here we have what we may call a *mechanical* combination of the earth and the body.   Suffer the falling body and the earth to dwindle in imagination to the size of atoms, and for the attraction of gravity substitute that of chemical affinity, which is the name given to the molecular attraction, we have then what is called a *chemical* combination.   The effect of the union in this case also is the development of heat, and from the amount of heat generated we can infer the intensity of the atomic pull.   Measured by ordinary mechanical standards, this is enormous.   Mix eight pounds of oxygen with one of hydrogen, and pass a spark through the mixture; the gases instantly combine, their atoms rushing over the little distances between them.   Take a weight of 47,000 pounds to an elevation of 1,000 feet above the earth's surface, and let it fall; the energy with which it would strike the earth would not exceed that of the eight pounds of oxygen atoms as they dash against one pound of hydrogen atoms to form water.

It is sometimes stated that the force of gravity is distinguished from all other forces by the fact of its resisting conversion into any other. Chemical affinity, it is said, can be converted into heat and light, and these again into magnetism and electricity. But gravity refuses to be so converted; it is a force which maintains itself under all circumstances, and is not capable of disappearing to give place to another. If by this is meant that a particle of matter can never be deprived of its weight, the assertion is correct; but the law which affirms the convertibility of natural forces was never meant, in the minds of those who understood it, to affirm that such a conversion as that here implied occurs in any case whatever. As regards convertibility into heat, gravity and chemical affinity stand on precisely the same footing. The *attraction* in the one case is as indestructible as in the other. Nobody affirms that when a stone rests upon the surface of the earth the mutual attraction of the earth and stone is abolished; nobody means to affirm that the mutual attraction of oxygen for hydrogen ceases after the atoms have combined to form water. What is meant in the case of chemical affinity is, that the pull of that affinity, acting through a certain space, imparts a motion of *translation* of the one atom towards the other. This motion of translation is *not* heat, nor is the force that produces it heat. But when the atoms strike and recoil, the motion of translation is converted into a motion of vibration, and this latter motion *is* heat. But the vibration, so far from causing the extinction of the original attraction, is in part carried on by that attraction. The atoms recoil in virtue of the elastic force which opposes actual contact, and in the recoil they are driven too far back. The original attraction then triumphs over the force of recoil, and urges the atoms once more together. Thus, like a pendulum, they oscillate, until their motion is imparted to the surrounding æther ; or, in other words, until their heat becomes *radiant* heat.

In this sense, and in this sense only, is chemical affinity converted into heat. There is, first of all, the attraction between the atoms ; there is, secondly, *space* between them. Across this space the attraction urges them. They collide, they recoil, they oscillate. There is a change in the form of the motion, but there is no real loss. It is so with the attraction of gravity. To produce motion here space must also intervene between the attracting bodies : when they strike motion is apparently destroyed, but in reality there is no destruction. Their atoms are suddenly urged together by the shock ; by their own perfect elasticity these atoms recoil ; and thus is set up the molecular oscillation which announces itself to the nerves as heat.

It was formerly universally supposed that by the collision of unelastic bodies force was destroyed. Men saw, for example, when two spheres of clay, or painter's putty, or lead, were urged together, that the motion possessed by the masses prior to impact was more or less annihilated. They believed in an absolute destruction of the force of impact. Until recent times, indeed, no difficulty was experienced in believing this, whereas, at present, the ideas of force and its destruction refuse to be united in most philosophic minds. In the collision of elastic bodies, on the contrary, it was observed that the motion with which they clashed together was in great part restored by the resiliency of the masses, the more perfect the elasticity the more complete being the restitution.

This led to the idea of perfectly elastic bodies—bodies competent to restore by their recoil the whole of the motion which they possessed before impact.

Hence the idea of the *conservation* of force, as opposed to the destruction of force, which was supposed to occur when inelastic bodies met in collision.

We now know that the principle of conservation holds equally good with elastic and unelastic bodies. Perfectly elastic bodies develope *no heat* on collision. They retain their motion afterwards, though its direction may be changed; and it is only when sensible motion is, in whole or in part, destroyed that heat is generated. This always occurs in unelastic collision, the heat developed being the exact equivalent of the motion extinguished. This heat virtually declares that the property of elasticity, denied to the masses, exists among their atoms, and by their recoil and oscillation the principle of conservation is vindicated.

But ambiguity in the use of the term ' force' has been for some time more and more creeping upon us. We called the attraction of gravity a force without any reference to motion. A body resting on a shelf is as much pulled by gravity as when after having been pushed off the shelf it falls towards the earth. We applied the term force also to that molecular attraction which we called chemical affinity. When, however, we spoke of the conservation of force in the case of elastic collision, we meant neither a pull nor a push, which, as just indicated, might be exerted upon inert matter, but we meant the *moving force*, if I may use the term, of the colliding masses.

What I have called moving force has a definite mechanical measure in the amount of work that it can perform. The simplest form of work is the raising of a weight. A man walking up-hill or up-stairs with a pound weight in his hand, to an elevation say of sixteen feet, performs a certain amount of work over and above the lifting of his own body. If he ascend to a height of thirty-two feet he does twice the work; if to a height of forty-eight feet, he does three times the work; if to sixty-four feet he does four times the work, and so on. If, moreover, he carries up two pounds instead of one, other things being equal, he does twice the work; if three, four, or five pounds, he does three, four, or five times the work. In fact it is plain that the work performed depends on two factors, the weight raised and the height to which it is raised. It is expressed by the product of these two factors.

But a body may be caused to reach a certain elevation in opposition to the force of gravity, without being actually carried up to the elevation. If a hodman, for example, wished to land a brick at an elevation of sixteen feet above the place where he stands, he would probably pitch it up to the bricklayer. He would thus impart, by a sudden effort, a velocity to the brick sufficient to raise it to the required height; the work accomplished by that effort being precisely the same as if he had slowly carried up the brick. The initial velocity which must be imparted in the case here assumed, is well known. To reach a height of sixteen feet, the brick must quit the man's hand with a velocity of thirty-two feet a second. It is needless to say that a body starting with any velocity, would, if wholly unopposed or unaided, continue to move *for ever* with the same velocity. But when, in the case before us, the body is thrown upwards, it moves in opposition to gravity, which incessantly retards its motion, and finally brings it to rest at an elevation of sixteen feet. If not here caught by the bricklayer, it would return to the hodman with an accelerated motion, and reach his hand with the precise velocity it possessed on quitting it.

Supposing the man competent to impart to the brick, at starting, a speed of sixty-four feet a second, or twice its former speed, would the amount of work performed in this effort be only twice what it was in the first instance? No; it would be four times that quantity. A body starting with twice the velocity of another, will rise to four times the height; in like manner, a three-fold velocity will give a nine-fold elevation, a fourfold velocity will give a sixteen-fold elevation, and so on. The height attained, then, or the work done, is not proportional to the velocity, but to the *square* of the velocity. As before, the work is also proportional to the weight elevated. Hence the work which any moving masses whatever are competent to perform, by the motion which they at any moment possess, is jointly proportional to the weight and the square of the velocity. Here, then, we have a second measure of work, in which we simply translate the idea of height into its equivalent idea of motion.

In mechanics, the product of the mass of a moving body into the square of its velocity, expresses what is called the *vis viva*, or living force. It is also sometimes called the 'mechanical effect.' If, for example, we point a cannon upwards, and start a ball with twice the velocity imparted by a second cannon, the ball will rise to four times the height. The speedier ball, if directed against a target, will also do four times the execution. Hence the importance of imparting a high velocity to projectiles in war. Having thus cleared our way to a perfectly clear conception of the *vis viva* of moving masses, we are prepared for the announcement that the heat generated by the collision of a falling body against the earth is proportional to the *vis viva* annihilated. In point of fact, it is not an annihilation at all, but a transference of *vis viva* from the mass, to its ultimate particles. This, as we now learn, is proportional to the square of the velocity. In the case, therefore, of two cannon balls of equal weight, if one strike a target with twice the velocity of the other, it will generate four times the heat ; if with three times the velocity it will generate nine times the heat, and so on.

Mr. Joule has shown that in falling from a height of 772 feet, a body will generate an amount of heat sufficient to raise its own weight of water one degree Fahrenheit in temperature. We have here the *mechanical equivalent* of heat. Now, a body falling from a height of 772 feet, has, upon striking the earth, a velocity of 223 feet a second ; and if this velocity were imparted to a body, by any other means, the quantity of heat generated by the stoppage of its motion would be that stated above. Six times that velocity, or 1,338 feet, would not be an inordinate one for a cannon ball as it quits the gun ; but if animated by six times the velocity, thirty-six times the heat will be generated by the stoppage of its motion. Hence a cannon ball moving with a velocity of 1,338 feet a second, would, by collision, generate an amount of heat competent to raise its own weight of water 36 degrees Fahrenheit in temperature. If composed of iron, and if all the heat generated were concentrated in the ball itself, its temperature would be raised about 360 degrees Fahrenheit; because one degree in the case of water is equivalent to about ten degrees in the case of iron. In artillery practice the heat generated is usually concentrated upon the front of the bolt, and on the portion of the target first struck. By this concentration the heat developed may become sufficiently intense to raise the dust of the metal to incandescence, a flash of light often accompanying collision with the target.

Let us now fix our attention for a moment on the gunpowder which urges the cannon ball. This is composed of combustible matter, which if burnt in the open air would yield a certain amount of heat. It will not yield this amount if it performs the work of urging a ball. The heat then generated by the gunpowder will fall short of that produced in the open air, by an amount equivalent to the *vis viva* of the ball; and this exact amount is restored by the ball on its collision with the target. In this perfect way are heat and mechanical motion connected.

Broadly enunciated, the principle of the conservation of force asserts that the quantity of force in the universe is as unalterable as the quantity of matter; that it is alike impossible to create force and to annihilate it. But in what sense are we to understand this assertion? It would be manifestly inapplicable to the force of gravity as Newton defined it; for this is a force varying inversely as the square of the distance, and to affirm the constancy of a varying force would be self-contradictory. Yet, when the question is properly understood, gravity forms no exception to the law of conservation. Following the method pursued by Helmholtz, I will here attempt an elementary exposition of this law, which, though destined in its applications to produce momentous changes in human thought, is not difficult of comprehension.

For the sake of simplicity we will consider a particle of matter, which we may call F, to be perfectly fixed, and a second movable particle, D, placed at a distance from F. We will assume that these two particles attract each other according to the Newtonian law. At a certain distance the attraction is of a certain definite amount, which might be determined by means of a spring balance. At half this distance the attraction would be augmented four times; at a third of the distance it would be augmented nine times; at one-fourth of the distance sixteen times, and so on. In every case the attraction might be measured by determining, with the spring balance, the amount of tension which is just sufficient to prevent D from moving towards F. Thus far we have nothing whatever to do with motion; we deal with statics, not with dynamics. We simply take into account the *distance* of D from F, and the pull exerted by gravity at that distance.

It is customary in mechanics to represent the magnitude of a force by a line of a certain length, a force of double magnitude being represented by a line of double length, and so on. Placing then the particle D at a distance from F, we can in imagination draw a straight line from D to F, and at D erect a perpendicular to this line, which shall represent the amount of the attraction exerted on D in this position. If D be at a very great distance from F the attraction will be very small, and the perpendicular consequently very short. Let us now suppose that at every point in the line joining F and D a perpendicular is erected proportional in length to the attraction exerted at that point ; we should thus obtain an infinite number of perpendiculars of gradually increasing length as D approaches F. Uniting the ends of all these perpendiculars, we should obtain a curve, and between this curve and the straight line joining F and D we should have an area containing all the perpendiculars placed side by side. Each one of this infinite series of perpendiculars representing an attraction, or tension as it is sometimes called, the area just referred to represents the total effort capable of being exerted by the tensions upon the particle D, during its passage from its first position up to F.

Up to the present point we have been dealing with tensions, and not with motion. Thus far *vis viva* has been entirely foreign to our contemplation of D and F. Let us now suppose D placed at a practically infinite distance from F ; here the pull of gravity would be nothing, and the perpendicular representing it would dwindle to a point. In this position the sum of the tensions capable of being exerted on D would be a maximum. Let D now begin to move in obedience to the attraction exerted upon it. Motion being once set up, the idea of *vis viva* arises. In moving towards F the particle D consumes, as it were, the tensions. Let us fix our attention on D at any point of the path over which it is moving. Between that point and F there is a quantity of unused tensions ; beyond that point the tensions have been all consumed, but we have in their place an equivalent quantity of *vis viva*. After D has passed any point, the tension previously in store at that point disappears, but not without having added, during the infinitely small duration of its action, a due amount of motion to that previously possessed by D. The nearer D approaches to F, the smaller is the sum of the tensions remaining, but the greater is the living force ; the farther D is from F, the greater is the sum of the unconsumed tensions, and the less is the living force. Now the principle of conservation affirms *not* the constancy of the value of the tensions of gravity, nor yet the constancy of the *vis viva*, taken separately, but the absolute constancy of the value of the sum of both. At the beginning the *vis viva* was zero and the tension area was a maximum ; close to F the *vis viva* is a maximum, while the tension area is zero. At every other point the work-producing power of the particle D consists in part of *vis viva* and in part of tensions.

If gravity, instead of being attraction, were repulsion, when the particles are in contact, the sum of the tensions between two material particles D and F would be a maximum, and the *vis viva* zero. If D, in obedience to the repulsion, moved away from F, *vis viva* would be generated ; and the farther D retreated from F the greater would be its *vis viva*, and the less the amount of tension still available for producing motion. Taking repulsion into account as well as attraction, the principle of the conservation of force affirms that the mechanical value of the *tensions* and *vires vivæ* of the material universe is a constant quantity. The universe, in short, possesses two kinds of property which are mutually convertible at an unvarying rate. The diminution of either carries with it the enhancement of the other, the total value of the property remaining unchanged.

The considerations that we have here applied to gravity apply equally to chemical affinity. In a mixture of oxygen and hydrogen the atoms exist apart, but by the application of proper means they may be caused to rush together across the space that separates them. While this space exists, and as long as the atoms have not begun to move towards each other, we have tensions and nothing else. During their motion towards each other the tensions, as in the case of gravity, are converted into *vis viva*. After they clash we have still *vis viva*, but in another form. It *was* translation, it *is* vibration. It *was* molecular transfer, it *is* heat. The same considerations apply to a mixture of hydrogen and chlorine. When these gases are mingled in the dark they remain separate, but if a sunbeam fall upon the mixture the atoms rush together with detonation. Here also we have tension converted into molecular translation, and molecular translation into heat and sound.

It is possible to reverse these processes, to unlock the embrace of the atoms and replace them in their first positions. But to accomplish this as much heat would be required as was generated by their union. Such reversals occur daily and hourly in nature. By the solar waves, the oxygen of water is divorced from its hydrogen in the leaves of plants. As molecular *vis viva* the waves disappear, but in so doing they re-endow the atoms of oxygen and hydrogen with tension. The atoms are thus enabled to re-combine, and when they do so they restore the precise amount of heat consumed in their separation. The same remarks apply to the compound of carbon and oxygen, called carbonic acid, which is exhaled from our lungs, produced by our fires, and found sparingly diffused everywhere throughout the air. In the leaves of plants the sunbeams also wrench these atoms asunder, and sacrifice themselves in the act ; but when the plants are

burnt the amount of heat consumed in their production is restored.

This, then, is the rhythmic play of nature as regards her forces. Throughout all her regions she oscillates from tension to *vis viva*, from *vis viva* to tension. We have the same play in the planetary system. The earth's orbit is an ellipse, one of the foci of which is occupied by the sun. Imagine the earth at the most distant part of the orbit. Her motion, and consequently her *vis viva*, is then a minimum. The planet rounds the curve, and begins to approach the sun. In front it has a store of tensions, which is gradually consumed, an equivalent amount of *vis viva* being generated. When nearest to the sun the motion, and consequently the *vis viva*, is a maximum. But here the available tensions have been used up. The earth rounds this portion of the curve and retreats from the sun. Tensions are now stored up, but *vis viva* is lost, to be again restored at the expense of the complementary force on the opposite side of the curve. Thus beats the heart of the universe, but without increase or diminution of its total stock of force.

I have thus far tried to steer clear amid confusion by fixing the mind of the reader upon things rather than upon names. But good names are essential; and here, as yet, we are not provided with such. We have had the force of gravity and living force—two utterly distinct things. We have had pulls and tensions; and we might have had the force of heat, the force of light, the force of magnetism, or the force of electricity—all of which terms have been employed more or less loosely by writers on physics. This confusion is happily avoided by the introduction of the term 'energy,' embracing under it both *tension* and *vis viva.* Energy is possessed by bodies already in motion; it is then actual, and we agree to call it *actual* or *dynamic energy.* It is our old *vis viva.* On the other hand, energy is possible to bodies not in motion, but which, in virtue of attraction or repulsion, possess a power of motion which would realise itself if all hindrances were removed. Looking, for example, at gravity, a body on the earth's surface in a position from which it cannot fall to a lower one possesses no energy. It has neither motion nor power of motion. But the same body suspended at a height above the earth has a power of motion though it may not have exercised it. Energy is possible to such a body, and we agree to call this *potential energy.* It embraces our old tensions. We, moreover, speak of the conservation of energy instead of the conservation of force; and say that the sum of the potential and dynamic energies of the material universe is a constant quantity.

A body cast upwards consumes the actual energy of projection, and lays up potential energy. When it reaches its utmost height all its actual energy is consumed, its potential energy being then a maximum. When it returns, there is a reconversion of the potential into the actual. A pendulum at the limit of its swing possesses potential energy; at the lowest point of its arc its energy is all actual. A patch of snow *resting* on a mountain slope has potential energy; loosened, and shooting down as an avalanche, it possesses dynamic energy. The pine-trees growing on the Alps have potential energy; but rushing down the *Holzrinne* of the woodcutters they possess actual energy. The same is true of the mountains themselves. As long as the rocks, which compose them can fall to a lower level, they possess potential energy, which is converted into actual when the frost ruptures their cohesion and hands them over to the action of gravity. The hammer of the great bell of Westminster, when raised before striking, possesses potential energy; when it falls, the energy becomes dynamic; and after the stroke, we have the rhythmic play of potential and dynamic in the vibrations of the bell. The same holds good for the molecular oscillations of a heated body. An atom is pressed against its neighbour, and recoils. But the ultimate amplitude of the recoil is soon attained, the motion of the atom in that direction is checked, and for an instant its energy is all potential. It is then drawn towards its neighbour with accelerated speed, thus, by attraction, converting its potential into dynamic energy. Its motion in this direction is also finally checked, and, for an instant, again its energy is all potential. It again retreats, converting, by repulsion, its potential into dynamic energy, till the latter attains a maximum, after which it is again changed into potential energy. Thus, what is true of the earth, as she swings to and fro in her yearly journey round the sun, is also true of her minutest atom. We have wheels within wheels, and rhythm within rhythm.

When a body is heated, a change of molecular arrangement always occurs, and to produce this change heat is consumed. Hence, a portion only of the heat communicated to the body remains as dynamic energy. Looking back on some of the statements made at the beginning of this article, now that our knowledge is more extensive, we see the necessity of qualifying them. When, for example, two bodies clash, heat is generated; but the heat, or molecular dynamic energy, developed at the moment of collision, is not the equivalent of the sensible dynamic energy destroyed. The true equivalent is this heat, plus the potential energy conferred upon the molecules by the placing of greater distances between them. This molecular potential energy is afterwards, on the cooling of the body, converted into heat.

Wherever two atoms capable of uniting together by their mutual attractions exist separately, they form a store of potential energy. Thus our woods, forests, and coal-fields on the one hand, and our atmospheric oxygen on the other, constitute a vast store of energy of this kind—vast, but far from infinite. We have, besides our coal-fields, bodies in the metallic condition more or less sparsely distributed in the earth's crust. These bodies can be oxydised, and hence are, so far as they go, stores of potential energy. But the attractions of the great mass of the earth's crust are already satisfied, and from them no further energy can possibly be obtained. Ages ago the elementary constituents of our rocks clashed together and produced the motion of heat, which was taken up by the æther and carried away through stellar space. It is lost for ever as far as we are concerned. In those ages the hot conflict of carbon, oxygen, and calcium produced the chalk and limestone hills which are now cold; and from this carbon, oxygen, and calcium no further energy can be derived. And so it is with almost all the other constituents of the earth's crust. They took their present form in obedience to molecular force; they turned their potential energy into dynamic, and gave it to the universe ages before man appeared upon this planet. For him a residue of potential energy remains, vast truly in relation to the life and wants of an individual, but exceedingly minute in comparison with the earth's primitive store.

To sum up.   The whole stock of *energy* or *working-power* in the world consists of *attractions, repulsions,* and *motions.*   If the attractions and repulsions are so circumstanced as to be able to produce motion, they are sources of working-power, but not otherwise.   As stated a moment ago, the attraction exerted between the earth and a body at a distance from the earth's surface is a source of working-power; because the body can be moved by the attraction, and in falling to the earth can perform work. When it rests upon the earth's surface it is *not* a source of power or energy, because it can fall no further.   But though it has ceased to be a source of *energy,* the attraction of gravity still acts as a *force,* which holds the earth and weight together.

The same remarks apply to attracting atoms and molecules.   As long as distance separates them, they can move across it in obedience to the attraction, and the motion thus produced may, by proper appliances, be caused to perform mechanical work.   When, for example, two atoms of hydrogen unite with one of oxygen, to form water, the atoms are first drawn towards each other— they move, they clash, and then by virtue of their resiliency, they recoil and *quiver.*   To this quivering motion we give the name of heat.   Now this atomic vibration is merely the redistribution of the motion produced by the chemical affinity; and this is the only sense in which chemical affinity can be said to be converted into heat. We must not imagine the chemical *attraction* destroyed, or converted into anything else.   For the atoms when mutually clasped to form a molecule of water, are held together by the very attraction which first drew them towards each other. That which has really been expended is the *pull* exerted through the space by which the distance between the atoms has been diminished.

If this be understood it will be at once seen that *gravity* may in this sense be said to be convertible into heat ; that it is in reality no more an outstanding and inconvertible agent, as it is sometimes stated to be, than chemical affinity. By the exertion of a certain pull through a certain space a body is caused to clash with a certain definite velocity against the earth. Heat is thereby developed, and this is the only sense in which gravity can be said to be converted into heat. In no case is the *force* which produces the motion annihilated or changed into anything else. The mutual *attraction* of the earth and weight exists when they are in contact as when they were separate ; but the ability of that attraction to employ itself in the production of motion does *not* exist.

The transformation, in this case, is easily followed by the mind's eye. First, the weight as a whole is set in motion by the attraction of gravity. This motion of the mass is arrested by collision with the earth, being broken up into molecular tremors, to which we give the name of heat.

And when we reverse the process, and employ those tremors of heat to raise a weight, as is done through the intermediation of an elastic fluid in the steam-engine, a certain definite portion of the molecular motion is destroyed in raising the weight. In this sense, and this sense only, can the heat be said to be converted into gravity, or more correctly, into potential energy of gravity. It is not that the destruction of the heat has created any *new* attraction, but simply that the old attraction has now a power conferred upon it, of exerting a certain definite pull in the interval between the starting-point of the falling weight and its collision with the earth.

When, therefore, writers on the conservation of energy speak of tensions being 'consumed' and 'generated,' they do not mean thereby that old attractions have been annihilated, and new ones brought into existence, but that, in the one case, the power of the attraction to produce motion has been diminished by the shortening of the distance between the attracting bodies, and that in the other case the power of producing motion has been augmented by the increase of the distance. These remarks apply to all bodies, whether they be sensible masses or molecules.

Of the inner quality that enables matter to attract matter we know nothing; and the law of conservation makes no statement regarding that quality. It takes the facts of attraction as they stand, and affirms only the constancy of *working-power*. That power may exist in the form of MOTION; or it may exist in the form of FORCE, *with distance to act through*. The former is dynamic energy, the latter is potential energy, the constancy of the sum of both being affirmed by the law of conservation. The *convertibility* of natural forces consists solely in transformations of dynamic into potential, and of potential into dynamic energy, which are incessantly going on. In no other sense has the convertibility of force, at present, any scientific meaning.

II.

THOUGHTS ON PRAYER AND NATURAL LAW.

AN EXTRACT.

[*Mountaineering in* 1861, *p.* 33.]

Aber im stillen Gemach entwirft bedeutende Zirkel
Sinnend der Weise.
Folgt durch die Lüfte dem Klang, folgt durch den Aether dem Strahl,
Sucht das vertraute Gesetz in des Zufalls grausenden Wundern,
Sucht den ruhenden Pol in der Erscheinungen Flucht.

SCHILLER.

II.

PRAYER AND NATURAL LAW.

THE aspects of nature are more varied and impressive in Alpine regions than elsewhere. The mountains in their setting of deep blue sky; the glow of firmament and peaks at sunrise and sunset; the formation and distribution of clouds; the descent of rain, hail, and snow; the stealthy slide of glaciers and the rush of avalanches and rivers; the fury of storms; thunder and lightning, with their occasional accompaniment of blazing woods;—all these things tend to excite the feelings and to bewilder the mind. In this entanglement of phenomena it seems hopeless to seek for law or orderly connection. And before the thought of law dawned upon the human mind men naturally referred these inexplicable effects to personal agency. The savage saw in the fall of a cataract the leap of a spirit, and the echoed thunder-peal was to him the hammer-clang of an exasperated god. Propitiation of these terrible powers was the consequence, and sacrifice was offered to the demons of earth and air.

But observation tends to chasten the emotions and to check those structural efforts of the intellect which have emotion for their base. One by one natural phenomena have been associated with their proximate causes; and the idea of direct personal volition mixing itself in the economy

of nature is retreating more and more. Many of us fear
this tendency; our faith and feelings are dear to us, and
we look with suspicion and dislike on any philosophy,
the apparent tendency of which is to dry up the soul.
Probably every change from ancient savagery to our present enlightenment excited, in a greater or less degree, a
fear of this kind. But the fact is, that we have not yet
determined whether the form under which they now
appear in the world is necessary to the life and warmth
of religious feeling. We may err in linking the imperishable with the transitory, and confound the living plant
with the decaying pole to which it clings. My object,
however, at present is not to argue, but to mark a tendency. We have ceased to propitiate the powers of
Nature—ceased even to pray for things in *manifest* contradiction to natural laws. In Protestant countries, at
least, I think it is conceded that the age of miracles is past.

The general question of miracles is at present in able
and accomplished hands; and were it not so, my polemical
acquirements are so limited, that I should not presume to
enter upon a discussion of this subject on its entire merits.
But there is one little outlying point, which attaches itself
to this question, on which a student of science, without
quitting the ground which strictly belongs to him, may
offer a remark.

At the auberge near the foot of the Rhone glacier, I
met in the summer of 1858, an athletic young priest, who,
after a solid breakfast, including a bottle of wine, informed me that he had come up to ' bless the mountains.'
This was the annual custom of the place. Year by year
the Highest was entreated, by official intercessors, to make
such meteorological arrangements as should ensure food

and shelter for the flocks and herds of the Valaisians. A diversion of the Rhone, or a deepening of the river's bed, would have been of incalculable benefit to the inhabitants of the valley at the time I now mention. But the priest would have shrunk from the idea of asking the Omnipotent to open a new channel for the river, or to cause a portion of it to flow over the Grimsel pass, and down the vale of Oberhasli to Brientz. This he would have deemed a *miracle*, and he did not come to ask the Creator to perform miracles, but to do something which he manifestly thought lay quite within the bounds of the natural and non-miraculous. A Protestant gentleman who was present at the time, smiled at this recital. He had no faith in the priest's blessing, still he deemed his prayer different in kind from a request to open a new river-cut, or to cause the water to flow up-hill.

In a similar manner we Protestants smile at the honest Tyrolese priest, who, when he feared the bursting of a glacier dam, offered the sacrifice of the mass upon the ice as a means of averting the calamity. That poor man did not expect to convert the ice into adamant, or to strengthen its texture so as to enable it to withstand the pressure of the water; nor did he expect that his sacrifice would cause the stream to roll back upon its source and relieve him, by a miracle, of its presence. But beyond the boundaries of his knowledge lay a region where rain was generated he knew not how. He was not so presumptuous as to expect a miracle, but he firmly believed that in yonder cloud-land matters could be so arranged, without trespass on the miraculous, that the stream which threatened him and his flock should be caused to shrink within its proper bounds.

D

Both these priests fashioned that which they did not understand to their respective wants and wishes. In their case imagination wrought, unconditioned by a knowledge of laws. A similar state of mind was long prevalent among mechanicians ; many of whom, and some of them extremely skilful ones, were occupied a century ago with the question of a perpetual motion. They aimed at constructing a machine which should execute work without the expenditure of power ; and many of them went mad in the pursuit of this object. The faith in such a consummation, involving as it did immense personal interest to the inventor, was extremely exciting, and every attempt to destroy this faith was met by bitter resentment on the part of those who held it. Gradually, however, as men became more and more acquainted with the true functions of machinery, the dream dissolved. The hope of getting work out of mere mechanical combinations disappeared ; but still there remained for the speculator a cloud-land denser than that which filled the imagination of the Tyrolese priest, and out of which he still hoped to evolve perpetual motion. There was the mystic store of chemic force, which nobody understood ; there were heat and light, electricity and magnetism, all competent to produce mechanical motions.[1] Here, then, is the mine in which we must seek our gem. A modified and more refined form of the ancient faith revived ; and, for aught I know, a remnant of sanguine designers may at the present moment be engaged on the problem which like-minded men in former years left unsolved.

And why should a perpetual motion, even under

---

[*] See Helmholtz—' Wechselwirkung der Naturkräfte.'

modern conditions, be impossible? The answer to this question is the statement of that great generalisation of modern science, which is known under the name of the Conservation of Energy. This principle asserts that no power can make its appearance in Nature without an equivalent expenditure of some other power; that natural agents are so related to each other as to be mutually convertible, but that no new agency is created. Light runs into heat; heat into electricity; electricity into magnetism; magnetism into mechanical force; and mechanical force again into light and heat. The Proteus changes, but he is ever the same; and his changes in Nature, supposing no miracle to supervene, are the expression, not of spontaneity, but of *physical necessity*. A perpetual motion, then, is deemed impossible, because it demands the creation of force, whereas the principle of Conservation is, no creation but infinite conversion.

It is an old remark that the law which moulds a tear also rounds a planet. In the application of law in Nature the terms great and small are unknown. Thus the principle referred to teaches us that the Italian wind gliding over the crest of the Matterhorn is as firmly ruled as the earth in its orbital revolution round the sun; and that the fall of its vapour into clouds is exactly as much a matter of necessity as the return of the seasons. The dispersion therefore of the slightest mist by the special volition of the Eternal, would be as much a miracle as the rolling of the Rhone over the Grimsel precipices and down Haslithal to Brientz.

It seems to me quite beyond the present power of science, to demonstrate that the Tyrolese priest, or his colleague of the Rhone valley, asked for an 'impossibility' in praying for good weather; but science *can* demonstrate the incompleteness of the knowledge of Nature which limited their prayers to this narrow ground; and she may lessen the number of instances in which we 'ask amiss,' by showing that we sometimes pray for the performance of a miracle when we do not intend it. She does assert, for example, that without a disturbance of natural law, quite as serious as the stoppage of an eclipse, or the rolling of the St. Lawrence up the Falls of Niagara, no act of humiliation, individual or national, could call one shower from heaven, or deflect towards us a single beam of the sun.

Those, therefore, who believe that the miraculous is still active in nature, may, with perfect consistency, join in our periodic prayers for fair weather and for rain : while those who hold that the age of miracles is past, will refuse to join in such petitions. And if these latter wish to fall back upon such a justification, they may fairly urge that the latest conclusions of science are in perfect accordance with the doctrine of the Master himself, which manifestly was that the distribution of natural phenomena is not affected by moral or religious causes. 'He maketh His sun to rise on the evil and on the good, and sendeth rain on the just and on the unjust.' Granting ' the power of Free Will in man,' so strongly claimed by Professor Mansel in his admirable defence of the belief in miracles, and assuming the efficacy of free prayer to produce changes in external nature, it necessarily follows that natural laws are more or less at the mercy of man's volition, and no conclusion founded on the assumed permanence of those laws would be worthy of confidence.

It is a wholesome sign for England that she numbers

among her clergy men wise enough to understand all this, and courageous enough to act up to their knowledge. Such men do service to the public character by encouraging a manly and intelligent conflict with the causes of disease and scarcity, instead of a delusive reliance on supernatural aid. But they have also a value beyond this local and temporary one. They prepare the public mind for changes, which though inevitable, could hardly, without such preparation, be wrought without violence. Iron is strong; still, water in crystallising will shiver an iron envelope, and the more unyielding the metal is, the worse for its safety. There are men amongst us who would encompass philosophic speculation by a rigid envelope, hoping thereby to restrain it, but in reality giving it explosive force. If we want an illustration of this we have only to look at modern Rome. In England, thanks to men of the stamp to which I have alluded, scope is gradually given to thought for changes of aggregation, and the envelope slowly alters its form in accordance with the necessities of the time.

The proximate origin of the foregoing slight article, and probably the remoter origin of the next following one, was this. Some years ago, a day of prayer and humiliation, on account of a bad harvest, was appointed by the proper religious authorities; but certain clergymen of the Church of England, doubting the wisdom of the demonstration, declined to join in the services of the day. For this act of nonconformity they were severely censured by some of their brethren. Rightly or wrongly, my sympathies were on the side of these men; and, to lend them a helping hand in their struggle against odds, I inserted the foregoing chapter in the little book mentioned on the title-page. Some time subsequently I received from a gentleman of great weight and distinction in the scientific world, and, I believe, of perfect orthodoxy in the religious one, a note directing my attention to an exceedingly thoughtful article on Prayer and Cholera in the 'Pall Mall Gazette.' My eminent correspondent deemed the article a fair answer to the remarks made by me in 1861. I, also, was struck by the temper and ability of the article, but I could not deem its arguments satisfactory, and in a short note to the editor of the 'Pall Mall Gazette' I ventured to state so much. This letter elicited some very able replies, and a second leading article was also devoted to the subject. In answer to all, I risked the publication of a second letter, and soon afterwards, by an extremely courteous note from the editor, the discussion was closed.

Though thus stopped locally, the discussion flowed in other directions. Sermons were preached, essays were published, articles were written, while a copious correspondence occupied the pages of some of the religious newspapers. It gave me sincere pleasure to notice that the discussion, save in a few cases where natural coarseness had the upper hand, was conducted with a minimum of vituperation. The severity shown was hardly more than sufficient to demonstrate earnestness, while gentlemanly feeling was too predominant to permit that earnestness to contract itself to bigotry or to clothe itself in abuse. It was probably the memory of this discussion which caused another excellent friend of mine to recommend to my perusal the exceedingly able work which in the next article I have endeavoured to review.

III.

MIRACLES AND SPECIAL PROVIDENCES.

A REVIEW.

[*Fortnightly Review, New Series, vol.* i. *p.* 645.]

'Mr. Mozley's book belongs to that class of writing of which Butler may be taken as the type. It is strong, genuine argument about difficult matters, fairly tracing what *is* difficult, fairly trying to grapple, not with what appears the gist and strong point of a question, but with what really at bottom is the knot of it. It is a book the reasoning of which may not satisfy everyone. . . . But we think it is a book for people who wish to see a great subject handled on a scale which befits it, and with a perception of its real elements. It is a book which will have attractions for those who like to see a powerful mind applying itself, without shrinking or holding back, without trick or reserve or show of any kind, as a wrestler closes body to body with his antagonist, to the strength of an adverse and powerful argument.'—*The Times, Tuesday, June 5, 1866.*

'We should add, that the faults of the work are wholly on the surface and in the arrangement; that the matter is as solid and as logical as that of any book within recent memory, and that it abounds in striking passages, of which we have scarcely been able even to give a sample. No future arguer against miracles can afford to pass it over.'—*Saturday Review, September* 15, 1866.

III.

MIRACLES AND SPECIAL PROVIDENCES.

IT is my privilege to enjoy the friendship of a select number of religious men, with whom I converse frankly upon theological subjects, expressing without disguise the notions and opinions I entertain regarding their tenets, and hearing in return these notions and opinions subjected to criticism. I have thus far found them liberal and loving men, patient in hearing, tolerant in reply, who know how to reconcile the duties of courtesy with the earnestness of debate. From one of these, nearly a year ago, I received a note, recommending strongly to my attention the volume of 'Bampton Lectures' for 1865, in which the question of miracles is treated by Mr. Mozley. Previous to receiving this note, I had in part made the acquaintance of the work, through the able and elaborate review of it which had appeared in the 'Times.' The combined effect of the letter and the review was to make the book the companion of my summer tour in the Alps. There, during the wet and snowy days which were only too prevalent last year, and during the days of rest interpolated between days of toil, I made myself more thoroughly conversant with Mr. Mozley's volume. I found it clear and strong—an intellectual tonic, as bracing and pleasant to my mind as the keen air of the mountains was to my body. From time to time I jotted down my thoughts regarding it, intending afterwards, if time permitted, to work them up into a coherent whole.   Other duties, however, interfered with the carrying out of this intention, and what I wrote last summer I now publish, not hoping within any reasonable time to be able to render my defence of scientific method more complete.

Mr. Mozley refers at the outset of his task to the movement against miracles which of late years has taken place, and which determined his choice of a subject. He acquits modern science of having had any great share in the production of this movement.   The objection against miracles, he says, does not arise from any minute knowledge of the laws of nature, but simply because they are opposed to that plain and obvious order of nature which everybody sees.   The present movement is, he thinks, to be ascribed to the greater earnestness and penetration of the present age.   Formerly miracles were accepted without question, because without reflection ; but the exercise of what Mr. Mozley calls the historic imagination is a characteristic of our own time.   Men are now accustomed to place before themselves vivid images of historic facts, and when a miracle rises to view, they halt before the astounding occurrence, and realising it with the same clearness as if it were now passing before their eyes, they ask themselves, ' Can this have taken place ? '   In some instances the effort to answer this question has led to a disbelief in miracles, in others to a strengthening of belief.   The end and aim of Mr. Mozley's lectures is to show that the strengthening of belief is the logical result which ought to follow from the examination of the facts.

Attempts have been made by religious men to bring the Scripture miracles within the scope of the order of nature, but all such attempts are rejected by Mr. Mozley as utterly futile and wide of the mark. Regarding miracles as a necessary accompaniment of a revelation, their evidential value in his eyes depends entirely upon their deviation from the order of nature. Thus deviating, they suggest and illustrate to him a power higher than nature, a 'personal will;' and they commend the person in whom this power is vested as a messenger from on high. Without these credentials such a messenger would have no right to demand belief, even though his assertions regarding his divine mission were backed by a holy life. Nor is it by miracles alone that the order of nature is, or may be, disturbed. The material universe is also the arena of 'special providences.' Under these two heads Mr. Mozley distributes the total preternatural. One form of the preternatural may shade into the other, as one colour passes into another in the rainbow; but while the line which divides the specially providential from the miraculous cannot be sharply drawn, their distinction broadly expressed is this, that while a special providence can only excite surmise more or less probable, it is 'the nature of a miracle to give proof, as distinguished from mere surmise of divine design.'

Mr. Mozley adduces various illustrations of what he regards to be special providences as distinguished from miracles. 'The death of Arius,' he says, 'was not miraculous, because the coincidence of the death of a heresiarch taking place when it was peculiarly advantageous to the orthodox faith . . . . was not such as to compel the inference of extraordinary Divine agency; but it was a special providence, because it carried a reasonable appearance of it. The miracle of the Thundering Legion was a special providence, but not a miracle, for the same reason, because the coincidence of an instantaneous fall of rain in answer to prayer carried some appearance, but not proof, of preternatural agency.' The eminent lecturer's remarks on this head brought to my recollection certain narratives published in Methodist magazines, which I used to read with avidity when a boy. The title of these chapters, if I remember aright, was ' The Providence of God asserted,' and in them the most extraordinary and exciting escapes from peril were recounted and ascribed to prayer, while equally wonderful instances of calamity were adduced as illustrations of Divine retribution. In such magazines, or elsewhere, I found recorded the case of the celebrated Samuel Hick, which, as it illustrates a whole class of special providences, approaching in conclusiveness to miracles, is worthy of mention here. It is related of this holy man—and I, for one, have no doubt of his holiness—that flour was lacking to make the sacramental bread. Grain was present, and a windmill was present, but there was no wind to grind the corn. With faith, undoubting Samuel Hick prayed to the Lord of the winds : the sails turned, the corn was ground, after which the wind ceased. According to the canon of the Bampton Lecturer, this, though carrying a strong appearance of an immediate exertion of Divine energy, lacks by a hair's-breadth the quality of a miracle. For the wind *might* have arisen, and *might* have ceased, in the ordinary course of nature. Hence the occurrence did not ' compel the inference of extraordinary Divine agency.' In like manner Mr. Mozley considers that ' the appearance of the cross to Constantine was a miracle, or a special providence, according to which account of it we adopt. As only a meteoric appearance in the shape of a cross it gave some token of preternatural agency, but not full evidence.'

In the Catholic canton of Switzerland where I now write, and still more among the pious Tyrolese, the mountains are dotted with shrines, containing offerings of all kinds, in acknowledgment of special mercies—legs, feet, arms, and hands of gold, silver, brass, and wood, according as worldly possessions enabled the grateful heart to express its indebtedness.   Most of these offerings are made to the Virgin Mary.   They are recognitions of ' special providences,' wrought through the instrumentality of the Mother of God.   Mr. Mozley's belief, that of the Methodist chronicler, and that of the Tyrolese peasant, are substantially the same.   Each of them assumes that Nature, instead of flowing ever onward in the uninterrupted rhythm of cause and effect, is mediately ruled by the free human will.   As regards *direct* action upon natural phenomena, man's will is confessedly powerless, but it is the trigger which, by its own free action, liberates the Divine power.   In this sense, and to this extent, man, of course, commands nature.

Did the existence of this belief depend solely upon the material benefits derived from it, it could not, in my opinion, last a decade.   As a purely objective fact we should soon see that the distribution of natural phenomena is unaffected by the merits or the demerits of man ; that the law of gravitation crushes the simple worshippers of Ottery St. Mary, while singing their hymns, just as surely as if they were engaged in a midnight brawl. The hold of this belief upon the human mind is not due to outward verification, but to the inner warmth, force, and elevation with which it is commonly associated. It is plain, however, that these feelings may exist under the most various forms. They are not limited to Church of England Protestantism—they are not even limited to Christianity. Though less refined, they are certainly not less strong, in the heart of the Methodist and the Tyrolese than in the heart of Mr. Mozley. Indeed, those feelings belong to the primal powers of man's nature. A 'sceptic' may have them. They find vent in the battle-cry of the Moslem. They take hue and form in the hunting-grounds of the red Indian; and raise all of them, as they raise the Christian, upon a wave of victory, above the terrors of the grave.

The character, then, of a miracle, as distinguished from a special providence, is that the former furnishes *proof*, while in the case of the latter we have only surmise. Dissolve the element of doubt, and the alleged fact passes from the one class of the preternatural into the other. In other words, if a special providence could be *proved* to be a special providence, it would cease to be a special providence and become a miracle. There is not the least cloudiness about Mr. Mozley's meaning here. A special providence is a doubtful miracle. Why, then, not use the correct phraseology? The term employed conveys no negative suggestion, whereas the negation of certainty is the peculiar characteristic of the thing intended to be expressed. There is an apparent unwillingness on the part of Mr. Mozley to call a special providence what his own definition makes it to be. Instead of speaking of it as a doubtful miracle, he calls it 'an invisible miracle.' He speaks of the point of contact of supernatural power with the chain of causation being so high up as to be wholly, or in part, out of sight, whereas the essence of a special providence is the uncertainty whether there is any contact at all, either high or low. By the use of an incorrect term, however, a grave danger is avoided. For the idea of doubt, if kept systematically before the mind, would soon be fatal to the special providence as a means of edification. The term employed, on the contrary, invites and encourages the trust which is necessary to supplement the evidence.

This inner trust, though at first rejected by Mr. Mozley in favour of external proof, is subsequently called upon to do momentous duty with regard to miracles. Whenever the evidence of the miraculous seems incommensurate with the fact which it has to establish, or rather when the fact is so amazing that hardly any evidence is sufficient to establish it, Mr. Mozley invokes ' the affections.' They must urge the reason to accept the conclusion from which unaided it recoils. The affections and emotions are eminently the court of appeal in matters of real religion, which is an affair of the heart, but they are not, I submit, the court in which to weigh allegations regarding the credibility of physical facts. These must be judged by the dry light of the intellect alone, appeals to the affections being reserved for cases where moral elevation, and not historic conviction, is the aim. It is, moreover, because the result, in the case under consideration, is deemed desirable that the affections are called upon to back it. If undesirable, they would, with equal right, be called upon to act the other way. Even to the disciplined scientific mind this would be a dangerous doctrine. A favourite theory—the desire to establish or avoid a certain result—can so warp the mind as to destroy its power of estimating facts. I have known men to work for years under a fascination of this kind, unable to extricate themselves from its fatal influence. They had certain data, but not, as it happened, enough. By a process exactly analogous to that invoked by Mr. Mozley they supplemented the data, and went wrong. From that hour their intellects were so blinded to the perception of adverse phenomena that they never reached truth. If, then, to the disciplined scientific mind, this incongruous mixture of proof and trust be fraught with danger, what must it be to the indiscriminate audience which Mr. Mozley addresses? In calling upon this agency he acts the part of Frankenstein. It is the monster thus evoked that we see stalking abroad, in the so-called spiritualistic phenomena of the present day. Again, I say, where the aim is to elevate the mind, to quicken the moral sense, to kindle the fire of religion in the soul, let the affections by all means be invoked ; but they must not be permitted to colour our reports, or to influence our acceptance of reports of occurrences in external nature. Testimony as to natural facts is usually worthless when wrapped in this atmosphere of the affections, the most earnest subjective truth being thus rendered perfectly compatible with the most astounding objective error.

There are questions in judging of which the affections or sympathies are often our best guides, the estimation of moral goodness being one of these. But at this precise point, where they are really of use, Mr. Mozley excludes the affections, and demands a miracle as a certificate of character. He will not accept any other evidence of the perfect goodness of Christ. ' No outward life or conduct,' he says, ' however irreproachable, could prove His perfect sinlessness, because goodness depends upon the inward motive, and the perfection of the inward motive is not proved by the outward act.' But surely the miracle is an outward act, and to pass from it to the inner motive imposes a greater strain upon logic than that involved in our ordinary methods of estimating men. There is, at least, moral congruity between the outward goodness and the inner life, but there is no such congruity between the miracle and the life within. The test of moral goodness laid down by Mr. Mozley is not the test of John, who says, ' He that doeth righteousness is righteous ;' nor is it the test of Jesus—' By their fruits ye shall know them ; do men gather grapes of thorns, or figs of thistles ?' But it *is* the test of another : ' If thou be the Son of God, command that these stones be made bread.' For my own part, I prefer the attitude of Fichte to that of Mr. Mozley. ' The Jesus of John,' says this noble and mighty thinker, ' knows no other God than the True God, in whom we all are, and live, and may be blessed, and out of whom there is only Death and Nothingness. And he appeals, and rightly appeals, in support of this truth, not to reasoning, but to the inward practical sense of truth in man, not even knowing any other proof than this inward testimony, " If any man will do the will of Him who sent me, he shall know of the doctrine whether it be of God." '

Accepting Mr. Mozley's test, with which alone I am now dealing, it is evident that, in the demonstration of moral goodness, the *quantity* of the miraculous comes into play. Had Christ, for example, limited himself to the conversion of water into wine, He would have fallen short of the performance of Jannes and Jambres, for it is

E

a smaller thing to convert one liquid into another than to convert a dead rod into a living serpent. But Jannes and Jambres, we are informed, were not good. Hence, if Mr. Mozley's test be a true one, a point must exist, on the one side, of which miraculous power demonstrates goodness, while on the other side it does not. How is this 'point of contrary flexure' to be determined? It must lie somewhere between the magicians and Moses, for within this space the power passed from the diabolical to the Divine. But how to mark the point of passage— how, out of a purely *quantitative* difference in the visible manifestation of power we are to infer a total inversion of quality—it is extremely difficult to see. Moses, we are informed, produced a large reptile, Jannes and Jambres produced a small one. I do not possess the intellectual faculty which would enable me to infer from those data either the goodness of the one or the badness of the other; and in the highest recorded manifestations of the miraculous I am equally at a loss. Let us not play fast and loose with the miraculous; either it is a demonstration of goodness in all cases or in none. If Mr. Mozley accepts Christ's goodness as transcendent, because He did such works as no other man did, he ought, logically speaking, to accept the works of those who, in His name, had cast out devils, as demonstrating a proportionate goodness on their part. But it is people of this class who are consigned to everlasting fire prepared for the devil and his angels. Such zeal as that of Mr. Mozley for miracles tends, I fear, to eat his religion up. The logical threatens to stifle the spiritual. The truly religious soul needs no miraculous proof of the goodness of Christ. The words addressed to Matthew at the receipt of custom required no miracle to produce obedience. It was by no stroke of the supernatural that Jesus caused those sent to seize him to go backward and fall to the ground. It was the sublime and holy effluence from within, which needed no prodigy to commend it to the reverence even of his foes.

As regards the function of miracles in the founding of a religion, Mr. Mozley institutes a comparison between the religion of Christ and that of Mahomet, and he derides the latter as ' irrational ' because it does not profess to adduce miracles in proof of its supernatural origin. But the religion of Mahomet, notwithstanding this drawback, has thriven in the world, and at one time it held sway over larger populations than Christianity itself. The spread and influence of Christianity are, however, brought forward by Mr. Mozley as ' a permanent, enormous, and incalculable practical result ' of Christian miracles ; and he actually makes use of this result to strengthen his plea for the miraculous. His logical warrant for this proceeding is not clear. It is the method of science, when a phenomenon presents itself, to the production of which several elements may contribute, to exclude them one by one, so as to arrive at length at the truly effective cause. Heat, for example, is associated with a phenomenon ; we exclude heat, but the phenomenon remains : hence, heat is not its cause. Magnetism is associated with a phenomenon ; we exclude magnetism, but the phenomenon remains : hence, magnetism is not its cause. Thus, also, when we seek the cause of the diffusion of a religion—whether it be due to miracles, or to the spiritual force of its founders— we exclude the miracles, and, finding the result unchanged, we infer that miracles are not the effective cause. This important experiment Mahometanism has made for us. It has lived and spread without miracles ; and to assert, in the face of this, that Christianity has spread *because* of miracles, is not more opposed to the spirit of science than to the common sense of mankind.

The incongruity of inferring moral goodness from miraculous power has been dwelt upon above ; in another particular also the strain put upon miracles by Mr. Mozley is, I think, more than they can bear. In consistency with his principles, it is difficult to see how he is to draw from the miracles of Christ any certain conclusion as to his Divine nature. He dwells very forcibly on what he calls ' the argument from experience,' in the demolition of which he takes evident delight. He destroys the argument, and repeats it for the mere pleasure of again and again knocking the breath out of it. Experience, he urges, can only deal with the past ; and the moment we attempt to project experience a hair's-breadth beyond the point it has at any moment reached, we are condemned by reason. It appears to me that when he infers from Christ's miracles a Divine and altogether superhuman energy, Mr. Mozley places himself precisely under this condemnation. For what is his logical ground for concluding that the miracles of the New Testament illustrate Divine power ? May they not be the result of expanded human power ? A miracle he defines as something impossible to man. But how does he know that the miracles of the New Testament are impossible to man ? Seek as he may he has absolutely no reason to adduce save this—that man has never hitherto accomplished such things. But does the fact that man *has* never raised the dead prove that he *can* never raise the dead? 'Assuredly not,' must be Mr. Mozley's reply; 'for this would be pushing experience beyond the limit it has now reached—which I pronounce unlawful.' Then a period *may* come when man will be able to raise the dead. If this be conceded—and I do not see how Mr. Mozley can avoid the concession—it destroys the *necessity* of inferring Christ's divinity from his miracles. He, it may be contended, antedated the humanity of the future; as a mighty tidal wave leaves high upon the beach a mark which by-and-by becomes the general level of the ocean. Turn the matter as you will, no other warrant will be found for the all-important conclusion that Christ's miracles demonstrate Divine power, than an argument which has been stigmatised by Mr. Mozley as ' a rope of sand '—the argument from experience.

The learned Bampton Lecturer would be in this position even if he had seen with his own eyes every miracle recorded in the New Testament. But he has *not* seen these miracles; and his intellectual plight is therefore worse. He accepts these miracles on testimony. Why does he believe that testimony? How does he know that it is not delusion; how is he sure that it is not even falsehood? He will answer that the writing bears the mark of sobriety and truth; and that in many cases the bearers of this message to mankind sealed it with their blood. Granted with all my heart; but whence the value of all this? Is it not solely derived from the fact that men, *as we know them*, do not sacrifice their lives in the attestation of that which they know to be untrue? Does not the entire value of the testimony of the apostles depend ultimately upon our experience of human nature?

It appears, therefore, that those who alleged to have seen the miracles based their inferences from what they saw on the argument from experience ; and that Mr. Mozley bases his belief in their testimony on the same argument. The weakness of his conclusion is augmented by this double insertion of a principle of belief to which he flatly denies rationality. His reasoning, in fact, cuts two ways—if it destroys our trust in the order of Nature, it far more effectually abolishes the basis on which Mr. Mozley seeks to found the Christian religion.

Over this argument from experience, which at bottom is *his* argument, Mr. Mozley rides rough-shod. There is a dash of scorn in the energy with which he tramples on it. Probably some previous writer had made too much of it, and thus invited his powerful assault. Finding the difficulty of belief in miracles to arise from their being in contradiction to the order of nature, he sets himself to examine the grounds of our belief in that order. With a vigour of logic rarely equalled, and with a confidence in its conclusions never surpassed, he disposes of this belief in a manner calculated to startle those who, without due examination, had come to the conclusion that the order of nature was secure.

What we mean, he says, by our belief in the order of nature, is the belief that the future will be like the past. There is not, according to Mr. Mozley, the slightest rational basis for this belief.

‘That any cause in nature is more permanent than its existing and known effects, extending further, and about to produce other and more instances besides what it has produced already, we have no evidence. Let us imagine,’ he continues, ‘the occurrence of a particular physical phenomenon for the first time. Upon that single occurrence we should have but the very faintest expectation of another. If it did occur again, once or twice, so far from counting on another occurrence, a cessation would occur as the most natural event to us. But let it continue one hundred times, and we should find no hesitation in inviting persons from a distance to see it; and if it occurred every day for years, its occurrence would be a certainty to us, its cessation a marvel. . . . . What ground of reason can we assign for an expectation that any part of the course of nature will be the next moment what it has been up to this moment, i.e. for our belief in the uniformity of nature? None. No demonstrative reason can be given, for the contrary to the recurrence of a fact of nature is no contradiction. No probable reason can be given, for all probable reasoning respecting the course of nature is founded *upon* this presumption of likeness, and therefore cannot be the foundation of it. No reason can be given for this belief. It is without a reason, It rests upon no rational grounds and can be traced to no rational principle.'

'Everything,' Mr. Mozley, however, adds, 'depends upon this belief, every provision we make for the future, every safeguard and caution we employ against it, all calculation, all adjustment of means to ends supposes this belief; and yet this belief has no more producible reason for it than a speculation of fancy. . . . . It is necessary, all-important for the purposes of life, but solely practical, and possesses no intellectual character. . . . . The proper function,' continues Mr. Mozley, ' of the inductive principle, the argument from experience, the belief in the order of nature—by whatever phrase we designate the same instinct—is to operate as a practical basis for the affairs of life and the carrying on of human society.' To sum up, the belief in the order of nature is general, but it is ' an unintelligent impulse, of which we can give no

rational account.' It is inserted in our constitution solely
to induce us to till our fields, to raise our winter fuel, and
thus to meet the future on the perfectly gratuitous supposition that that future will be like the past.

' Thus step by step,' says Mr. Mozley, with the emphasis
of a man who feels his position to be a strong one, ' has
philosophy loosened the connection of the order of nature
with the ground of reason, befriending in exact proportion
as it has done this the principle of miracles.' For ' this
belief not having itself a foundation in reason, the ground
is gone upon which it could be maintained that miracles,
as opposed to the order of nature, are opposed to reason.'
When we regard this belief in connection with science,
' in which connection it receives a more imposing name,
and is called the inductive principle,' the result is the
same. ' The inductive principle is only this unreasoning
impulse applied to a scientifically ascertained fact. . . . .
Science has led up to the fact, but there it stops, and for
converting this fact into a law, a totally unscientific principle comes into play, the same as that which generalises
the commonest observation of nature.'

The eloquent pleader of the cause of miracles passes
over without a word the results of scientific investigation
as proving anything rational regarding the principles or
methods by which such results have been achieved. Here,
as before, he declines the test, ' By their fruits shall ye
know them.' Perhaps the best way of proceeding will
be to give one or two examples of the mode in which
men of science apply the unintelligent impulse with
which Mr. Mozley credits them, and which shall show by
illustration the surreptitious character of the method by
which they climb from the region of facts to that of laws.

It was known before the sixteenth century that, the end of an open tube being dipped into water, on drawing an air-tight piston up the tube the water follows the piston, and this fact had been turned to account in the construction of the common pump. The effect was explained at the time by the maxim, ' Nature abhors a vacuum.' It was not known that there was any limit to the height to which the water would ascend, until, on one occasion, the gardeners of Florence, while attempting to raise the water a very great elevation, found that the column ceased at a height of thirty-two feet. Beyond this all the skill of the pump-maker could not get it to rise. The fact was brought to the notice of Galileo, and he, soured by a world which had not treated his science over kindly, is said to have twitted the philosophy of the time by remarking that nature evidently abhorred a vacuum only to a height of thirty-two feet. But Galileo did not solve the problem. It was taken up by his pupil Torricelli, who pondered it, and while he did so various thoughts regarding it arose in his mind. It occurred to him that the water might be forced up in the tube by a pressure applied to the surface of the water outside. But where, under the actual circumstances, was such a pressure to be found ? After much reflection, it flashed upon Torricelli that the atmosphere might possibly exert the pressure ; that the impalpable air might possess weight, and that a column of water thirty-two feet high might be of the exact weight necessary to hold the pressure of the atmosphere in equilibrium.

There is much in this process of pondering and its results which it is impossible to analyse. It is by a kind of inspiration that we rise from the wise and sedulous contemplation of facts to the principles on which they depend. The mind is, as it were, a photographic plate, which is gradually cleansed by the effort to think rightly, and which when so cleansed, and not before, receives impressions from the light of truth. This passage from facts to principles is called induction, which in its highest form is inspiration; but, to make it sure, the inward sight must be shown to be in accordance with outward fact. To prove or disprove the induction, we must resort to deduction and experiment.

Torricelli reasoned thus—If a column of water thirty-two feet high holds the pressure of the atmosphere in equilibrium, a shorter column of a heavier liquid ought to do the same. Now, mercury is thirteen times heavier than water; hence, if my induction be correct, the atmosphere ought to be able to sustain only thirty inches of mercury. Here, then, is a deduction which can be immediately submitted to experiment. Torricelli took a glass tube a yard or so in length, closed at one end and open at the other, and filling it with mercury, he stopped the open end with his thumb and inverted it in a basin filled with the liquid metal. One can imagine the feeling with which Torricelli removed his thumb, and the delight he experienced when he found that his thought had forestalled a fact never before revealed to human eyes. The column sank, but ceased to sink at a height of thirty inches, leaving the Torricellian vacuum overhead. From that hour the theory of the pump was established.

The celebrated Pascal followed Torricelli with a still further deduction. He reasoned thus—If the mercurial can be supported by the atmosphere, the higher we ascend in the air the lower the column ought to sink, for the less will be the weight of the air overhead. He ascended the Puy de Dome, carrying with him a barometric column, and found that as he ascended the mountain the column sank, and that as he descended the column rose.

Between the time here referred to and the present, millions of experiments have been made upon this subject. Every village pump is an apparatus for such experiments. In thousands of instances, moreover, pumps have refused to work; but on examination it has infallibly been found that the well was dry, that the pump required priming, or that some other defect in the apparatus accounted for the anomalous action. In every case of the kind the skill of the pump-maker has been found to be the true remedy. In no case has the pressure of the atmosphere ceased; constancy, as regards the lifting of pump-water, has been hitherto the demonstrated rule of nature. So also as regards Pascal's experiment. His experience has been the universal experience ever since. Men have climbed mountains, and gone up in balloons; but no deviation from Pascal's result has ever been observed. Barometers, like pumps, have refused to act; but instead of indicating any suspension of the operations of nature, or any interference on the part of its Author with atmospheric pressure, examination has in every instance fixed the anomaly upon the instruments themselves. It is this welding, then, of rigid logic to verifying fact that Mr. Mozley refers to an 'unreasoning impulse.'

Let us now briefly consider the case of Newton. Before his time men had occupied themselves with the problem of the solar system. Kepler had deduced, from a

vast mass of observations, the general expressions of
planetary motion known as 'Kepler's Laws.' It had
been observed that a magnet attracts iron ; and by one
of those flashes of inspiration which reveal to the human
mind the vast in the minute, the general in the particular,
it occurred to Kepler, that the force by which bodies fall
to the earth might also be an attraction.   Newton pondered all these things.   He had a great power of pondering.   He could look into the darkest subject until it
became entirely luminous.   How this light arises we cannot explain ; but, as a matter of fact, it does arise.   Let
me remark here, that this power of pondering facts is one
with which the ancients could be but imperfectly acquainted.   They found the uncontrolled exercise of the
imagination too pleasant to expend much time in gathering and brooding over facts.   Hence it is that when those
whose education has been derived from the ancients
speak of 'the reason of man,' they are apt to omit from
their conception of reason one of its greatest powers.
Well, Newton slowly marshalled his thoughts, or rather
they came to him while he 'intended his mind,' rising one
after another like a series of intellectual births out of
chaos.   He made this idea of attraction his own.   But to
apply the idea to the solar system, it was necessary to
know the magnitude of the attraction and the law of its
variation with the distance.   His conceptions first of all
passed from the action of the earth as a whole, to that of
its constituent particles, the integration of which composes
the whole.   And persistent thought brought more and
more clearly out the final divination, that every particle
of matter attracts every other particle by a force which
varies inversely as the square of the distance between the

particles. This is Newton's celebrated law of inverse squares. Here we have the flower and outcome of his induction; and how to verify it, or to disprove it, was the next question. The first step of Newton in this direction was to prove, mathematically, that if this law of attraction be the true one; if the earth be constituted of particles which obey this law; then the action of a sphere equal to the earth in size on a body outside of it, is the same as that which would be exerted if the whole mass of the sphere were contracted to a point at its centre. Practically speaking, then, the centre of the earth is the point from which distances must be measured to bodies attracted by the earth. This was the first-fruit of his deduction.

From experiments executed before his time, Newton knew the amount of the earth's attraction at the earth's surface, or at a distance of 4,000 miles from its centre. His object now was to measure the attraction at a greater distance, and thus to determine the law of its diminution. But how was he to find a body at a sufficient distance? He had no balloon, and even if he had, he knew that any height which he could attain would be too small to enable him to solve his problem. What did he do? He fixed his thoughts upon the moon;—a body at a distance of 240,000 miles, or sixty times the earth's radius from the earth's centre. He virtually weighed the moon, and found that weight to be $\frac{1}{3600}$th of what it would be at the earth's surface. This is exactly what his theory required. I will not dwell here upon the pause of Newton after his first calculations, or speak of his self-denial in withholding them, because they did not quite agree with the observations then at his command. Newton's action in this matter is the normal action of the scientific mind. If it were otherwise—if scientific men were not accustomed to demand verification—if they were satisfied with the imperfect while the perfect is attainable, their science, instead of being, as it is, a fortress of adamant, would be a house of clay, ill-fitted to bear the buffetings of the theologic storms to which it has been from time to time, and is at present exposed.

Thus we see that Newton, like Torricelli, first pondered his facts, illuminated them with persistent thought, and finally divined the character of the force of gravitation. But having thus travelled inward to the principle, he had to reverse his steps, carry the principle outward, and justify it by demonstrating its fitness to external nature. This he did by determining the attraction of the earth and moon.

And here, in passing, I would notice a point which is well worthy of attention. Kepler had deduced his laws from observation. As far back as those observations extended, the planetary motions had obeyed these laws; and neither Kepler nor Newton entertained a doubt as to their continuing to obey them. Year after year, as the ages rolled, they believed that those laws would continue to illustrate themselves in the heavens. But this was not sufficient. The scientific mind can find no repose in the mere registration of sequence in nature. The further question intrudes itself with resistless might: whence comes the sequence? What is it that binds the consequent with its antecedent in nature? The truly scientific intellect never can attain rest until it reaches the *forces* by which the observed succession is produced. It was thus with Torricelli; it was thus with Newton ; it is thus pre-eminently with the real scientific man of to-day. In common with the most ignorant, he shares the belief that spring will succeed winter, that summer will succeed spring, that autumn will succeed summer, and that winter will succeed autumn. But he knows still further—and this knowledge is essential to his intellectual repose—that this succession, besides being permanent, is, under the circumstances, *necessary* ; that the gravitating force exerted between the sun, and a revolving sphere with an axis inclined to the plane of its orbit, *must* produce the observed succession of the seasons. Not until this relation between forces and phenomena has been established is the law of reason rendered concentric with the law of nature, and not until this is effected does the mind of the scientific philosopher rest in peace.

The expectation of likeness, then, in the procession of phenomena is not that on which the scientific mind founds its belief in the order of nature. If the force be *permanent* the phenomena are *necessary*, whether they resemble or do not resemble anything that has gone before. Hence, in judging of the order of nature, our enquiries eventually relate to the permanence of force. From Galileo to Newton, from Newton to our own time, eager eyes have been scanning the heavens, and clear heads have been pondering the phenomena of the solar system. The same eyes and minds have been also observing, experimenting, and reflecting on the action of gravity at the surface of the earth. Nothing has occurred to indicate that the operation of the law has for a moment been suspended ; nothing has ever intimated that nature has been crossed by spontaneous action, or that a state of things at any time existed which could not be rigorously deduced from the preceding state. Given the distribution of matter and the forces in operation in the time of Galileo, the competent mathematician of that day could predict what is now occurring in our own. We calculate eclipses before they have occurred and find them true to the second. We determine the dates of those that have occurred in the early times of history, and find calculation and history at peace. Anomalies and perturbations in the planets have been over and over again observed, but these, instead of demonstrating any inconstancy on the part of natural law, have invariably been reduced to consequences of that law. Instead of referring the perturbations of Uranus to any interference on the part of the Author of nature with the law of gravitation, the question which the astronomer proposed to himself was, 'how, in accordance with this law, can the perturbation be produced?' Guided by a principle, he was enabled to fix the point of space in which, if a mass of matter were placed, the observed perturbations would follow. We know the result. The practical astronomer turned his telescope towards the region which the intellect of the theoretic astronomer had already explored, and the planet now named Neptune was found in its predicted place. A very respectable outcome, it will be admitted, of an impulse which 'rests upon no rational grounds, and can be traced to no rational principle;' which possesses 'no intellectual character;' which 'philosophy' has uprooted from 'the ground of reason,' and fixed in that 'large irrational department' discovered for it, by Mr. Mozley, in the hitherto unexplored wildernesses of the human mind.

The proper function of the inductive principle, or the belief in the order of nature, says Mr. Mozley, is 'to act as a practical basis for the affairs of life, and the carrying on of human society.' But what, it may be asked, has the planet Neptune, or the belts of Jupiter, or the whiteness about the poles of Mars, to do with the affairs of society? How is society affected by the fact that the sun's atmosphere contains sodium, or that the nebula of Orion contains hydrogen gas? Nineteen-twentieths of the force employed in the exercise of the inductive principle, which, reiterates Mr. Mozley, is 'purely practical,' have been expended upon subjects as unpractical as these. What practical interest has society in the fact that the spots on the sun have a decennial period, and that when a magnet is closely watched for half a century, it is found to perform small motions which synchronise with the appearance and disappearance of the solar spots? And yet, I doubt not, Sir Edward Sabine would deem a life of intellectual toil amply rewarded by being priveleged to solve, at its close, these infinitesimal motions.

The inductive principle is founded in man's desire to know—a desire arising from his position among phenomena which are reducible to order by his intellect. The material universe is the complement of the intellect, and without the study of its laws reason would never have awoke to its higher forms of self-consciousness at all. It is the non-ego, through and by which the ego is endowed with self-discernment. We hold it to be an exercise of reason to explore the meaning of a universe to which we stand in this relation, and the work we have accomplished is the proper commentary on the methods we

F

have pursued. Before these methods were adopted the unbridled imagination roamed through Nature, putting in the place of law the figments of superstitious dread. For thousands of years witchcraft, and magic, and miracles, and special providences, and Mr. Mozley's 'distinctive reason of man,' had the world to themselves. They made worse than nothing of it—*worse*, I say, because they let and hindered those who might have made something of it. Hence it is that during a single lifetime of this era of 'unintelligent impulse,' the progress in natural knowledge is all but infinite as compared with that of the ages which preceded ours.

The believers in magic and miracles of a couple of centuries ago had all the strength of Mr. Mozley's present logic on their side. They had done for themselves what he rejoices in having so effectually done for us— cleared the ground of the belief in the order of nature, and declared magic, miracles, and witchcraft to be matters for ordinary evidence to decide. 'The principle of miracles' thus 'befriended' had free scope, and we know the result. Lacking that rock-barrier of natural knowledge which we, laymen of England, now possess, keen jurists and cultivated men were hurried on to deeds, the bare recital of which makes the blood run cold. Skilled in all the rules of human evidence, and versed in all the arts of cross-examination, these men, nevertheless, went systematically astray, and committed the deadliest wrongs against humanity. And why? Because they could not put nature into the witness-box, and question her; of her voiceless 'testimony' they knew nothing. In all cases between man and man, their judgment was to be relied on; but in all cases between man and nature they were blind leaders of the blind.[1]

Mr. Mozley concedes that it would be no great result for miracles to be accepted by the ignorant and super-stitious, ' because it is easy to satisfy those who do not enquire.' But he does consider it ' a great result' that they have been accepted by the *educated.* In what sense educated? Like those statesmen, jurists, and church dignitaries whose education was unable to save them from the frightful errors glanced at above? Not even in this sense; for the great mass of Mr. Mozley's educated people had no legal training, and must have been abso-lutely defenceless against delusions which could set even that training at nought. Like nine-tenths of our clergy at the present day, they were versed in the literature of Greece, Rome, and Judea; but as regards a knowledge of nature, which is here the one thing needful, they were ' noble savages,' and nothing more. In the case of miracles, then, it behoves us to understand the weight of the negative, before we assign a value to the positive; to comprehend the protest of nature before we attempt to measure, with it, the assertions of men. We have only to open our eyes to see what honest, and even intel-lectual, men and women are capable of in the way of evidence in this nineteenth century of the Christian era, and in latitude fifty-two degrees north. The experience

[1] ' In 1664 two women were hung in Suffolk, under a sentence of Sir Matthew Hale, who took the opportunity of declaring that the reality of witchcraft was unquestionable; " for first, the Scriptures had affirmed so much; and secondly, the wisdom of all nations had provided laws against such persons, which is an argument of their confidence of such a crime." Sir Thomas Browne, who was a great physician as well as a great writer, was called as a witness, and swore " that he was clearly of opinion that the persons were bewitched." '—Lecky's *History of Rationalism,* vol. i. p. 120.

thus gained ought, I imagine, to influence our opinion regarding the testimony of people inhabiting a sunnier clime, with a richer imagination, and without a particle of that restraint which the discoveries of physical science have imposed upon mankind.

Having thus submitted Mr. Mozley's views to the examination which they challenged at the hands of a student of the order of nature, I am unwilling to quit his book without expressing my high admiration and respect for his ability.   His failure, as I consider it to be, must, I think, await all attempts, however able, to deal with the material universe by logic and imagination, unaided by experiment and observation.   With regard to the style of the book, I willingly subscribe to the description with which the ' Times ' winds up its able and appreciative review.   ' It is marked throughout with the most serious and earnest conviction, but is without a single word from first to last of asperity or insinuation against opponents, and this not from any deficiency of feeling as to the importance of the issue, but from a deliberate and resolutely maintained self-control, and from an over-ruling, ever-present sense of the duty, on themes like these, of a more than judicial calmness.' [1]

[To the argument regarding the quantity of the miraculous, introduced at page 49, Mr. Mozley has done me the honour of publishing a Reply in the seventh volume of the ' Contemporary Review.'—J. T. 1871.]

[1] See Appendix at the end of the book.

IV.

MATTER AND FORCE.

A LECTURE TO THE WORKING MEN OF DUNDEE

5th September, 1867.

' Heard are the voices,
 Heard are the sages,
 The worlds and the ages,
 " Choose well, your choice is
 Brief and yet endless.

' " Here eyes do regard you
 In eternity's stillness ;
 Here is all fulness
 Ye brave to reward you,
 Work and despair not." '
                      GOETHE.

IV.

MATTER AND FORCE.

It is the custom of the Professors in the Royal School of Mines in London to give courses of evening lectures every year to working men. Each course is duly advertised, and at a certain hour the working men assemble to purchase tickets for the course. The lecture-room holds 600 people; and tickets to this amount are disposed of as quickly as they can be handed to those who apply for them. So desirous are the working men of London to attend these lectures, that the persons who fail to obtain tickets always bear a large proportion to those who succeed. Indeed, if the lecture-room could hold 2,000 instead of 600, I do not doubt that every one of its benches would be occupied on these occasions. It is, moreover, worthy of remark that the lectures are but rarely of a character which could help the working man in his daily pursuits. The knowledge acquired is hardly ever of a nature which admits of being turned into money. It is a pure desire for knowledge, as a thing good in itself, and without regard to its practical application, which animates these men. They wish to know more of the wonderful universe around them; their minds desire this knowledge as naturally as their bodies desire food and drink, and to satisfy this intellectual want they come to the School of Mines.

It is also my privilege to lecture to another audience
in London, composed in part of the aristocracy of rank,
while the audience just referred to is composed wholly
of the aristocracy of labour.   As regards attention and
courtesy to the lecturer, neither of these audiences has
anything to learn of the other ; neither can claim superiority over the other.   I do not, however, think that
it would be quite correct to take those persons who flock
to the School of Mines as average samples of their
class ; they are probably picked men—the aristocracy of
labour, as I have just called them.   At all events, their
conduct demonstrates that the essential qualities of a
gentleman are confined to no class, and they have often
raised in my mind the wish that the gentlemen of all
classes, artisans as well as lords, could, by some process
of selection, be sifted from the general mass of the com-
munity, and caused to know each other better.

When pressed some months ago by the Council of the
British Association to give an evening lecture to the working men of Dundee, my experience of the working men of
London naturally rose to my mind ; and, though heavily
weighted with other duties, I could not bring myself to
decline the request of the Council.   Hitherto, the evening
discourses of the Association have been delivered before
its members and associates alone.   But after the meeting
at Nottingham, last year, where the working men, at
their own request, were addressed by our late President, Mr. Grove, and by my excellent friend, Professor
Huxley, the idea arose of incorporating with all subsequent meetings of the Association an address to the
working men of the town in which the meeting is held.
A resolution to that effect was sent to the Committee of

Recommendations; the Committee supported the resolution; the Council of the Association ratified the decision of the Committee; and here I am to carry out to the best of my ability their united wishes.

Whether it be a consequence of long-continued development, or an endowment conferred once for all on man at his creation, we find him here gifted with a mind, curious to know the causes of things, and surrounded by objects which excite its questionings, and raise the desire for an explanation. It is related of a young Prince of one of the Pacific Islands, that when he first saw himself in a looking-glass, he ran round the glass to see who was standing at the back. And thus it is with the general human intellect, as regards the phenomena of the external world. It wishes to get behind and learn the causes and connections of these phenomena. What is the sun, what is the earth, what should we see if we came to the edge of the earth and looked over? What is the meaning of thunder and lightning, of hail, rain, storm, and snow? Such questions presented themselves to early men, and by and by it was discovered that this desire for knowledge was not implanted in vain. After many trials it became evident the man's capacities were, so to speak, the complement of nature's facts, and that, within certain limits, the secret of the universe was open to the human understanding. It was found that the mind of man had the power of penetrating far beyond the boundaries of his five senses; that the things which are seen in the material world depend for their action upon things unseen; in short, that besides the phenomena which address the senses, there are laws and principles and processes which do not address the senses at all, but which must be, and can be, spiritually discerned.

There are two things which form, so to say, the substance of all scientific thought. The entire play of the scientific intellect is confined to the combination and resolution of the ideas of *matter* and *force*. Newton, it is said, saw an apple fall. To the common mind this presented no difficulty and excited no question. Not so with Newton. He observed the fact; but one side of his great intellectual nature was left unsatisfied by the mere act of observation. He sought after the principle which ruled the fact. Whether this anecdote be true or not, it illustrates how the ordinary operations of nature, which most people take for granted as perfectly plain and simple, are often those which most puzzle the scientific man. To the conception of the matter of the apple, Newton added that of the force that moved it. The falling of the apple was due to an attraction exerted mutually between it and the earth. He applied the idea of this force to suns and planets and moons, and showed that all their motions were necessary consequences of this attraction.

Newton, you know, was preceded by a grand fellow named John Kepler—a true working man—who, by analysing the astronomical observations of his master, Tycho Brahe, had actually found that the planets moved as they are now known to move. As a matter of fact, Kepler knew as much about the motion of the planets as Newton did; in fact, Kepler taught Newton and the world generally the facts of planetary motion. But this was not enough. The question arose—Why should the facts be so? This was the great question for Newton, and it was the solution of this question which renders his name and fame immortal. He proved that the planetary motions were what observation made them to be, because every particle of matter in the solar system attracts every other particle by a force which varies as the inverse square of the distance between the particles. He showed that the moon fell towards the earth, and that the planets fell towards the sun, through the operation of the same force that pulls an apple from its tree. This all pervading force, which forms the solder of the material universe, and the conception of which was necessary to Newton's intellectual peace, is called the force of gravitation.

All force may be ultimately reduced to a push or a pull in a straight line ; but its manifestations are various, and sometimes so complex as entirely to disguise its elementary constituents. Its different manifestations have received different names. Here, for example, is a magnet freely suspended. I bring the end of a second magnet near one of the ends of the suspended one—attraction is the consequence. I reverse the position of one of the magnets—repulsion follows. This display of power is called magnetic force. In the case of gravitation we have a simple attraction, in the case of magnetism attraction and repulsion always go together. Thus magnetism is a double force, or, as it is usually called, a polar force. I present a bit of common iron to the magnet, the iron itself becomes a temporary magnet, and it now possesses the power of attracting other iron. And if several pieces of iron be presented at the same time, not only will the magnet act on them, but they will also act upon each other.

This leads me to an experiment which will give you some idea of how bodies arrange themselves under the operation of a polar force. Underneath this plate of glass is placed a small magnet, and by an optical arrangement comprising a powerful lamp, a magnified image of the magnet is now cast upon the screen before you. I scatter iron filings over the glass. You already notice a certain arrangement of the particles of iron. Their free action is, however, hampered by friction. I therefore tap the glass, liberate the particles, which, as I tap, arrange themselves in these beautiful curves. This experiment is intended to make clear to you how a definite arrangement of particles—a kind of incipient structure—may result from the operation of a polar force. We shall by-and-bye see far more wonderful exhibitions of the same structural action when we come to deal with the force of crystallisation.

The magnetic force has here acted upon particles of matter visible to the eye. But, as already stated, there are numerous processes in nature which entirely elude the eye of the body, and must be figured by the eye of the mind. The processes of chemistry are examples of these. Long thinking and experimenting on the materials which compose our world has led philosophers to conclude that matter is composed of atoms from which, whether separate or in combination, the whole material world is built up. The air we breathe, for example, is mainly a mixture of the atoms of two distinct substances, called oxygen and nitrogen. The water we drink is also composed of two distinct substances, called oxygen and hydrogen. But it differs from the air in this particular, that in water the oxygen and hydrogen are not *mechanically* mixed, but *chemically* combined. In fact, the atoms of oxygen and those of hydrogen exert enormous attractions on each other, so that when brought into sufficient proximity they rush together with an almost incredible force to form a chemical compound. But powerful as is the force with which these atoms lock themselves together, we have the means of tearing them asunder, and the agent by which we accomplish this may here receive a few moments' attention.

Into a vessel containing acidulated water I dip these two strips of metal, the one being zinc and the other platinum, not permitting them to touch each other in the liquid. I now connect the two upper ends of the strips by a piece of copper wire. The wire is apparently unchanged, but it is not so in reality. It is now the channel of what, for want of a better name, we call an electric current—a power generated and maintained by the chemical action going on in the vessel of acidulated water. What the inner change of the wire is we do not know, but we do know that a change has occurred, by the external effects produced by the wire. Let me show you one or two of these effects. And here it is convenient to operate with greater power than can be obtained from a single small pair of strips of metal, and a single vessel of acidulated water. Before you is a series of ten vessels, each with its pair of metals, and I wish to get the added force of all ten. This arrangement is called a voltaic battery. I take a piece of copper wire in my hand, and plunge it among these iron filings ; they refuse to cling to it ; the wire has no power over the filings. I now employ the self-same wire to connect the two ends of the battery, and subject it to the same test. The iron filings now crowd round the wire and cling to it. This is one of the effects of the electric current now traversing the wire. I interrupt the current, and the filings immediately fall ; the power of attraction continues only so long as the wire connects the two ends of the battery.

Here is a piece of similar wire, overspun with cotton, to prevent the contact of its various parts. It is formed into a coil, which at present has no power over these iron nails ; but I now make the coil part of the wire which connects the two ends of the voltaic battery. No visible change has occurred in the coil, but it is no longer what it was. By the attractive force with which it has become suddenly endowed, it now empties this tool-box of its nails. I twist a covered copper wire round this common poker. At present the poker is powerless over these iron nails ; but when we connect with the wire surrounding the poker the two ends of the voltaic battery, the poker is instantly transformed into a strong magnet. Here, again, are two flat spirals suspended facing each other. They are about six inches apart. By turning this handle in a certain direction a current is sent through both spirals. When this is done they clash suddenly together, being drawn together by their mutual attraction. By turning the handle in another direction, I reverse what is called the direction of the current in one of the spirals, and now they fly asunder, being driven apart by their mutual repulsion. All these effects are due to the power which we name an electric current, and which we figure as flowing through the wire when the voltaic circuit is complete.

I have said that no visible change occurs in the wire when the current passes through it. Still a change over and above what you have seen really does take place. Lay of those spirals, and you will find them warm. Let me exalt this warmth so as to render it visible to you. In front of the table is a thin platinum wire six feet long. On sending a current from a battery of fifty pairs of plates through this wire it glows, as you see, vividly red. I shorten the wire; more electricity now flows through it, and its light becomes more intense. It is now bright yellow; and now it is a dazzling white. This light is so strong that, though the wire is not much thicker than a bristle, it appears to those on the nearest benches as thick as a quill; while to those at a distance it appears as thick as a man's finger. This effect, which we call irradiation, is always produced by a very strong light. It is this same electric current that furnished us with the powerful light employed in one of our first experiments. The lamp then made use of is provided with these coke rods; and when the electric current passes between them we obtain a light almost as brilliant as that of the sun.

And now let us return to the point at which the electric current was introduced—the point, namely, where the tearing asunder of the locked atoms of a chemical compound was spoken of. The agent by which we effect this is also the electric current; and I hope to make its action visible to you all. Into this small cell, containing water, dip two thin wires. By means of a solar microscope and the powerful light of our electric lamp, a magnified image of this cell is thrown upon the screen before you. You see plainly the images of the wires. And now I send from a second small battery which rests upon this table an electric current from wire to wire. Bubbles of gas rise immediately from each of them, and these are the two gases of which the water is composed. The oxygen is always liberated on the one wire, the hydrogen on the other. The two gases may be collected separately; in fact, they have been thus collected in these jars. A lighted taper placed in one jar inflames the gas, which proves it to be hydrogen; a burning ember of wood placed in the other jar instantly bursts into vivid combustion, which proves the gas in the jar to be oxygen. I place upon my hand a soap bubble filled with a mixture of both gases in the exact proportions in which they exist in water. Applying a taper to the bubble, a loud explosion is heard. The gases have rushed together with detonation, but without injury to my hand, and the water from which they were extracted is the result of the re-union.

I wish you to see with the utmost possible clearness what has here taken place. First, then, you are to remember that to form water the proportions by weight of oxygen and hydrogen are as eight to one. Eight ounces of oxygen, for example, unite with one of hydrogen to form nine ounces of water. But if, instead of comparing weights, we compare volumes, two volumes of hydrogen unite with one of oxygen to form water. Now, these volumes, and not the weights, express the proportions in which the atoms of hydrogen unite with those of oxygen. In the act of combination two atoms of hydrogen combine with one of oxygen to form what we call the *molecule* of water. Every such molecule is a group of three atoms, two of which are hydrogen and one oxygen.

One consequence of the rushing together of the atoms is the development of heat. What is this heat? How are we to figure it before our minds? I do not despair of being able to give you a tolerably distinct answer to this question. Here are two ivory balls suspended from the same point of support by two short strings. I draw them thus apart and then liberate them. They clash together, but, by virtue of their elasticity, they quickly recoil from each other, and a sharp vibratory rattle succeeds their collision. This experiment will enable you to figure to your mind a pair of clashing atoms. We have, in the first place, a motion of the one atom towards the other—a motion of translation, as it is usually called. But when the atoms come sufficiently near each other, elastic repulsion sets in, the motion of translation is stopped and converted into a motion of vibration. To this vibratory motion we give the name of heat. Thus, three things are to be kept before the mind—first, the atoms themselves; secondly, the force with which they attract each other; and thirdly, the motion consequent upon the exertion of that force. This motion must be figured first as a motion of translation, and then as a motion of vibration; and it is not until the motion reaches the vibratory stage that we give it the name of heat. It is this motion imparted to the nerves that produces the sensation of heat.

It would be useless to attempt a more detailed description of this molecular motion. After the atoms have been thrown into this state of agitation, very complicated motions must ensue from their incessant collision. There must be a wild whirling about among the molecules. For some time after the act of combination this action is so violent as to prevent the molecules from coming together. The water is maintained for a time in a state of vapour. But as the vapour cools, or in other words loses its motion, the water molecules coalesce to form a liquid. And now we are approaching a new and wonderful display of force.　No one who had only seen water in its vaporous or liquid form could imagine the existence of the forces now to be referred to; for as long as the substance remains in a liquid or vaporous condition, the play of these forces is altogether masked and hidden. But let the heat be gradually withdrawn, the antagonist to their union being removed, the molecules prepare for new arrangements and combinations.　Like the particles of iron in our magnetic experiment, the water molecules are endowed with attractive and repulsive poles, and they arrange themselves together in accordance with these attractions and repulsions.　Solid crystals of water are thus formed, to which we give the familiar name of ice.　To the eye of science these ice crystals are as precious as the diamond — as purely formed, as delicately built.　Where no disturbing causes intervene, there is no disorder in this crystalline architecture.　By their own constructive power molecule builds itself on to molecule with a precision far greater than that attainable by the hands of man.　We are apt to overlook the wonderful when it becomes common.　Imagine the bricks and stones of this town of Dundee endowed with locomotive power.　Imagine them attracting and repelling each other, and arranging themselves in consequence of these attractions and repulsions to form streets and houses and Kinnaird Halls; would not that be wonderful?　Hardly less wonderful is the play of force by which the molecules of water build themselves into the sheets of crystal which every winter roof your ponds and lakes.

　If I could show you the actual progress of this molecular architecture, its beauty would delight and astonish you. A reversal of the process may be actually shown. The molecules of a piece of ice may be taken asunder before your eyes, and from the manner in which they separate, you may to some extent infer the manner in which they aggregate. When a beam is sent from our electric lamp through a plate of glass, a portion of the beam is intercepted, and the glass is warmed by the portion thus retained within it. When the beam is sent through a plate of ice, a portion of the beam is also absorbed; but instead of warming the ice, the intercepted heat melts it internally. It is to the delicate silent action of this beam within the ice that I now wish to direct your attention. Upon the screen is thrown a magnified image of the slab of ice : the light of the beam passes freely through the ice without melting it, and enables us to form the image, but the heat of the beam is in great part intercepted by the ice, and that heat now applies itself to the work of internal liquefaction. Observe those stars breaking out over the white surface, and expanding in size as the action of the beam continues. These stars are liquefied ice, and each of them, you observe, has six rays. They still more closely resemble flowers, each of six petals. Under the action of the heat the molecules of the ice fall asunder, so as to leave behind them these exquisite forms. We have here the process of crystallisation reversed. In this fashion, and in strict accordance with this hexangular type, every ice molecule takes its place upon our ponds and lakes during the frosts of winter. To use the language of an American poet, ' the atoms march in tune,' moving to the music of law, which thus renders the commonest substance in nature a miracle of beauty.

It is the function of science, not as some think to divest this universe of its wonder and its mystery, but, as in the case here before us, to point out the wonder and the mystery of common things. Those fern-like forms, which on a frosty morning overspread your window panes, illustrate the action of the same force. Breathe upon such a pane before the fires are lighted, and reduce the solid crystalline film to the liquid condition; then watch its subsequent appearance. You will see it all the better if you look at it through a common magnifying glass. After you have ceased breathing, the film, abandoned to the action of its own forces, appears for a moment to be alive. Lines of motion run through it; molecule closes with molecule, until finally the whole film passes from the state of liquidity, through this state of motion, to its final crystalline repose.

I can show you something similar. Over a piece of perfectly clean glass I pour a little water in which a crystal has been dissolved. A film of the solution clings to the glass, and this film will now be caused to crystallise before your eyes. By means of a microscope and a lamp, an image of the plate of glass is thrown upon the screen. The beam of the lamp, besides illuminating the glass, also heats it; evaporation sets in, and, at a certain moment, when the solution has become supersaturated, splendid branches of crystals shoot out over the screen. A dozen square feet of surface are now covered by those beautiful forms. With another solution we obtain crystalline spears, feathered right and left by other spears. From distant nuclei in the middle of the field of view the spears shoot with magical rapidity in all directions. The film of water on a window pane on a frosty morning exhibits effects quite as wonderful as these. Latent in this formless solution, latent in every drop of water, lies this marvellous structural power, which only requires the withdrawal of opposing forces to bring it into action.

Our next experiment on crystallisation you will probably consider more startling even than these. The clear liquid now held up before you is a solution of nitrate of silver—a compound of silver and nitric acid. When an electric current is sent through this liquid the silver is severed from the acid, as the hydrogen was separated from the oxygen in a former experiment; and I would ask you to observe how the metal behaves when its molecules are thus successively set free. The image of the cell, and of the two wires which dip into the liquid of the cell, are now clearly shown upon the screen. Let us close the circuit, and send the current through the liquid. From one of the wires a beautiful silver tree commences immediately to sprout. Branches of the metal are thrown out, and umbrageous foliage loads the branches. You have here a growth apparently as wonderful as that of any vegetable perfected in a minute before your eyes. Substituting for the nitrate of silver acetate of lead, which is a compound of lead and acetic acid, the electric current severs the lead from the acid, and there you see the metal slowly branching into these exquisite metallic ferns, the fronds of which, as they become too heavy, break from their roots and fall to the bottom of the cell.

These experiments show that the common matter of our earth—'brute matter,' as Dr. Young pleases to call it— when its atoms and molecules are permitted to bring their forces into free play, arranges itself, under the operation of these forces, into forms which rival in beauty those of the vegetable world. And what is the vegetable world itself but the result of the complex play of these molecular forces? Here, as elsewhere throughout nature, if matter moves it is force that moves it, and if a certain structure, vegetable or mineral, is produced, it is through the operation of the forces exerted between the atoms and molecules. These atoms and molecules resemble little magnets with mutually attractive and mutually repellent poles. The attracting poles unite, the repellent poles retreat, and vegetable, as well as mineral, forms are the final expression of this complicated molecular action.

In the formation of our lead and silver trees, we needed an agent to wrest the lead and the silver from the acids with which they were combined. A similar agent is required in the vegetable world. The solid matter of which our lead and silver trees were formed was, in the first instance, disguised in a transparent liquid ; the solid matter of which our woods and forests are composed is also, for the most part, disguised in a transparent gas, which is mixed in small quantities with the air of our atmosphere. This gas is formed by the union of carbon and oxygen, and is called carbonic acid gas. Two atoms of oxygen and one of carbon unite to form the molecule of carbonic acid which, as I have said, is the material from which wood and vegetable tissues are mainly derived. The carbonic acid of the air being subjected to an action somewhat analogous to that of the electric current in the case of our lead and silver solutions, has its carbon liberated and deposited as woody fibre. The watery vapour of the air is subjected to similar action ; its hydrogen is liberated from its oxygen, and lies down side by side with the carbon in the tissues of the tree.

The oxygen in both cases is permitted to wander away into the atmosphere. But what is it which thus tears the carbon and the hydrogen from the strong embrace of the oxygen? What is it in nature that plays the part of the electric current in our experiments? The rays of the sun. The leaves of the plants absorb both the carbonic acid and the aqueous vapour of the air; these leaves answer to the cells in which our decompositions by the electric current took place. In the leaves the solar rays decompose both the carbonic acid and the water, permitting the oxygen in both cases to escape into the air, and allowing the carbon and the hydrogen to follow the bent of their own forces. And just as the molecular attractions of the silver and the lead found expression in the production of those beautiful branching forms seen in our experiments, so do the molecular attractions of the liberated carbon and hydrogen find expression in the architecture of grasses, plants, and trees.

In the fall of a cataract and the rush of the wind we have examples of mechanical power. In the combinations of chemistry and in the formation of crystals and vegetables we have examples of molecular power. But before proceeding further I should like to make clear to you the present condition of the surface of our globe with reference to power generally. You have learned how the atoms of oxygen and hydrogen rush together to form water. I have not thought it necessary to dwell upon the mighty mechanical energy of their act of combination, but, in passing, I would say that the clashing together of 1 lb. of hydrogen and 8 lbs. of oxygen to form 9 lbs. of aqueous vapour, is greater than the clash of a weight of 1,000 tons falling from a height of 20 feet against the earth. Now, in order that the atoms of oxygen and hydrogen should rise by their mutual attractions to the velocity corresponding to this enormous mechanical effect, a certain distance must exist between the particles. It is in rushing over this that the velocity is attained.

This idea of distance between the attracting atoms is of the highest importance in our conception of the system of the world. For the world may be divided into two kinds of matter; or rather the matter of the world may be classified under two distinct heads—namely, of atoms and molecules which have already rushed together and thus satisfied their mutual attractions, and of atoms and molecules which have not yet rushed together, and whose mutual attractions are, therefore, as yet unsatisfied. Now, as regards motive power, the working of machinery, or the performance of mechanical work generally, by means of the materials of the earth's crust, we are entirely dependent on those atoms and molecules whose attractions are as yet unsatisfied. Those attractions can produce motion, because sufficient distance intervenes between the attracting molecules, and it is this molecular motion that we utilise in our machines. Thus we can get power out of oxygen and hydrogen by the act of their union, but once they are combined, and once the motion consequent on their combination has been expended, no further power can be got out of the mutual attraction of oxygen and hydrogen. As dynamic agents they are dead. If we examine the materials of which the earth's crust is composed, we find them to consist for the most part of substances whose atoms have already closed in chemical union—whose mutual attractions are satisfied. Granite, for instance, is a widely-diffused substance, but granite

consists, in great part, of silicon, oxygen, potassium, calcium, and aluminum, the atoms of which substances met
long ago in chemical combination, and are therefore
dead. Limestone is also a widely-diffused substance. It
is composed of carbon, oxygen, and a metal called calcium. But the atoms of those substances closed long ago
in chemical union, and are therefore dead. And in this
way we might go over the whole of the materials of the
earth's crust, and satisfy ourselves that though they were
sources of power in ages past, and long before any being
appeared on the surface of the earth capable of turning
their power to account, they are sources of power no
longer. And here we might halt for a moment to remark
on that tendency, so prevalent in the world, to regard
everything as made for human use. Those who entertain
this notion hold, I think, an overweening opinion of their
own importance in the system of nature. Flowers bloomed
before men saw them, and the quantity of power wasted
before man could utilise it is all but infinite compared
with what now remains to be applied. The healthy attitude of mind with reference to this subject is that of
the poet, who, when asked whence came the rhodora,
replied—

> Why thou wert there, O rival of the rose !
> I never thought to ask, I never knew,
> But in my simple ignorance supposed
> The self-same power that brought me there brought you.[1]

A few exceptions to this general state of union of the
particles of the earth's crust—all-important to us, but
trivial in comparison to the total store of which they are
the residue—still remain. They constitute our main

[1] Emerson.

sources of motive power. By far the most important of these are our beds of coal, composed chiefly of carbon, which has not yet closed in chemical union with oxygen. Distance still intervenes between the atoms of carbon and those of oxygen, across which the atoms may be impelled by their mutual attractions ; and we can do nothing more than utilise the motion produced by this attraction. Once the carbon and the oxygen have rushed together, so as to form carbonic acid, their mutual attractions are satisfied, and, while they continue in this condition, as dynamic agents they are dead. A pound of coal produces by its combination with oxygen an amount of heat which, if mechanically applied, would raise a weight of 100 lbs. to a height of twenty miles above the earth's surface. Conversely, 100 lbs. falling from a height of twenty miles, and striking against the earth, would generate an amount of heat equal to that developed by the combustion of a pound of coal. Wherever work is done by heat, heat disappears. A gun which fires a ball is less heated than one which fires blank cartridge. The quantity of heat communicated to the boiler of a working steam-engine is greater than that which could be obtained from the re-condensation of the steam after it had done its work ; and the amount of work performed is the exact equivalent of the amount of heat missing. We dig annually nearly 100 millions of tons of coal from our pits. The amount of mechanical force represented by this quantity of coal seems perfectly fabulous. The combustion of a single pound of coal, supposing it to take place in a minute, would be equivalent to the work of 300 horses ; and if we suppose 120 millions of horses working day and night with unimpaired strength, for a year, their united energies would enable them to perform an amount of work just equivalent to the heat to be derived from the annual produce of our coal-fields. Our woods and forests are also sources of mechanical energy, because they also have the power of uniting with the atmospheric oxygen, and the molecular motion produced in the act of union may be turned to mechanical account. Passing from dead matter to living matter, we find that the source of motive power here referred to is also the source of muscular power. A horse can perform work, and so can a man, but this work is at bottom the molecular work of the elements of the food and the oxygen of the air. We inhale this vital gas, and bring it into sufficiently close proximity with the carbon and the hydrogen of the food. They unite in obedience to their mutual attractions, and their motion towards each other, properly turned to account by the wonderful mechanism of the body, becomes muscular motion.

One fundamental thought pervades all these statements : there is one tap root from which they all spring. This is the ancient maxim that out of nothing nothing comes ; that neither in the organic world nor in the inorganic is power produced without the expenditure of other power ; that neither in the plant nor in the animal is there a creation of force or motion. Trees grow, and so do men and horses ; and here we have new power incessantly introduced upon the earth. But its source, as I have already stated, is the sun. For he it is who separates the carbon from the oxygen of the carbonic acid, and thus enables them to recombine. Whether they recombine in the furnace of the steam-engine or in the animal body, the origin of the power they produce is the same. In this sense we are all ' souls of fire and children of the sun.' But, as remarked by Helmholtz, we must be content to share our celestial pedigree with the meanest living things. The frog, and the toad, and those terrible creatures the monkey and the gorilla, draw their power from the same source as man.

Some estimable persons, here present, very possibly shrink from accepting these statements; they may be frightened by their apparent tendency towards what is called materialism—a word which, to many minds, expresses something very dreadful. But it ought to be known and avowed that the physical philosopher, as such, must be a pure materialist. His enquiries deal with matter and force, and with them alone. The action which he has to investigate is necessary action; not spontaneous action—the transformation, not the creation, of matter and force. And whatever be the forms which matter and force may assume, whether in the organic world or the inorganic, whether in the coal-beds and forests of the earth, or in the brains and muscles of men, the physical philosopher will make good his right to investigate them. It is perfectly vain to attempt to stop enquiry as to the actual and possible actions of matter and force. Depend upon it, if a chemist by bringing the proper materials together, in a retort or crucible, could make a baby, he would do it. There is no law, moral or physical, forbidding him to do it—his enquiries in this direction are limited solely by his own capacity and the laws of matter and force. At the present moment there are, no doubt, persons experi-

menting on the possibility of producing what we call life
out of inorganic materials. Let them pursue their studies
in peace; it is only by such trials that they will learn
the limits of their powers.

But while I thus make the largest demand for freedom
of investigation—while I as a man of science feel a
natural pride in scientific achievement, while I regard
science as the most powerful instrument of intellectual
culture, as well as the most powerful ministrant to the
material wants of men; if you ask me whether science
has solved, or is likely in our day to solve, the problem
of this universe, I must shake my head in doubt. You
remember the first Napoleon's question, when the *savans*
who accompanied him to Egypt discussed in his presence
the origin of the universe, and solved it to their own
apparent satisfaction. He looked aloft to the starry
heavens, and said, ' It is all very well, gentlemen ; but
who made all these?' That question still remains un-
answered, and science makes no attempt to answer it.
As far as I can see, there is no quality in the human
intellect which is fit to be applied to the solution of the
problem. It entirely transcends us. The mind of man
may be compared to a musical instrument with a certain
range of notes, beyond which in both directions we have
an infinitude of silence. The phenomena of matter and
force lie within our intellectual range, and as far as they
reach we will at all hazards push our enquiries. But
behind, and above, and around all, the real mystery of
this universe lies unsolved, and, as far as we are concerned, is incapable of solution. Fashion this mystery
as you will, with that I have nothing to do. But be
careful that your conception of it be not an unworthy

one. Invest that conception with your highest and holiest thought, but be careful of pretending to know more about it than is given to man to know. Be careful, above all things, of professing to see in the phenomena of the material world the evidences of Divine pleasure or displeasure. Doubt those who would deduce from the fall of the tower of Siloam the anger of the Lord against those who were crushed Doubt those equally who pretend to see in cholera, cattle-plague, and bad harvests, evidences of Divine anger. Doubt those spiritual guides who in Scotland have lately propounded the monstrous theory that the depreciation of railway scrip is a consequence of railway travelling on a Sunday. Let them not, as far as you are concerned, label and libel the system of nature with their ignorant hypotheses. Well might the mightiest of living Scotchmen, that hero of the intellect who might have been a hero in the field, that strong and earnest soul who has made every soul of like nature in these islands his debtor—looking from the solitudes of thought into this highest of questions, well, I say, might your noble old Carlyle scornfully retort on such interpreters of the ways of God to men :—

> The Builder of this universe was wise,
>   He formed all souls, all systems, planets, particles;
> The plan he formed his worlds and Æons by,
>   Was—Heavens !—was thy small nine-and-thirty articles !

V.

ADDRESS TO THE STUDENTS OF UNIVERSITY COLLEGE, LONDON

ON THE DISTRIBUTION OF PRIZES IN THE FACULTY OF ARTS.

Session 1868—69.

'Self-reverence, self-knowledge, self-control,
These three alone lead life to sovereign power.
Yet not for power (power of herself
Would come uncalled for), but to live by law,
Acting the law we live by without fear ;
And, because right is right, to follow right
Were wisdom in the scorn of consequence.'

TENNYSON.

V.

AN ADDRESS TO STUDENTS.

THERE is an idea regarding the nature of man which modern philosophy has sought, and is still seeking, to raise into clearness, the idea, namely, of secular growth. Man is not a thing of yesterday; nor do I imagine that the slightest controversial tinge is imported into this address when I say that he is not a thing of 6,000 years ago. Whether he came originally from stocks or stones, from nebulous gas or solar fire, I know not; if he had any such origin the process of his transformation is as inscrutable to you and to me as that of the grand old legend, according to which ' the Lord God formed man of the dust of the ground, and breathed into his nostrils the breath of life; and man became a living soul.' But however obscure man's origin may be, his growth is not to be denied. Here a little and there a little added through the ages have slowly transformed him from what he was into what he is. The doctrine has been held that the mind of the child is like a sheet of white paper, on which by education we can write what characters we please. This doctrine assuredly needs qualification and correction. In physics, when an external force is applied to a body with a view of affecting its inner texture, if we wish to predict the result, we must know whether the external force conspires with or opposes the internal forces of the body itself; and in bringing the influence of education to bear upon the new-born man his inner powers must be also taken into account. He comes to us as a bundle of inherited capacities and tendencies, labelled ' from the indefinite past to the indefinite future ; ' and he makes his transit from the one to the other through the education of the present time. The object of that education is, or ought to be, to provide wise exercise for his capacities, wise direction for his tendencies, and through this exercise and this direction to furnish his mind with such knowledge as may contribute to the usefulness, the beauty, and the nobleness of his life.

How is this discipline to be secured, this knowledge imparted ? Two rival methods now solicit attention,— the one organised and equipped, the labour of centuries having been expended in bringing it to its present state of perfection ; the other, more or less chaotic, but becoming daily less so, and giving signs of enormous power, both as a source of knowledge and as a means of discipline. These two methods are the classical and the scientific method. I wish they were not rivals ; it is only bigotry and short-sightedness that make them so ; for assuredly it is possible to give both of them fair play. Though hardly authorised to express any opinion whatever upon the subject, I nevertheless hold the opinion that the proper study of a language is an intellectual discipline of the highest kind. If I except discussions on the comparative merits of popery and protestantism, English grammar was the most important discipline of my boyhood. The piercing through the involved and inverted sentences of ' Paradise Lost ' ; the linking of the verb to its often distant nominative, of the relative to its distant antecedent, of the agent to the object of the

transitive verb, of the preposition to the noun or pro-
noun which it governed,—the study of variations in
mood and tense, the transformations often necessary
to bring out the true grammatical structure of a sentence,—all this was to my young mind a discipline of
the highest value, and indeed a source of unflagging delight. How I rejoiced when I found a great author trip-
ping, and was fairly able to pin him to a corner from
which there was no escape! As I speak, some of the
sentences which exercised me when a boy rise to my recollection. 'He that hath ears to hear let him hear.'
That was one of them, where the 'He' is left, as it were,
floating in mid air without any verb to support it. I
speak thus of English because it was of real value to me.
I do not speak of other languages because their educa-
tional value for me was almost insensible. But knowing
the value of English so well, I should be the last to deny,
or even to doubt, the high discipline involved in the
proper study of Latin and Greek.

That study, moreover, has other merits and recommendations which have been already slightly touched upon.
It is organised and systematised by long-continued use.
It is an instrument wielded by some of the best intellects
of the country in the education of youth ; and it can
point to results in the achievements of our foremost
men. What, then, has science to offer which is in
the least degree likely to compete with such a system?
I cannot better reply than by recurring to the grand old
story from which I have already quoted. Speaking of
the world and all that therein is, of the sky and the stars
around it, the ancient writer says, 'And God saw all that
he had made, and behold it was very good.' It is the

body of things thus described which science offers to the study of man. There is a very renowned argument much prized and much quoted by theologians, in which the universe is compared to a watch. Let us deal practically with this comparison. Supposing a watchmaker, having completed his instrument, to be so satisfied with his work as to call it very good, what would you understand him to mean? You would not suppose that he referred to the dial-plate in front and the chasing of the case behind, so much as to the wheels and pinions, the springs and jewelled pivots of the works within, those qualities and powers, in short, which enable the watch to perform accurately its work as a keeper of time. With regard to the knowledge of such a watch he would be a mere ignoramus who would content himself with outward inspection. I do not wish to say one severe word here to-day, but I fear that many of those who are very loud in their praise of the works of the Lord know them only in this outside and superficial way. It is the inner works of the universe which science reverently uncovers ; it is the study of these that she recommends as a discipline worthy of all acceptation.

The ultimate problem of physics is to reduce matter by analysis to its lowest condition of divisibility, and force to its simplest manifestations, and then by synthesis to construct from these elements the world as it stands. We are still a long way from the final solution of this problem ; and when the solution comes, it will be one more of spiritual insight than of actual observation. But though we are still a long way from this complete intellectual mastery of nature, we have conquered vast regions of it, have learned their polities and the play of their powers. We live upon a ball of matter 8,000 miles in diameter, swathed by an atmosphere of unknown height. This ball has been molten by heat, chilled to a solid, and sculptured by water; it is made up of substances possessing distinctive properties and modes of action, properties which have an immediate bearing upon the continuance of man in health, and on his recovery from disease, on which moreover depend all the arts of industrial life. These properties and modes of action offer problems to the intellect, some profitable to the child, and others sufficient to tax the highest powers of the philosopher. Our native sphere turns on its axis and revolves in space. It is one of a band which do the same. It is illuminated by a sun which, though nearly a hundred millions of miles distant, can be brought virtually into our closets and there subjected to examination. It has its winds and clouds, its rain and frost, its light, heat, sound, electricity, and magnetism. And it has its vast kingdoms of animals and vegetables. To a most amazing extent the human mind has conquered these things, and revealed the logic which runs through them. Were they facts only, without logical relationship, science might, as a means of discipline, suffer in comparison with language. But the whole body of phenomena is instinct with law; the facts are hung on principles, and the value of physical science as a means of discipline consists in the motion of the intellect, both inductively and deductively, along the lines of law marked out by phenomena. As regards that discipline to which I have already referred as derivable from the study of languages,—that, and more, are involved in the study of physical science. Indeed, I believe it would be possible so to limit and arrange the study of a portion of physics as to render the mental exercise involved in it almost qualitatively the same as that involved in the unravelling of a language.

I have thus far limited myself to the purely intellectual side of this question. But man is not all intellect. If he were so, science would, I believe, be his proper nutriment. But he feels as well as thinks; he is receptive of the sublime and the beautiful as well as of the true. Indeed, I believe that even the intellectual action of a complete man is, consciously or unconsciously, sustained by an under current of the emotions. It is vain, I think, to attempt to separate moral and emotional nature from intellectual nature. Let a man but observe himself, and he will, if I mistake not, find that in nine cases out of ten, moral or immoral considerations, as the case may be, are the motive force which pushes his intellect into action. The reading of the works of two men, neither of them imbued with the spirit of modern science, neither of them, indeed, friendly to that spirit, has placed me here to-day. These men are the English Carlyle and the American Emerson. I must ever remember with gratitude that through three long cold German winters Carlyle placed me in my tub, even when ice was on its surface, at five o'clock every morning; not slavishly, but cheerfully, meeting each day's studies with a resolute will, determined whether victor or vanquished not to shrink from difficulty. I never should have gone through Analytical Geometry and the Calculus had it not been for those men. I never should have become a physical investigator, and hence without them I should not have been here to-day. They told me what I ought to do in a way that caused me to do it, and all my consequent intellectual action is to be traced to this purely moral source. To Carlyle and Emerson I ought to add Fichte, the greatest representative of pure idealism. These three unscientific men made me a practical scientific worker. They called out 'act!' I hearkened to the summons, taking the liberty, however, of determining for myself the direction which effort was to take.

And I may now cry 'act!' but the potency of action must be yours. I may pull the trigger, but if the gun be not charged there is no result. We are creators in the intellectual world as little as in the physical. We may remove obstacles, and render latent capacities active, but we cannot suddenly change the nature of man. The 'new birth' itself implies the pre-existence of the new character which requires not to be created but brought forth. You cannot by any amount of missionary labour suddenly transform the savage into the civilised Christian. The improvement of man is *secular*—not the work of an hour or of a day. But though indubitably bound by our organisations, no man knows what the potentialities of any human mind may be, which require only release to be brought into action. Let me illustrate this point. There are in the mineral world certain crystals, certain forms, for instance, of fluor-spar, which have lain darkly in the earth for ages, but which nevertheless have a potency of light locked up within them. In their case the potential has never become actual—the light is in fact held back by a molecular detent. When these crystals are warmed, the detent is lifted, and an outflow of light immediately begins. I know not how many of you may be in the condition of this fluor-spar. For aught I know, every one of you may be in this condition, requiring but the proper agent to be applied—the proper word to be spoken—to remove a detent, and to render you conscious of light within yourselves and sources of light to others.

The circle of human nature, then, is not complete without the arc of feeling and emotion. The lilies of the field have a value for us beyond their botanical ones—a certain lightening of the heart accompanies the declaration that 'Solomon in all his glory was not arrayed like one of these.' The sound of the village bell which comes mellowed from the valley to the traveller upon the hill, has a value beyond its acoustical one. The setting sun when it mantles with the bloom of roses the alpine snows, has a value beyond its optical one. The starry heavens, as you know, had for Immanuel Kant a value beyond their astronomical one. Round about the intellect sweeps the horizon of emotions from which all our noblest impulses are derived. I think it very desirable to keep this horizon open ; not to permit either priest or philosopher to draw down his shutters between you and it. And here the dead languages, which are sure to be beaten by science in the purely intellectual fight, have an irresistible claim. They supplement the work of science by exalting and refining the æsthetic faculty, and must on this account be cherished by all who desire to see human culture complete. There must be a reason for the fascination which these languages have so long exercised upon the most powerful and elevated minds—a fascination which will probably continue for men of Greek and Roman mould to the end of time.

In connection with this question of the emotions one very obvious danger besets many of the more earnest spirits of our day—the danger of *haste* in endeavouring to give the feelings repose. We are distracted by systems of theology and philosophy which were taught to us when young, and which now excite in us a hunger and a thirst for knowledge not proved to be attainable. There are periods when the judgment ought to remain in suspense, the data on which a decision might be based being absent. This discipline of suspending the judgment is a common one in science, but not so common as it ought to be elsewhere. I walked down Regent Street some time ago with a man of great gifts and acquirements, discussing with him various theological questions. I could not accept his views of the origin and destiny of the universe, nor was I prepared to enunciate any definite views of my own. He turned to me at length and said, ' You surely must have a theory of the universe.' That I should in one way or another have solved this mystery of mysteries seemed to my friend a matter of course. ' I have not even a theory of magnetism ' was my reply. We ought to learn to wait, and pause before closing with the advances of those expounders of the ways of God to men, who offer us intellectual peace at the modest cost of intellectual life.

The teachers of the world ought to be its best men, and for the present at all events such men must learn self-trust. They must learn more and more to do without external aid ; save such aid as comes from the contemplation of a universe, which, though it baffles the intellect, can elevate the heart. But they must learn to feel the mystery of that universe without attempting to give it a rigid form, personal or otherwise. By the fullness and freshness of their own lives and utterances they must awaken life in others. The position of science is already assured, but I think the poet also will have a great part to play in the future of the world. To him it is given for a long time to come to fill those shores which the recession of the theologic tide has left exposed; to him, when he rightly understands his mission, and does not flinch from the tonic discipline which it assuredly demands, we have a right to look for that heightening and brightening of life which so many of us need. He ought to be the interpreter of that power which as

'Jehovah, Jove, or Lord,'

has hitherto filled and strengthened the human heart.

Let me utter one practical word in conclusion—take care of your health. There have been men who by wise attention to this point might have risen to any eminence —might have made great discoveries, written great poems, commanded armies, or ruled states, but who by unwise neglect of this point have come to nothing. Imagine Hercules as oarsman in a rotten boat; what can he do there but by the very force of his stroke expedite the ruin of his craft. Take care then of the timbers of your boat, and avoid all practices likely to introduce either wet or dry rot amongst them. And this is not to be accomplished by desultory or intermittent efforts of the will, but by the formation of *habits*. The will no doubt has sometimes to put forth its strength in order to strangle or crush the special temptation. But the formation of right habits is essential to your permanent security. They diminish your chance of falling when assailed, and they augment your chance of recovery when overthrown.

VI.

SCOPE AND LIMIT OF SCIENTIFIC MATERIALISM.

AN ADDRESS.

DELIVERED IN THE MATHEMATICAL AND PHYSICAL SECTION OF THE
BRITISH ASSOCIATION IN NORWICH.

19th August, 1868.

'As I proceeded I found my philosopher altogether forsaking mind or any other principle of order, and having recourse to air and æther, and water, and other eccentricities. I might compare him to a person who began by maintaining generally that mind is the cause of the actions of Socrates, but who when he endeavoured to explain the cause of my several actions in detail, went on to show that I sit here because my body is made up of bones and muscles; and the bones he would say are hard and have ligaments which divide them, and the muscles are elastic, and they cover the bones, which have also a covering or environment of flesh and skin which contains them; and as the bones are lifted at their joints by the contraction or relaxation of the muscles, I am able to bend my limbs, and this is why I am sitting here in a curved posture;—that is what he would say, and he would have a similar explanation of my talking to you, which he would attribute to sound, and air, and hearing, and he would assign ten thousand other causes of the same sort, forgetting to mention the true cause, which is that the Athenians have thought fit to condemn me, and accordingly I have thought it better and more right to remain here and undergo my sentence; for I am inclined to think that these muscles and bones of mine would have gone off to Megara or Bœotia—by the dog of Egypt they would, if they had been guided by their own idea of what was best, and if I had not chosen as the better and nobler part, instead of playing truant and running away, to undergo any punishment which the State inflicts.'—PLATO, *Jowett's Translation.*

VI.

SCIENTIFIC MATERIALISM.

THE CELEBRATED FICHTE, in his lectures on the ' Vocation of the Scholar,' insisted on a culture which should not be one-sided, but all-sided. The scholar's intellect was to expand spherically and not in a single direction only. In one direction, however, Fichte required that the scholar should apply himself directly to nature, become a creator of knowledge, and thus repay by original labours of his own the immense debt he owed to the labours of others. It was these which enabled him to supplement the knowledge derived from his own researches, so as to render his culture rounded and not one-sided.

As regards science Fichte's idea is to some extent illustrated by the constitution and the labours of the British Association. We have here a body of men engaged in the pursuit of Natural Knowledge, but variously engaged. While sympathising with each of its departments, and supplementing his culture by knowledge drawn from all of them, each student amongst us selects one subject for the exercise of his own original faculty—one line along which he may carry the light of his private intelligence a little way into the darkness by which all knowledge is surrounded. Thus, the geologist deals with the rocks ; the biologist with the conditions and phenomena of life ; the astronomer with stellar masses and motions; the mathematician with the relations of space and number; the chemist pursues his atoms, while the physical investigator has his own large field in optical, thermal, electrical, acoustical, and other phenomena. The British Association then, as a whole, faces physical nature on all sides and pushes knowledge centrifugally outwards, the sum of its labours constituting what Fichte might call the *sphere* of natural knowledge. In the meetings of the Association it is found necessary to resolve this sphere into its component parts, which take concrete form under the respective letters of our Sections.

This is the Mathematical and Physical Section. Mathematics and physics have been long accustomed to coalesce. For, no matter how subtle a natural phenomenon may be, whether we observe it in the region of sense, or follow it into that of imagination, it is in the long run reducible to mechanical laws. But the mechanical data once guessed or given, methematics become all-powerful as an instrument of deduction. The command of geometry over the relations of space, the far-reaching power which organised symbolic reasoning confers, are potent both as means of physical discovery, and of reaping the entire fruits of discovery. Indeed, without mathematics, expressed or implied, our knowledge of physical science would be friable in the extreme.

Side by side with the mathematical method we have the method of experiment. Here, from a starting-point furnished by his own researches or those of others, the investigator proceeds by combining intuition and verification. He ponders the knowledge he possesses and tries to push it further, he guesses and checks his guess, he conjectures and confirms or explodes his conjecture. These guesses and conjectures are by no means leaps in the dark ; for knowledge once gained casts a faint light beyond its own immediate boundaries. There is no discovery so limited as not to illuminate something beyond itself. The force of intellectual penetration into this penumbral region which surrounds actual knowledge is not, as some seem to think, dependent upon method, but upon the genius of the investigator. There is, however, no genius so gifted as not to need control and verification. The profoundest minds know best that Nature's ways are not at all times their ways, and that the brightest flashes in the world of thought are incomplete until they have been proved to have their counterparts in the world of fact. Thus the vocation of the true experimentalist may be defined as the continued exercise of spiritual insight, and its incessant correction and realisation. His experiments constitute a body, of which his purified intuitions are, as it were, the soul.

Partly through mathematical and partly through experimental research, physical science has of late years assumed a momentous position in the world. Both in a material and in an intellectual point of view it has produced, and it is destined to produce, immense changes—vast social ameliorations, and vast alterations in the popular conception of the origin, rule, and governance of natural things. By science, in the physical world, miracles are wrought, while philosophy is forsaking its ancient metaphysical channels and pursuing others which have been opened or indicated by scientific research. This must become more and more the case as philosophical writers become more deeply imbued with the methods of science, better acquainted with the facts which scientific men have won, and with the great theories which they have elaborated.

If you look at the face of a watch, you see the hour- and minute-hands, and possibly also a second-hand, moving over the graduated dial. Why do these hands move? and why are their relative motions such as they are observed to be? These questions cannot be answered without opening the watch, mastering its various parts, and ascertaining their relationship to each other. When this is done, we find that the observed motion of the hands follows of necessity from the inner mechanism of the watch, when acted upon by the force invested in the spring.

The motion of the hands may be called a phenomenon of art, but the case is similar with the phenomena of nature. These also have their inner mechanism, and their store of force to set that mechanism going. The ultimate problem of physical science is to reveal this mechanism, to discern this store, and to show that from the combined action of both the phenomena of which they constitute the basis must of necessity flow.

I thought an attempt to give you even a brief and sketchy illustration of the manner in which scientific thinkers regard this problem would not be uninteresting to you on the present occasion; more especially as it will give me occasion to say a word or two on the tendencies and limits of modern science; to point out the region which men of science claim as their own, and where it is mere waste of time to oppose their advance, and also to define, if possible, the bourne between this and that other region to which the questionings and yearnings of the scientific intellect are directed in vain.

But here your tolerance will be needed. It was the American Emerson, I think, who said that it is hardly possible to state any truth strongly without apparent injustice to some other truth. Truth is often of a dual character, taking the form of a magnet with two poles; and many of the differences which agitate the thinking part of mankind are to be traced to the exclusiveness with which partisan reasoners dwell upon one half of the duality in forgetfulness of the other. The proper course appears to be to state both halves strongly, and allow each its fair share in the formation of the resultant conviction. But this waiting for the statement of the two sides of a question implies patience. It implies a resolution to suppress indignation if the statement of the one half should clash with our convictions, and to repress equally undue elation if the half-statement should happen to chime in with our views. It implies a determination to wait calmly for the statement of the whole, before we pronounce judgment in the form of either acquiescence or dissent.

This premised, and, I trust, accepted, let us enter upon our task. There have been writers who affirmed that the pyramids of Egypt were the productions of nature; and in his early youth Alexander von Humboldt wrote a learned essay with the express object of refuting this notion. We now regard the pyramids as the work of men's hands, aided probably by machinery of which no record remains. We picture to ourselves the swarming workers toiling at those vast erections, lifting the inert stones, and, guided by the volition, the skill, and possibly at times by the whip of the architect, placing them in their proper positions. The blocks in this case were

I

moved and posited by a power external to themselves, and the final form of the pyramid expressed the thought of its human builder.

Let us pass from this illustration of constructive power to another of a different kind. When a solution of common salt is slowly evaporated, the water which holds the salt in solution disappears, but the salt itself remains behind. At a certain stage of concentration the salt can no longer retain the liquid form; its particles, or molecules, as they are called, begin to deposit themselves as minute solids, so minute, indeed, as to defy all microscopic power. As evaporation continues solidification goes on, and we finally obtain, through the clustering together of innumerable molecules, a finite crystalline mass of a definite form. What is this form? It sometimes seems a mimicry of the architecture of Egypt. We have little pyramids built by the salt, terrace above terrace from base to apex, forming a series of steps resembling those up which the Egyptian traveller is dragged by his guides. The human mind is as little disposed to look unquestioning at these pyramidal salt-crystals as to look at the pyramids of Egypt without enquiring whence they came. How, then, are those salt-pyramids built up?

Guided by analogy, you may, if you like, suppose that swarming among the constituent molecules of the salt, there is an invisible population, controlled and coerced by some invisible master, and placing the atomic blocks in their positions. This, however, is not the scientific idea, nor do I think your good sense will accept it as a likely one. The scientific idea is that the molecules act upon each other without the intervention of slave labour; that they attract each other and repel each other at certain definite points, or poles, and in certain definite directions; and that the pyramidal form is the result of this play of attraction and repulsion. While, then, the blocks of Egypt were laid down by a power external to themselves, these molecular blocks of salt are self-posited, being fixed in their places by the forces with which they act upon each other.

I take common salt as an illustration because it is so familiar to us all; but any other crystalline substance would answer my purpose equally well. Everywhere, in fact, throughout inorganic nature, we have this formative power, as Fichte would call it—this structural energy ready to come into play, and build the ultimate particles of matter into definite shapes. The ice of our winters and of our polar regions is its handywork, and so equally are the quartz, felspar, and mica of our rocks. Our chalk-beds are for the most part composed of minute shells, which are also the product of structural energy; but behind the shell, as a whole, lies a more remote and subtle formative act. These shells are built up of little crystals of calc-spar, and to form these crystals the structural force had to deal with the intangible molecules of carbonate of lime. This tendency on the part of matter to organise itself, to grow into shape, to assume definite forms in obedience to the definite action of force, is, as I have said, all pervading. It is in the ground on which you tread, in the water you drink, in the air you breathe. Incipient life, as it were, manifests itself throughout the whole of what we call inorganic nature.

The forms of the minerals resulting from this play of polar forces are various, and exhibit different degrees of complexity. Men of science avail themselves of all

possible means of exploring their molecular architecture.
For this purpose they employ in turn as agents of ex-
ploration, light, heat, magnetism, electricity, and sound.
Polarised light is especially useful and powerful here.   A
beam of such light, when sent in among the molecules of
a crystal, is acted on by them, and from this action we
infer with more or less of clearness the manner in which
the molecules are arranged.   That differences, for example, exist between the inner structure of rocksalt and
crystallised sugar or sugar-candy, is thus strikingly revealed.   These actions often display themselves in chromatic phenomena of great splendour, the play of molecular force being so regulated as to remove some of the
coloured constituents of white light, and to leave others
with increased intensity behind.

And now let us pass from what we are accustomed to
regard as a dead mineral to a living grain of corn. When
it is examined by polarised light, chromatic phenomena
similar to those noticed in crystals are observed. And
why?   Because the architecture of the grain resembles the
architecture of the crystal. In the grain also the molecules
are set in definite positions, and in accordance with their
arrangement they act upon the light.   But what has built
together the molecules of the corn?   I have already said
regarding crystalline architecture that you may, if you
please, consider the atoms and molecules to be placed in
position by a power external to themselves.   The same
hypothesis is open to you now.   But if in the case of
crystals you have rejected this notion of an external
architect, I think you are bound to reject it now, and to
conclude that the molecules of the corn are self-posited by
the forces with which they act upon each other. It would

be poor philosophy to invoke an external agent in the one case and to reject it in the other.

Instead of cutting our grain of corn into slices and subjecting it to the action of polarised light, let us place it in the earth and subject it to a certain degree of warmth. In other words, let the molecules, both of the corn and of the surrounding earth, be kept in that state of agitation which we call warmth. Under these circumstances, the grain and the substances which surround it interact, and a definite molecular architecture is the result. A bud is formed ; this bud reaches the surface, where it is exposed to the sun's rays, which are also to be regarded as a kind of vibratory motion. And as the motion of common heat with which the grain and the substances surrounding it were first endowed, enabled the grain and these substances to exercise their attractions and repulsions, and thus to coalesce in definite forms, so the specific motion of the sun's rays now enables the green bud to feed upon the carbonic acid and the aqueous vapour of the air. The bud appropriates those constituents of both for which it has an elective attraction, and permits the other constituent to resume its place in the air. Thus the architecture is carried on. Forces are active at the root, forces are active in the blade, the matter of the earth and the matter of the atmosphere are drawn towards the root and blade, and the plant augments in size. We have in succession the bud, the stalk, the ear, the full corn in the ear ; the cycle of molecular action being completed by the production of grains similar to that with which the process began.

Now there is nothing in this process which necessarily eludes the conceptive or imagining power of the purely human mind. An intellect the same in kind as our own would, if only sufficiently expanded, be able to follow the whole process from beginning to end. It would see every molecule placed in its position by the specific attractions and repulsions exerted between it and other molecules, the whole process and its consummation being an instance of the play of molecular force. Given the grain and its environment, the purely human intellect might, if sufficiently expanded, trace out *à priori* every step of the process of growth, and by the application of purely mechanical principles demonstrate that the cycle must end, as it is seen to end, in the reproduction of forms like that with which it began. A similar necessity rules here to that which rules the planets in their circuits round the sun.

You will notice that I am stating my truth strongly, as at the beginning we agreed it should be stated. But I must go still further, and affirm that in the eye of science *the animal body* is just as much the product of molecular force as the stalk and ear of corn, or as the crystal of salt or sugar. Many of the parts of the body are obviously mechanical. Take the human heart, for example, with its system of valves, or take the exquisite mechanism of the eye or hand. Animal heat, moreover, is the same in kind as the heat of a fire, being produced by the same chemical process. Animal motion, too, is as directly derived from the food of the animal, as the motion of Trevethyck's walking-engine from the fuel in its furnace. As regards matter, the animal body creates nothing; as regards force, it creates nothing. Which of you by taking thought can add one cubit to his stature? All that has been said, then, regarding the plant may be restated with regard to the animal. Every particle that enters into the composition of a muscle, a nerve, or a bone, has been placed in its position by molecular force. And unless the existence of law in these matters be denied, and the element of caprice introduced, we must conclude that, given the relation of any molecule of the body to its environment, its position in the body might be determined mathematically. Our difficulty is not with the *quality* of the problem, but with its *complexity* ; and this difficulty might be met by the simple expansion of the faculties which we now possess. Given this expansion, with the necessary molecular data, and the chick might be deduced as rigorously and as logically from the egg as the existence of Neptune from the disturbances of Uranus, or as conical refraction from the undulatory theory of light.

You see I am not mincing matters, but avowing nakedly what many scientific thinkers more or less distinctly believe. The formation of a crystal, a plant, or an animal, is in their eyes a purely mechanical problem, which differs from the problems of ordinary mechanics in the smallness of the masses and the complexity of the processes involved. Here you have one half of our dual truth ; let us now glance at the other half. Associated with this wonderful mechanism of the animal body we have phenomena no less certain than those of physics, but between which and the mechanism we discern no necessary connection. A man, for example, can say *I feel, I think, I love* ; but how does *consciousness* infuse itself into the problem ? The human brain is said to be the organ of thought and feeling ; when we are hurt the brain feels it, when we ponder it is the brain that thinks, when our passions or affections are excited it is through the instrumentality of the brain. Let us endeavour to be a little more precise here. I hardly imagine there exists a profound scientific thinker, who has reflected upon the subject, unwilling to admit the extreme probability of the hypothesis, that for every fact of consciousness, whether in the domain of sense, of thought, or of emotion, a definite molecular condition of motion or structure is set up in the brain ; or who would be disposed even to deny that if the motion or structure be induced by internal causes instead of external, the effect on consciousness will be the same ? Let any nerve, for example, be thrown by morbid action into the precise state of motion which would be communicated to it by the pulses of a heated body, surely that nerve will declare itself hot—the mind will accept the subjective intimation exactly as if it were objective. The retina may be excited by purely mechanical means. A blow on the eye causes a luminous flash, and the mere pressure of the finger on the external ball produces a star of light, which Newton compared to the circles on a peacock's tail. Disease makes people see visions and dream dreams ; but, in all such cases, could we examine the organs implicated, we should, on philosophical grounds, expect to find them in that precise molecular condition which the real objects, if present, would superinduce.

The relation of physics to consciousness being thus invariable, it follows that, given the state of the brain, the corresponding thought or feeling might be inferred ; or given the thought or feeling, the corresponding state of the brain might be inferred. But how inferred ?   It would be at bottom not a case of logical inference at all, but of empirical association. You may reply that many of the inferences of science are of this character ; the inference, for example, that an electric current of a given direction will deflect a magnetic needle in a definite way ; but the cases differ in this, that the passage from the current to the needle, if not demonstrable, is thinkable, and that we entertain no doubt as to the final mechanical solution of the problem. But the passage from the physics of the brain to the corresponding facts of consciousness is unthinkable. Granted that a definite thought, and a definite molecular action in the brain occur simultaneously; we do not possess the intellectual organ, nor apparently any rudiment of the organ, which would enable us to pass, by a process of reasoning, from the one to the other. They appear together, but we do not know why. Were our minds and senses so expanded, strengthened, and illuminated as to enable us to see and feel the very molecules of the brain ; were we capable of following all their motions, all their groupings, all their electric discharges, if such there be ; and were we intimately acquainted with the corresponding states of thought and feeling, we should be as far as ever from the solution of the problem, ' How are these physical processes connected with the facts of consciousness ?' The chasm between the two classes of phenomena would still remain intellectually impassable. Let the consciousness of *love*, for example, be associated with a right-handed spiral motion of the molecules of the brain, and the consciousness of *hate* with a left-handed spiral motion. We should then know when we love that the motion is in one direction, and when we hate that the motion is in the other; but the ' WHY ? ' would remain as unanswerable as before.

In affirming that the growth of the body is mechanical, and that thought, as exercised by us, has its correlative in the physics of the brain, I think the position of the

' Materialist' is stated, as far as that position is a tenable one.   I think the materialist will be able finally to maintain this position against all attacks; but I do not think, in the present condition of the human mind, that he can pass beyond this position.   I do not think he is entitled to say that his molecular groupings and his molecular motions *explain* everything.   In reality they explain nothing.   The utmost he can affirm is the association of two classes of phenomena, of whose real bond of union he is in absolute ignorance.   The problem of the connection of body and soul is as insoluble in its modern form as it was in the prescientific ages.   Phosphorus is known to enter into the composition of the human brain, and a trenchant German writer has exclaimed, ' Ohne Phosphor, kein Gedanke.'   That may or may not be the case; but even if we knew it to be the case, the knowledge would not lighten our darkness.   On both sides of the zone here assigned to the materialist he is equally helpless.   If you ask him whence is this 'Matter' of which we have been discoursing, who or what divided it into molecules, who or what impressed upon them this necessity of running into organic forms, he has no answer.   Science is mute in reply to these questions.   But if the materialist is confounded and science rendered dumb, who else is prepared with a solution?   To whom has this arm of the Lord been revealed?   Let us lower our heads and acknowledge our ignorance, priest and philosopher, one and all.

Perhaps the mystery may resolve itself into knowledge at some future day.   The process of things upon this earth has been one of amelioration.   It is a long way from the Iguanodon and his contemporaries to the President and Members of the British Association. And whether we regard the improvement from the scientific or from the theological point of view, as the result of progressive development, or as the result of successive exhibitions of creative energy, neither view entitles us to assume that man's present faculties end the series,—that the process of amelioration stops at him. A time may therefore come when this ultra-scientific region by which we are now enfolded may offer itself to terrestrial, if not to human investigation. Two-thirds of the rays emitted by the sun fail to arouse in the eye the sense of vision. The rays exist, but the visual organ requisite for their translation into light does not exist. And so from this region of darkness and mystery which surrounds us, rays may now be darting which require but the development of the proper intellectual organs to translate them into knowledge as far surpassing ours as ours surpasses that of the wallowing reptiles which once held possession of this planet. Meanwhile the mystery is not without its uses. It certainly may be made a power in the human soul; but it is a power which has feeling, not knowledge, for its base. It may be, and will be, and I hope is turned to account, both in steadying and strengthening the intellect, and in rescuing man from that littleness to which, in the struggle for existence, or for precedence in the world, he is continually prone.

Musings on the Matterhorn, 27th July, 1868.

'Hacked and hurt by time, the aspect of the mountain from its higher crags saddened me. Hitherto the impression it made was that of savage strength; here we had inexorable decay. But this notion of decay implied a reference to a period when the Matterhorn was in the full strength of mountainhood. Thought naturally ran back to its remoter origin and sculpture. Nor did thought halt there, but wandered on through molten worlds to that nebulous haze which philosophers have regarded, and with good reason, as the proximate source of all material things. I tried to look at this universal cloud, containing within itself the prediction of all that has since occurred; I tried to imagine it as the seat of those forces whose action was to issue in solar and stellar systems, and all that they involve. Did that formless fog contain potentially the *sadness* with which I regarded the Matterhorn? Did the *thought* which now ran back to it simply return to its primeval home? If so, had we not better recast our definitions of matter and force; for if life and thought be the very flower of both, any definition which omits life and thought must be inadequate, if not untrue. Are questions like these warranted? Why not? If the final goal of man has not been yet attained; if his development has not been yet arrested, who can say that such yearnings and questionings are not necessary to the opening of a finer vision, to the budding and the growth of diviner powers? When I look at the heavens and the earth, at my own body, at my strength and weakness of mind, even at these ponderings, and ask myself, is there no being or thing in the universe that knows more about these matters than I do; what is my answer? Supposing our theologic schemes of creation, condemnation, and redemption to be dissipated; and the warmth of denial which they excite, and which, as a motive force, can match the warmth of affirmation, dissipated at the same time; would the undeflected human mind return to the meridian of absolute neutrality as regards these ultra-physical questions? Is such a position one of stable equilibrium? The channels of thought being already formed, such are the questions without replies, which could run athwart consciousness during a ten minutes' halt upon the weathered point of the Matterhorn.'

VII.

ON THE

SCIENTIFIC USE OF THE IMAGINATION.

A DISCOURSE.

DELIVERED BEFORE THE BRITISH ASSOCIATION AT LIVERPOOL.

16th September, 1870.

'If thou would'st know the mystic song
  Chaunted when the sphere was young,
  Aloft, abroad, the pæan swells,
  O wise man, hear'st thou half it tells?
  To the open ear it sings
  The early genesis of things;
  Of tendency through endless ages
  Of star-dust and star-pilgrimages,
  Of rounded worlds, of space and time,
  Of the old floods' subsiding slime,
  Of chemic matter, force and form,
  Of poles and powers, cold, wet, and warm.
  The rushing metamorphosis
  Dissolving all that fixture is,
  Melts things that be to things that seem,
  And solid nature to a dream.'

EMERSON.

'Was wär' ein Gott der nur von aussen stiesse
  Im Kreis das All am Finger laufen liesse!
  Ihm ziemt's, die Welt im Innern zu bewegen,
  Natur in Sich, Sich in Natur zu hegen.'

GOETHE.

VII.

SCIENTIFIC USE OF THE IMAGINATION.

' Lastly, physical investigation more than anything besides helps to teach us the actual value and right use of the Imagination—of that wondrous faculty, which, left to ramble uncontrolled, leads us astray into a wilderness of perplexities and errors, a land of mists and shadows ; but which properly controlled by experience and reflection, becomes the noblest attribute of man : the source of poetic genius, the instrument of discovery in Science, without the aid of which Newton would never have invented fluxions, nor Davy have decomposed the earths and alkalies, nor would Columbus have found another Continent.'—Address to the Royal Society by its President Sir Benjamin Brodie, November 30, 1859.

I CARRIED with me to the Alps this year the heavy burden of this evening's work. In the way of new investigation I had nothing complete enough to be brought before you ; so all that remained to me was to fall back upon such residues as I could find in the depths of consciousness, and out of them to spin the fibre and weave the web of this discourse. Save from memory I had no direct aid upon the mountains ; but to spur up the emotions, on which so much depends, as well as to nourish indirectly the intellect and will, I took with me two volumes of poetry, Goethe's ' Farbenlehre,' and the work on ' Logic ' recently published by Mr. Alexander Bain.[1] The spur, I am sorry to say, was no match for the integument of dulness it had to pierce. In Goethe, so

---

[1] One of my critics remarks, that he does not see the wit of calling Goethe's 'Farbenlehre' and Bain's 'Logic,' 'two volumes of poetry.' Nor do I.

glorious otherwise, I chiefly noticed the self-inflicted hurts of genius, as it broke itself in vain against the philosophy of Newton. For a time, Mr. Bain became my principal companion. I found him learned and practical, shining generally with a dry light, but exhibiting at times a flush of emotional strength, which proved that even logicians share the common fire of humanity. He interested me most when he became the mirror of my own condition. Neither intellectually nor socially is it good for man to be alone, and the griefs of thought are more patiently borne when we find that they have been experienced by another. From certain passages in his book I could infer that Mr. Bain was no stranger to such sorrows. Take this passage as an illustration. Speaking of the ebb of intellectual force, which we all from time to time experience, Mr. Bain says, 'The uncertainty where to look for the next opening of discovery brings the pain of conflict and the debility of indecision.' These words have in them the true ring of personal experience. The action of the investigator is periodic. He grapples with a subject of enquiry, wrestles with it, overcomes it, exhausts, it may be, both himself and it for the time being. He breathes a space, and then renews the struggle in another field. Now this period of halting between two investigations is not always one of pure repose. It is often a period of doubt and discomfort, of gloom and ennui. 'The uncertainty where to look for the next opening of discovery brings the pain of conflict and the debility of indecision.' Such was my precise condition in the Alps this year; in a score of words Mr. Bain has here sketched my mental diagnosis; and it was under these evil circumstances that I had to equip myself for the hour and the ordeal that are now come.

Gladly, however, as I should have seen this duty in other hands, I could by no means shrink from it. Disloyalty would have been worse than failure. In some fashion or other—feebly or strongly, meanly or manfully, on the higher levels of thought, or on the flats of commonplace—the task had to be accomplished. I looked in various directions for help and furtherance; but without me for a time I saw only ' antres vast,' and within me ' deserts idle.' My case resembled that of a sick doctor who had forgotten his art and sorely needed the prescription of a friend. Mr. Bain wrote one for me. He said, ' Your present knowledge must forge the links of connection between what has been already achieved and what is now required.'[1] In these words he admonished me to review the past and recover from it the broken ends of former investigations. I tried to do so. Previous to going to Switzerland I had been thinking much of light and heat, of magnetism and electricity, of organic germs, atoms, molecules, spontaneous generation, comets, and skies. With one or another of these I now sought to reform an alliance, and finally succeeded in establishing a kind of cohesion between thought and Light. The wish grew within me to trace, and to enable you to trace, some of the more occult operations of this agent. I wished, if possible, to take you behind the drop-scene of the senses, and to show you the hidden mechanism of optical action. For I take it to be well worth the while of the scientific teacher to take some pains, and even

---

[1] Induction, page 422.

great pains, to make those whom he addresses copartners
of his thoughts. To clear his own mind in the first place
of all haze and vagueness, and then to project into
language which shall leave no mistake as to his meaning
—which shall leave even his errors naked—the definite
ideas he has shaped. A great deal is, I think, possible to
scientific exposition conducted in this way. It is possible,
I believe, even before an audience like the present, to
uncover to some extent the unseen things of nature; and
thus to give not only to professed students, but to others
with the necessary bias, industry, and capacity, an intelligent interest in the operations of science. Time and
labour are necessary to this result, but science is the
gainer from the public sympathy thus created.

How then are those hidden things to be revealed?
How, for example, are we to lay hold of the physical
basis of light, since, like that of life itself, it lies entirely
without the domain of the senses? Philosophers may
be right in affirming that we cannot transcend experience; but we can, at all events, carry it a long way
from its origin. We can also magnify, diminish, qualify,
and combine experiences, so as to render them fit for
purposes entirely new. We are gifted with the power of
Imagination—combining what the Germans call An-
schauungsgabe and Einbildungskraft—and by this power
we can lighten the darkness which surrounds the world
of the senses. There are tories even in science who
regard imagination as a faculty to be feared and avoided
rather than employed. They had observed its action in
weak vessels, and were unduly impressed by its disasters.
But they might with equal justice point to exploded
boilers as an argument against the use of steam. Bounded

and conditioned by cooperant Reason, imagination becomes the mightiest instrument of the physical discoverer. Newton's passage from a falling apple to a falling moon was, at the outset, a leap of the imagination. When William Thomson tries to place the ultimate particles of matter between his compass points, and to apply to them a scale of millimetres, he is powerfully aided by this faculty. And in much that has been recently said about protoplasm and life, we have the outgoings of the imagination guided and controlled by the known analogies of science. In fact, without this power, our knowledge of nature would be a mere tabulation of co-existences and sequences. We should still believe in the succession of day and night, of summer and winter ; but the soul of Force would be dislodged from our universe ; causal relations would disappear, and with them that science which is now binding the parts of nature to an organic whole.

I should like to illustrate by a few simple instances the use that scientific men have already made of this power of imagination, and to indicate afterwards some of the further uses that they are likely to make of it. Let us begin with the rudimentary experiences. Observe the falling of heavy rain-drops into a tranquil pond. Each drop as it strikes the water becomes a centre of disturbance, from which a series of ring-ripples expand outwards. Gravity and inertia are the agents by which this wave-motion is produced, and a rough experiment will suffice to show that the rate of propagation does not amount to a foot a second. A series of slight mechanical shocks is experienced by a body plunged in the water as the wavelets reach it in succession. But a finer motion is at the same time set up and propagated. If the head and ears be immersed in the water, as in an experiment of Franklin's, the shock of the drop is communicated to the auditory nerve—the *tick* of the drop is heard. Now this sonorous impulse is propagated, not at the rate of a foot a second, but at the rate of 4,700 feet a second. In this case it is not the gravity, but the *elasticity* of the water that is the urging force. Every liquid particle pushed against its neighbour delivers up its motion with extreme rapidity, and the pulse is propagated as a thrill. The incompressibility of water, as illustrated by the famous Florentine experiment, is a measure of its elasticity, and to the possession of this property in so high a degree the rapid transmission of a sound-pulse through water is to be ascribed.

But water, as you know, is not necessary to the conduction of sound; air is its most common vehicle. And you know that when the air possesses the particular density and elasticity corresponding to the temperature of freezing water the velocity of sound in it is 1,090 feet a second. It is almost exactly one-fourth of the velocity in water; the reason being that though the greater weight of the water tends to diminish the velocity, the enormous molecular elasticity of the liquid far more than atones for the disadvantage due to weight. By various contrivances we can compel the vibrations of the air to declare themselves; we know the length and frequency of sonorous waves, and we have also obtained great mastery over the various methods by which the air is thrown into vibration. We know the phenomena and laws of vibrating rods, of organ-pipes, strings, membranes, plates, and bells. We can abolish one sound by another.

We know the physical meaning of music and noise, of harmony and discord. In short, as regards sounds we have a very clear notion of the external physical processes which correspond to our sensations.

In these phenomena of sound we travel a very little way from downright sensible experience. Still the imagination is to some extent exercised. The bodily eye, for example, cannot see the condensations and rarefactions of the waves of sound. We construct them in thought, and we believe as firmly in their existence as in that of the air itself. But now our experience has to be carried into a new region, where a new use is to be made of it. Having mastered the cause and mechanism of sound, we desire to know the cause and mechanism of light. We wish to extend our enquiries from the auditory nerve to the optic nerve. There is in the human intellect a power of expansion—I might almost call it a power of creation—which is brought into play by the simple brooding upon facts. The legend of the Spirit brooding over chaos may have originated in a knowledge of this power. In the case now before us it has manifested itself by transplanting into space, for the purposes of light, an adequately modified form of the mechanism of sound. We know intimately whereon the velocity of sound depends. When we lessen the density of a medium and preserve its elasticity constant we augment the velocity. When we heighten the elasticity and keep the density constant we also augment the velocity. A small density, therefore, and a great elasticity, are the two things necessary to rapid propagation. Now light is known to move with the astounding velocity of 185,000 miles a second. How is such a velocity to be obtained?

By boldly diffusing in space a medium of the requisite
tenuity and elasticity.

Let us make such a medium our starting point, en-
dowing it with one or two other necessary qualities ; let
us handle it in accordance with strict mechanical laws ;
let us give to every step of our deduction the surety of
the syllogism ; let us carry it thus forth from the world of
imagination into the world of sense, and see whether the
final outcrop of the deduction be not the very phenomena
of light which ordinary knowledge and skilled experiment
reveal. If in all the multiplied varieties of these phenomena, including those of the most remote and entangled
description, this fundamental conception always brings us
face to face with the truth ; if no contradiction to our
deductions from it be found in external nature, but
on all sides agreement and verification ; if, moreover,
as in the case of Conical Refraction and in other cases,
it has actually forced upon our attention phenomena
which no eye had previously seen, and which no mind
had previously imagined, such a conception, which never
disappoints us, but always lands us on the solid shores of
fact, must, we think, be something more than a mere
figment of the scientific fancy. In forming it that com-
posite and creative unity in which reason and imagination
are together blent, has, we believe, led us into a world
not less real than that of the senses, and of which the
world of sense itself is the suggestion and justification.

Far be it from me, however, to wish to fix you immov-
ably in this or in any other theoretic conception. With
all our belief of it, it will be well to keep the theory
plastic and capable of change. You may, moreover, urge
that although the phenomena occur *as if* the medium

existed, the absolute demonstration of its existence is still wanting.   Far be it from me to deny to this reasoning such validity as it may fairly claim.   Let us endeavour by means of analogy to form a fair estimate of its force. You believe that in society you are surrounded by reasonable beings like yourself.   You are perhaps as firmly convinced of this as of anything.   What is your warrant for this conviction?   Simply and solely this, your fellow-creatures behave as if they were reasonable; the hypothesis, for it is nothing more, accounts for the facts.   To take an eminent example: you believe that our President is a reasonable being.   Why?   There is no known method of superposition by which any one of us can apply himself intellectually to another so as to demonstrate coincidence as regards the possession of reason.   If, therefore, you hold our President to be reasonable, it is because he behaves *as if* he were reasonable.   As in the case of the æther, beyond the ' *as if* ' you cannot go.   Nay I should not wonder if a close comparison of the data on which both inferences rest, caused many respectable persons to conclude that the æther had the best of it.

This universal medium, this light-æther as it is called, is a vehicle, not an origin of wave-motion.   It receives and transmits, but it does not create.   Whence does it derive the motions it conveys?   For the most part from luminous bodies.   By this motion of a luminous body I do not mean its sensible motion, such as the flicker of a candle, or the shooting out of red prominences from the limb of the sun.   I mean an intestine motion of the atoms or molecules of the luminous body.   But here a certain reserve is necessary.   Many chemists of the present day refuse to speak of atoms and molecules as real things.

Their caution leads them to stop short of the clear, sharp, mechanically intelligible atomic theory enunciated by Dalton, or any form of that theory, and to make the doctrine of multiple proportions their intellectual bourne. I respect the caution, though I think it is here misplaced. The chemists who recoil from these notions of atoms and molecules accept without hesitation the Undulatory Theory of Light. Like you and me they one and all believe in an æther and its light-producing waves. Let us consider what this belief involves. Bring your imaginations once more into play and figure a series of soundwaves passing through air. Follow them up to their origin, and what do you there find? A definite, tangible, vibrating body. It may be the vocal chords of a human being, it may be an organ-pipe, or it may be a stretched string. Follow in the same manner a train of æther waves to their source; remembering at the same time that your æther is matter, dense, elastic, and capable of motions subject to and determined by mechanical laws. What then do you expect to find as the source of a series of æther waves? Ask your imagination if it will accept a vibrating multiple proportion—a numerical ratio in a state of oscillation? I do not think it will. You cannot crown the edifice by this abstraction. The scientific imagination, which is here authoritative, demands as the origin and cause of a series of æther waves a particle of vibrating matter quite as definite, though it may be excessively minute, as that which gives origin to a musical sound. Such a particle we name an atom or a molecule. I think the seeking intellect when focussed so as to give definition without penumbral haze, is sure to realise this image at the last.

With the view of preserving thought continuous throughout this discourse, and of preventing either failure of knowledge or of memory from causing any rent in our picture, I here propose to run rapidly over a bit of ground which is probably familiar to most of you, but which I am anxious to make familiar to you all. The waves generated in the æther by the swinging atoms of luminous bodies are of different lengths and amplitudes. The amplitude is the width of swing of the individual particles of the wave. In water-waves it is the height of the crest above the trough, while the length of the wave is the distance between two consecutive crests. The aggregate of waves emitted by the sun may be broadly divided into two classes: the one class competent, the other incompetent, to excite vision. But the light-producing waves differ markedly among themselves in size, form, and force. The length of the largest of these waves is about twice that of the smallest, but the amplitude of the largest is probably a hundred times that of the smallest. Now the force or energy of the wave, which, expressed with reference to sensation, means the intensity of the light, is proportional to the square of the amplitude. Hence the amplitude being one-hundredfold, the energy of the largest light-giving waves would be ten-thousandfold that of the smallest. This is not improbable. I use these figures not with a view to numerical accuracy, but to give you definite ideas of the differences that probably exist among the light-giving waves. And if we take the whole range of solar radiation into account—its non-visual as well as its visual waves—I think it probable that the force or energy of the largest wave is a million times that of the smallest.

Turned into their equivalents of sensation, the different

light-waves produce different colours. Red, for example, is produced by the largest waves, violet by the smallest, while green is produced by a wave of intermediate length and amplitude. On entering from air into more highly refracting substances, such as glass or water, or the sulphide of carbon, all the waves are retarded, but the smallest ones most. This furnishes a means of separating the different classes of waves from each other; in other words, of analysing the light. Sent through a refracting prism, the waves of the sun are turned aside in different degrees from their direct course, the red least, the violet most. They are virtually pulled asunder, and they paint upon a white screen placed to receive them 'the solar spectrum.' Strictly speaking, the spectrum embraces an infinity of colours, but the limits of language and of our powers of distinction cause it to be divided into seven segments : red, orange, yellow, blue, indigo, violet. These are the seven primary or prismatic colours.

Separately, or mixed in various proportions, the solar waves yield all the colours observed in nature and employed in art. Collectively, they give us the impression of whiteness. Pure unsifted solar light is white ; and if all the wave-constituents of such light be reduced in the same proportion, the light, though diminished in intensity, will still be white. The whiteness of Alpine snow with the sun shining upon it, is barely tolerable to the eye. The same snow under an overcast firmament is still white. Such a firmament enfeebles the light by reflection, and when we lift ourselves above a cloud-field—to an Alpine summit, for instance, or to the top of Snowdon—and see, in the proper direction, the sun shining on the clouds, they appear dazzlingly white. Ordinary clouds, in fact, divide the solar light impinging on them into two parts—a reflected part and a transmitted part, in each of which the proportions of wave-motion which produce the impression of whiteness are sensibly preserved.

It will be understood that the conditions of whiteness would fail if all the waves were diminished *equally*, or by the same absolute quantity. They must be reduced *proportionately*, instead of equally. If by the act of reflection the waves of red light are split into exact halves, then, to preserve the light white, the waves of yellow, orange, green, and blue must also be split into exact halves. In short, the reduction must take place, not by absolutely equal quantities, but by equal fractional parts. In white light the preponderance as regards energy of the larger over the smaller waves must always be immense. Were the case otherwise, the physiological correlative, *blue*, of the smaller waves would have the upper hand in our sensations.

My wish to render our mental images complete, causes me to dwell briefly upon these known points, and the same wish will cause me to linger a little longer among others. But here I am disturbed by my reflections. When I consider the effect of dinner upon the nervous system, and the relation of that system to the intellectual powers I am now invoking—when I remember that the universal experience of mankind has fixed upon certain definite elements of perfection in an after-dinner speech, and when I think how conspicuous by their absence these elements are on the present occasion, the thought is not comforting to a man who wishes to stand well with his fellow-creatures in general, and with the members of the British Association in particular. My condition might well resemble that of the æther, which is scientifically defined as an assemblage of vibrations. And the worst of it is that unless you reverse the general verdict regarding the effect of dinner, and prove in your own persons that a uniform experience need not continue uniform—which will be a great point gained for some people—these tremors of mine are likely to become more and more painful. But I call to mind the comforting words of an inspired though uncanonical writer, who admonishes us in the Apocrypha that fear is a bad counsellor. Let me then cast him out, and let me trustfully assume that you will one and all postpone that balmy sleep, of which dinner might under the circumstances be regarded as the indissoluble antecedent, and that you will manfully and womanfully prolong your investigations of the æther and its waves into regions which have been hitherto crossed by the pioneers of science alone.

Not only are the waves of æther reflected by clouds, by solids, and by liquids, but when they pass from light air to dense, or from dense air to light, a portion of the wave-motion is always reflected. Now our atmosphere changes continually in density from top to bottom. It will help our conceptions if we regard it as made up of a series of thin concentric layers, or shells of air, each shell being of the same density throughout, and a small and sudden change of density occurring in passing from shell to shell. Light would be reflected at the limiting surfaces of all these shells, and their action would be practically the same as that of the real atmosphere. And now I would ask your imagination to picture this act of reflection. What must become of the reflected light? The atmospheric layers turn their convex surfaces towards the sun, they are so many convex mirrors of feeble power, and you will immediately perceive that the light regularly reflected from these surfaces cannot reach the earth at all, but is dispersed in space.

But though the sun's light is not reflected in this fashion from the aërial layers to the earth, there is indubitable evidence to show that the light of our firmament is reflected light. Proofs of the most cogent description could be here adduced; but we need only consider that we receive light at the same time from all parts of the hemisphere of heaven. The light of the firmament comes to us across the direction of the solar rays, and even against the direction of the solar rays; and this lateral and opposing rush of wave-motion can only be due to the rebound of the waves from the air itself, or from something suspended in the air. It is also evident that, unlike the action of clouds, the solar light is not reflected by the sky in the proportions which produce white. The sky is blue, which indicates a deficiency on part of the larger waves. In accounting for the colour of the sky, the first question suggested by the analogy would undoubtedly be, Is not the air blue? The blueness of the air has in fact been given as a solution of the blueness of the sky. But reason, basing itself on observation, asks in reply, How, if the air be blue, can the light of sunrise and sunset, which travels through vast distances of air, be yellow, orange, or even red? The passage of white solar light through a blue medium could by no possibility redden the light. The hypothesis of a blue air is therefore untenable. In fact the agent, whatever it is, which sends us the light of the sky, exercises in so doing a dichroitic action.   The light reflected is blue, the light transmitted is orange or red.   A marked distinction is thus exhibited between the matter of the sky and that of an ordinary cloud, which exercises no such dichroitic action.

By the force of imagination and reason combined we may penetrate this mystery also.   The cloud takes no note of size on the part of the waves of æther, but reflects them all alike.   It exercises no selective action.   Now the cause of this may be that the cloud particles are so large in comparison with the size of the waves of æther as to reflect them all indifferently.   A broad cliff reflects an Atlantic roller as easily as a ripple produced by a sea-bird's wing ; and in the presence of large reflecting surfaces, the existing differences of magnitude among the waves of æther may disappear.   But supposing the reflecting particles, instead of being very large, to be very small, in comparison with the size of the waves.   In this case, instead of the whole wave being fronted and in great part thrown back, a small portion only is shivered off.   The great mass of the wave passes over such a particle without reflection.   Scatter then a handful of such minute foreign particles in our atmosphere, and set imagination to watch their action upon the solar waves. Waves of all sizes impinge upon the particles, and you see at every collision a portion of the impinging wave struck off.   All the waves of the spectrum, from the extreme red to the extreme violet, are thus acted upon.   But in what proportions will the waves be scattered ?   A clear picture will enable us to anticipate the experimental answer.   Remembering that the red waves are to the blue much in the relation of billows to ripples, let us consider whether those extremely small particles are competent to scatter all the waves in the same proportion. If they be not—and a little reflection will make it clear to you that they are not—the production of colour must be an incident of the scattering. Largeness is a thing of relation ; and the smaller the wave, the greater is the relative size of any particle on which the wave impinges, and the greater also the ratio of the scattered portion to the total wave. A pebble placed in the way of the ring-ripples produced by our heavy rain-drops on a tranquil pond will throw back a large fraction of the ripple incident upon it, while the fractional part of a larger wave thrown back by the same pebble might be infinitesimal. Now we have already made it clear to our minds that to preserve the solar light white, its constituent proportions must not be altered ; but in the act of division performed by these very small particles we see that the proportions *are* altered ; an undue fraction of the smaller waves is scattered by the particles, and, as a consequence, in the scattered light, blue will be the predominant colour. The other colours of the spectrum must, to some extent, be associated with the blue. They are not absent but deficient. We ought, in fact, to have them all, but in diminishing proportions, from the violet to the red.

We have here presented a case to the imagination, and, assuming the undulatory theory to be a reality, we have, I think, fairly reasoned our way to the conclusion, that were particles, small in comparison to the size of the æther waves, sown in our atmosphere, the light scattered by those particles would be exactly such as we observe in our azure skies. When this light is analysed, all the colours of the spectrum are found ; but they are found in the proportions indicated by our conclusion.

Let us now turn our attention to the light which passes unscattered among the particles. How must it be finally affected ? By its successive collisions with the particles the white light is more and more robbed of ,its shorter waves ; it therefore loses more and more of its due proportion of blue. The result may be anticipated. The transmitted light, where short distances are involved, will appear yellowish. But as the sun sinks towards the horizon the atmospheric distances increase, and consequently the number of the scattering particles. They abstract in succession the violet, the indigo, the blue, and even disturb the proportions of green. The transmitted light under such circumstances must pass from yellow through orange to red. This also is exactly what we find in nature. Thus, while the reflected light gives us at noon the deep azure of the Alpine skies, the transmitted light gives us at sunset the warm crimson of the Alpine snows. The phenomena certainly occur *as if* our atmosphere were a medium rendered slightly turbid by the mechanical suspension of exceedingly small foreign particles.

Here, as before, we .encounter our sceptical ' *as if.*' It is one of the parasites of science, ever at hand, and ready to plant itself and sprout, if it can, on the weak points of our philosophy. But a strong constitution defies the parasite, and in our case, as we question the phenomena, probability grows like growing health, until in the end the malady of doubt is completely extirpated. The first question that naturally arises is—Can small particles be really proved to act in the manner indicated ? No doubt of it. Each one of you can submit the question to an experimental test. Water will not dissolve resin, but spirit will; and when spirit which holds resin in solution is dropped into water, the resin immediately separates in solid particles, which render the water milky. The coarseness of this precipitate depends on the quantity of the dissolved resin. You can cause it to separate in thick clots or in exceedingly fine particles. Professor Brücke has given us the proportions which produce particles particularly suited to our present purpose. One gramme of clean mastic is dissolved in eighty-seven grammes of absolute alcohol, and the transparent solution is allowed to drop into a beaker containing clear water kept briskly stirred. An exceedingly fine precipitate is thus formed, which declares its presence by its action upon light. Placing a dark surface behind the beaker, and permitting the light to fall into it from the top or front, the medium is seen to be distinctly blue. It is not perhaps so perfect a blue as I have seen on exceptional days, this year, among the Alps, but it is a very fair sky-blue. A trace of soap in water gives a tint of blue. London, and I fear Liverpool milk, makes an approximation to the same colour through the operation of the same cause; and Helmholtz has irreverently disclosed the fact that the deepest blue eye is simply a turbid medium.

The action of turbid media upon light was illustrated by Goethe, who, though unacquainted with the undulatory theory, was led by his experiments to regard the firmament as an illuminated turbid medium with the darkness of space behind it. He describes glasses showing a bright yellow by transmitted, and a beautiful blue by reflected light. Professor Stokes, who was probably the first to discern the real nature of the action of small particles on the waves of æther, describes a glass of a similar kind.[1] Capital specimens of such glass are to be found at Salviati's in St. James's Street. What artists call 'chill' is no doubt an effect of this description. Through the action of minute particles, the browns of a picture often present the appearance of the bloom of a plum. By rubbing the varnish with a silk handkerchief optical continuity is established, and the chill disappears. Some years ago I witnessed Mr. Hirst experimenting at Zermatt on the turbid water of the Visp, which was charged with the finely divided matter ground down by the glaciers. When kept still for a day or so, the grosser matter sank, but the finer matter remained suspended, and gave a distinctly blue tinge to the water. The blueness of certain Alpine lakes has been shown to be in part due to this cause. Prof. Roscoe has noticed several striking cases of a similar kind. In a very remarkable paper the late Principal Forbes showed that steam issuing from the safety-valve of a locomotive, when favourably observed, exhibits at a certain stage of its condensation the colours of the sky. It is blue by reflected light, and orange or red by transmitted light. The same effect, as pointed out by Goethe, is to some extent exhibited by peat smoke. More than ten years ago I amused myself at Killarney by observing on a calm day the straight smoke-columns rising from the cabin chimneys. It was easy to project

[1] This glass, by reflected light, had a colour 'strongly resembling that of a decoction of horse-chestnut bark.' Curiously enough Goethe refers to this very decoction: 'Man nehme einen Streifen frischer Rinde von der Rosskastanie, man stecke denselben in ein Glas Wasser, und in der kürzesten Zeit werden wir das vollkommenste Himmelblau entstehen sehen.'—Goethe's *Werke*, b. xxix. p. 24.

the lower portion of a column against a dark pine, and its upper portion against a bright cloud. The smoke in the former case was blue, being seen mainly by reflected light ; in the latter case it was reddish, being seen mainly by transmitted light. Such smoke was not in exactly the condition to give us the glow of the Alps, but it was a step in this direction. Brücke's fine precipitate above referred to looks yellowish by transmitted light, but by duly strengthening the precipitate you may render the white light of noon as ruby-coloured as the sun when seen through Liverpool smoke, or upon Alpine horizons. I do not, however, point to the gross smoke arising from coal as an illustration of the action of small particles, because such smoke soon absorbs and destroys the waves of blue instead of sending them to the eyes of the observer.

These multifarious facts, and numberless others which cannot now be referred to, are explained by reference to the single principle, that where the scattering particles are small in comparison to the size of the waves we have in the reflected light a greater proportion of the smaller waves, and in the transmitted light a greater proportion of the larger waves, than existed in the original white light. The physiological consequence is that in the one light blue is predominant, and in the other light orange or red. And now let us push our enquiries forward. Our best microscopes can readily reveal objects not more than $\frac{1}{50000}$th of an inch in diameter. This is less than the length of a wave of red light. Indeed a first-rate microscope would enable us to discern objects not exceeding in diameter the length of the smallest waves of the visible spectrum. By the microscope therefore we can submit our particles to an experimental test. If they are as large as the light-waves they will infallibly be seen : and if they are not seen it is because they are smaller. I placed in the hands of our President a bottle containing Brücke's particles in greater number and coarseness than those examined by Brücke himself. The liquid was a milky blue, and Mr. Huxley applied to it his highest microscopic power. He satisfied me at the time that had particles of even $\frac{1}{100000}$th of an inch in diameter existed in the liquid they could not have escaped detection. But no particles were seen. Under the microscope the turbid liquid was not to be distinguished from distilled water. Brücke, I may say, also found the particles to be of ultra-microscopic magnitude.

But we have it in our power to imitate far more closely than we have hitherto done the natural conditions of this problem. We can generate in air, as many of you know, artificial skies, and prove their perfect identity with the natural one, as regards the exhibition of a number of wholly unexpected phenomena. By a continuous process of growth, moreover, we are able to connect sky-matter, if I may use the term, with molecular matter on the one side, and with molar matter, or matter in sensible masses, on the other. In illustration of this, I will take an experiment described by M. Morren of Marseilles at the last meeting of the British Association. Sulphur and oxygen combine to form sulphurous acid gas. It is this choking gas that is smelt when a sulphur match is burnt in air. Two atoms of oxygen and one of sulphur constitute the molecule of sulphurous acid. Now it has been recently shown in a great number of instances that waves of æther issuing from a strong source, such as the sun or the electric light, are competent to shake asunder the atoms of gaseous molecules. A chemist would call this 'decomposition' by light; but it behoves us, who are examining the power and function of the imagination, to keep constantly before us the physical images which underlie our terms. Therefore I say, sharply and definitely, that the components of the molecules of sulphurous acid are shaken asunder by the æther waves. Enclosing the substance in a suitable vessel, placing it in a dark room, and sending through it a powerful beam of light, we at first see nothing: the vessel containing the gas is as empty as a vacuum. Soon, however, along the track of the beam a beautiful sky-blue colour is observed, which is due to the liberated particles of sulphur. For a time the blue grows more intense ; it then becomes whitish ; and from a whitish blue it passes to a more or less perfect white. If the action be continued long enough, we end by filling the tube with a dense cloud of sulphur particles, which by the application of proper means may be rendered visible.

Here then our æther waves untie the bond of chemical affinity, and liberate a body—sulphur—which at ordinary temperatures is a solid, and which therefore soon becomes an object of the senses. We have first of all the free atoms of sulphur, which are both invisible and incompetent to stir the retina sensibly with scattered light. But these atoms gradually coalesce and form particles, which grow larger by continual accretion until after a minute or two they appear as sky-matter. In this condition they are invisible themselves, but competent to send an amount of wave-motion to the retina sufficient to produce the firmamental blue. The particles continue, or may be caused to continue, in this condition for a considerable time, during which no microscope can cope with them. But they continually grow larger, and pass by insensible gradations into the state of *cloud,* when they can no longer elude the armed eye. Thus without solution of continuity we start with matter in the molecule, and end with matter in the mass, sky-matter being the middle term of the series of transformations.

Instead of sulphurous acid, we might choose from a dozen other substances, and produce the same effect with any of them. In the case of some—probably in the case of all—it is possible to preserve matter in the skyey condition for fifteen or twenty minutes under the continual operation of the light. During these fifteen or twenty minutes the particles are constantly growing larger, without ever. exceeding the size requisite to the production of the celestial blue. Now when two vessels are placed before you, each containing sky-matter, it is possible to state with great distinctness which vessel contains the largest particles. The retina is very sensitive to differences of light, when, as here, the eye is in comparative darkness, and when the quantities of wave-motion thrown against the retina are small. The larger particles declare themselves by the greater whiteness of their scattered light. Call now to mind the observation, or effort at observation, made by our President, when he failed to distinguish the particles of mastic in Brücke's medium, and when you have done so follow me. I permitted a beam of light to act upon a certain vapour. In two minutes the azure appeared, but at the end of fifteen minutes it had not ceased to be azure. After fifteen minutes, for example, its colour, and some other phenomena, pronounced it to be a blue of distinctly smaller particles than those sought for in vain by Mr. Huxley. These particles, as already stated, must have been less than $\frac{1}{100000}$th of an inch in diameter. And now I want you to submit to your imagination the following question : Here are particles which have been growing continually for fifteen minutes, and at the end of that time are demonstrably smaller than those which defied the microscope of Mr. Huxley :—*what must have been the size of these particles at the beginning of their growth?* What notion can you form of the magnitude of such particles? The distances of stellar space give us simply a bewildering sense of vastness without leaving any distinct impression on the mind, and the magnitudes with which we have here to do bewilder us equally in the opposite direction. We are dealing with infinitesimals compared with which the test objects of the microscope are literally immense.

From their perviousness to stellar light, and other considerations, Sir John Herschel drew some startling conclusions regarding the density and weight of comets. You know that these extraordinary and mysterious bodies sometimes throw out tails 100,000,000 of miles in length, and 50,000 miles in diameter. The diameter of our earth is 8,000 miles. Both it and the sky, and a good portion of space beyond the sky, would certainly be included in a sphere 10,000 miles across. Let us fill a hollow sphere of this diameter with cometary matter, and make it our unit of measure. To produce a comet's tail of the size just mentioned about 300,000 such measures would have to be emptied into space. Now suppose the whole of this stuff to be swept together, and suitably compressed, what do you suppose its volume would be ?　Sir John Herschel would probably tell you that the whole mass might be carted away at a single effort by one of your dray-horses.　In fact, I do not know that he would require more than a small fraction of a horse-power to remove the cometary dust.　After this you will hardly regard as monstrous a notion I have sometimes entertained concerning the quantity of matter in our sky.　Suppose a shell to surround the earth at a height above the surface which would place it beyond the grosser matter that hangs in the lower regions of the air —say at the height of the Matterhorn or Mont Blanc. Outside this shell we have the deep blue firmament.　Let the atmospheric space beyond the shell be swept clean, and let the sky-matter be properly gathered up.　What is its probable amount ?　I have sometimes thought that a lady's portmanteau would contain it all.　I have thought that even a gentleman's portmanteau—possibly his snuff-box—might take it in.　And whether the actual sky be capable of this amount of condensation or not, I entertain no doubt that a sky quite as vast as ours, and as good in appearance, could be formed from a quantity of matter which might be held in the hollow of the hand.

Small in mass, the vastness in point of number of the particles of our sky may be inferred from the continuity of its light.　It is not in broken patches, nor at scattered points that the heavenly azure is revealed.　To the observer on the summit of Mont Blanc the blue is as uniform and coherent as if it formed the surface of the most close-grained solid.　A marble dome would not exhibit a stricter continuity.　And Mr. Glaisher will inform you that if our hypothetical shell were lifted to twice the height of Mont Blanc above the earth's surface, we should still have the azure overhead. Everywhere through the atmosphere those sky-particles are strewn. They fill the Alpine valleys, spreading like a delicate gauze in front of the slopes of pine. They sometimes so swathe the peaks with light as to abolish their definition. This year I have seen the Weisshorn thus dissolved in opalescent air. By proper instruments the glare thrown from the sky-particles against the retina may be quenched, and then the mountain which it obliterated starts into sudden definition. Its extinction in front of a dark mountain resembles exactly the withdrawal of a veil. It is the light then taking possession of the eye, and not the particles acting as opaque bodies, that interfere with the definition. By day this light quenches the stars; even by moonlight it is able to exclude from vision all stars between the fifth and the eleventh magnitude. It may be likened to a noise, and the stellar radiance to a whisper drowned by the noise.

What is the nature of the particles which shed this light? The celebrated De la Rive ascribes the haze of the Alps in fine weather to floating organic germs. Now the possible existence of germs in such profusion has been held up as an absurdity. It has been affirmed that they would darken the air, and on the assumed impossibility of their existence in the requisite numbers, without invasion of the solar light, a powerful argument has been based by believers in spontaneous generation. Similar arguments have been used by the opponents of the germ theory of epidemic disease, who have triumphantly challenged an appeal to the microscope and the chemist's balance to decide the question. Such arguments are absolutely valueless. Without committing myself in the least to De la Rive's notion, without offering

any objection here to the doctrine of spontaneous generation, without expressing any adherence to the germ theory
of disease, I would simply draw attention to the fact that
in the atmosphere we have particles which defy both the
microscope and the balance, which do not darken the air,
and which exist, nevertheless, in multitudes sufficient to
reduce to insignificance the Israelitish hyperbole regarding the sands upon the seashore.

The varying judgments of men on these and other
questions may perhaps be, to some extent, accounted for
by that doctrine of Relativity which plays so important a
part in philosophy. This doctrine affirms that the impressions made upon us by any circumstance, or combination
of circumstances, depend upon our previous state. Two
travellers upon the same peak, the one having ascended
to it from the plain, the other having descended to it
from a higher elevation, will be differently affected by the
scene around them. To the one nature is expanding, to
the other it is contracting, and feelings are sure to differ
which have two such different antecedent states. In our
scientific judgments the law of relativity may also play an
important part. To two men, one educated in the school
of the senses, who has mainly occupied himself with observation, and the other educated in the school of imagination as well, and exercised in the conceptions of atoms
and molecules to which we have so frequently referred, a
bit of matter, say $\frac{1}{50000}$th of an inch in diameter, will
present itself differently. The one descends to it from his
molar heights, the other climbs to it from his molecular
low-lands. To the one it appears small, to the other
large. So also as regards the appreciation of the most
minute forms of life revealed by the microscope. To one

of these men they naturally appear conterminous with the ultimate particles of matter, and he readily figures the molecules from which they directly spring; with him there is but a step from the atom to the organism. The other discerns numberless organic gradations between both. Compared with his atoms, the smallest vibrios and bacteria of the microscopic field are as behemoth and leviathan. The law of relativity may to some extent explain the different attitudes of these two men with regard to the question of spontaneous generation. An amount of evidence which satisfies the one entirely fails to satisfy the other; and while to the one the last bold defence and startling expansion of the doctrine will appear perfectly conclusive, to the other it will present itself as imposing a profitless labour of demolition on subsequent investigators.[1]

I trust, Mr. President, that you—whom untoward circumstances have made a biologist, but who still keep alive your sympathy with that class of enquiries which nature intended you to pursue and adorn—will excuse me to your brethren if I say that some of them seem to form an inadequate estimate of the distance which separates the microscopic from the molecular limit, and that, as a consequence, they sometimes employ a phraseology which is calculated to mislead. When, for example, the contents of a cell are described as perfectly homogeneous, as absolutely structureless, because the microscope fails to distinguish any structure, then I think the microscope begins to play a mischievous part. A little consideration will make it plain to all of you that the microscope can have no voice in the real question of germ structure.

---

[1] A resolute scrutiny of the experiments, recently executed with reference to this question, is sure to yield instructive results.

Distilled water is more perfectly homogeneous than the contents of any possible organic germ. What causes the liquid to cease contracting at 39° Fahr., and to expand until it freezes? It is a structural process of which the microscope can take no note, nor is it likely to do so by any conceivable extension of its powers. Place this distilled water in the field of an electro-magnet, and bring a microscope to bear upon it. Will any change be observed when the magnet is excited? Absolutely none; and still profound and complex changes have occurred. First of all, the particles of water are rendered diamagnetically polar; and secondly, in virtue of the structure impressed upon it by the magnetic strain of its molecules, the liquid twists a ray of light in a fashion perfectly determinate both as to quantity and direction. It would be immensely interesting to both you and me if one whom I hoped to see here present,[1] who has brought his brilliant imagination to bear upon this subject, could make us see as he sees the entangled molecular processes involved in the rotation of the plane of polarisation by magnetic force. While dealing with this question, he lived in a world of matter and of motion to which the microscope has no passport, and in which it can offer no aid. The cases in which similar conditions hold are simply numberless. Have the diamond, the amethyst, and the countless other crystals formed in the laboratories of nature and of man no structure? Assuredly they have; but what can the microscope make of it? Nothing. It cannot be too distinctly borne in mind that between the microscope limit and the true molecular

---

[1] Sir William Thomson.

limit there is room for infinite permutations and combinations. It is in this region that the poles of the atoms are arranged, that tendency is given to their powers, so that when these poles and powers have free action and proper stimulus in a suitable environment, they determine first the germ and afterwards the complete organism. This first marshalling of the atoms on which all subsequent action depends baffles a keener power than that of the microscope. Through pure excess of complexity, and long before observation can have any voice in the matter, the most highly trained intellect, the most refined and disciplined imagination, retires in bewilderment from the contemplation of the problem. We are struck dumb by an astonishment which no microscope can relieve, doubting not only the power of our instrument, but even whether we ourselves possess the intellectual elements which will ever enable us to grapple with the ultimate structural energies of nature.

But the speculative faculty, of which imagination forms so large a part, will nevertheless wander into regions where the hope of certainty would seem to be entirely shut out. We think that though the detailed analysis may be, and may for ever remain, beyond us, general notions may be attainable. At all events, it is plain that beyond the present outposts of microscopic enquiry lies an immense field for the exercise of the speculative power. It is only, however, the privileged spirits who know how to use their liberty without abusing it, who are able to surround imagination by the firm frontiers of reason, that are likely to work with any profit here. But freedom to them is of such paramount importance that, for the sake of securing it, a good deal of wildness on the part of weaker brethren may be overlooked. In more senses than one Mr. Darwin has drawn heavily upon the scientific tolerance of his age. He has drawn heavily upon *time* in his development of species, and he has drawn adventurously upon *matter* in his theory of pangenesis. According to this theory, a germ already microscopic is a world of minor germs. Not only is the organism as a whole wrapped up in the germ, but every organ of the organism has there its special seed. This, I say, is an adventurous draft on the power of matter to divide itself and distribute its forces. But, unless we are perfectly sure that he is overstepping the bounds of reason, that he is unwittingly sinning against observed fact or demonstrated law—for a mind like that of Darwin can never sin wittingly against either fact or law—we ought, I think, to be cautious in limiting his intellectual horizon. If there be the least doubt in the matter, it ought to be given in favour of the freedom of such a mind. To it a vast possibility is in itself a dynamic power, though the possibility may never be drawn upon. It gives me pleasure to think that the facts and reasonings of this discourse tend rather towards the justification of Mr. Darwin than towards his condemnation, that they tend rather to augment than to diminish the cubic space demanded by this soaring speculator; for they seem to show the perfect competence of matter and force, as regards divisibility and distribution, to bear the heaviest strain that he has hitherto imposed upon them.

In the case of Mr. Darwin, observation, imagination, and reason combined have run back with wonderful sagacity and success over a certain length of the line of biological succession. Guided by analogy, in his ' Origin of Species,' he placed at the root of life a primordial germ, from which he conceived the amazing richness and variety of the life that now is upon the earth's surface might be deduced. If this hypothesis were true, it would not be final. The human imagination would infallibly look behind the germ, and, however hopeless the attempt, would enquire into the history of its genesis. In this dim twilight of conjecture the searcher welcomes every gleam, and seeks to augment his light by indirect incidences. He studies the methods of nature in the ages and the worlds within his reach, in order to shape the course of speculation in the antecedent ages and worlds. And though the certainty possessed by experimental enquiry is here shut out, the imagination is not left entirely without guidance. From the examination of the solar system, Kant and Laplace came to the conclusion that its various bodies once formed parts of the same undislocated mass; that matter in a nebulous form preceded matter in a dense form; that as the ages rolled away, heat was wasted, condensation followed, planets were detached, and that finally the chief portion of the fiery cloud reached, by self-compression, the magnitude and density of our sun. The earth itself offers evidence of a fiery origin; and in our day the hypothesis of Kant and Laplace receives the independent countenance of spectrum analysis, which proves the same substances to be common to the earth and sun.

Accepting some such view of the construction of our system as probable, a desire immediately arises to connect the present life of our planet with the past. We wish to know something of our remotest ancestry. On its first detachment from the central mass, life, as we understand it, could hardly have been present on the earth. How then did it come there? The thing to be encouraged here is a reverent freedom—a freedom preceded by the hard discipline which checks licentiousness in speculation—while the thing to be repressed, both in science and out of it, is dogmatism. And here I am in the hands of the meeting—willing to end, but ready to go on. I have no right to intrude upon you, unasked, the unformed notions which are floating like clouds, or gathering to more solid consistency in the modern speculative scientific mind. But if you wish me to speak plainly, honestly, and undisputatiously, I am willing to do so. On the present occasion—

> You are ordained to call, and I to come.

Two views, then, offer themselves to us. Life was present potentially in matter when in the nebulous form, and was unfolded from it by the way of natural development, or it is a principle inserted into matter at a later date. With regard to the question of time, the views of men have changed remarkably in our day and generation; and I must say as regards courage also, and a manful willingness to engage in open contest, with fair weapons, a great change has also occurred. The clergy of England—at all events the clergy of London—have nerve enough to listen to the strongest views which anyone amongst us would care to utter; and they invite, if they do not challenge, men of the most decided opinions to state and stand by those opinions in open court. Let the hardiest theory be stated only in the language current among gentlemen, and they look it in the face; smiting the theory, if they do not like it, not with theologic fulmination, but with honest secular strength. With the country clergy I am told the case is different. It is right that I should say this, because the clergy of London have more than once offered me the chance of meeting them in open, honourable discussion.

Two or three years ago, in an ancient London College, I listened to such a discussion at the end of a remarkable lecture by a very remarkable man. Three or four hundred clergymen were present at the lecture. The orator began with the civilisation of Egypt in the time of Joseph; pointing out that the very perfect organisation of the kingdom, and the possession of chariots, in one of which Joseph rode, indicated a long antecedent period of civilisation. He then passed on to the mud of the Nile, its rate of augmentation, its present thickness, and the remains of human handywork found therein; thence to the rocks which bound the Nile valley, and which teem with organic remains. Thus in his own clear and admirable way he caused the idea of the world's age to expand itself indefinitely before the mind of his audience, and he contrasted this with the age usually assigned to the world. During his discourse he seemed to be swimming against a stream; he manifestly thought that he was opposing a general conviction. He expected resistance; so did I. But it was all a mistake : there was no adverse current, no opposing conviction, no resistance, merely here and there a half-humorous, but unsuccessful attempt to entangle him in his talk. The meeting agreed with all that had been said regarding the antiquity of the earth and of its life. They had, indeed, known it all long ago, and they good-humouredly rallied the lecturer

M

for coming amongst them with so stale a story. It was quite plain that this large body of clergymen, who were I should say, the finest samples of their class, had entirely given up the ancient landmarks, and transported the conception of life's origin to an indefinitely distant past.

This leads us to the gist of our present enquiry, which is this:—Does life belong to what we call matter, or is it an independent principle inserted into matter at some suitable epoch—say when the physical conditions became such as to permit of the development of life? Let us put the question with all the reverence due to a faith and culture in which we all were cradled—a faith and culture, moreover, which are the undeniable historic antecedents of our present enlightenment. I say, let us put the question reverently, but let us also put it clearly and definitely. There are the strongest grounds for believing that during a certain period of its history the earth was not, nor was it fit to be, the theatre of life. Whether this was ever a nebulous period, or merely a molten period, does not much matter; and if we revert to the nebulous condition, it is because the probabilities are really on its side. Our question is this:—Did creative energy pause until the nebulous matter had condensed, until the earth had been detached, until the solar fire had so far withdrawn from the earth's vicinity as to permit a crust to gather round the planet? Did it wait until the air was isolated, until the seas were formed, until evaporation, condensation, and the descent of rain had begun, until the eroding forces of the atmosphere had weathered and decomposed the molten rocks so as to form soils, until the sun's rays had become so tempered by distance and by waste as to be chemically fit for the decompositions necessary to vegetable life? Having waited through those Æons until the proper conditions had set in, did it send the fiat forth, ' Let Life be!'? These questions define a hypothesis not without its difficulties, but the dignity of which was demonstrated by the nobleness of the men whom it sustained.

Modern scientific thought is called upon to decide between this hypothesis and another: and public thought generally will afterwards be called upon to do the same. You may, however, rest secure in the belief that the hypothesis just sketched can never be stormed, and that it is sure, if it yield at all, to yield to a prolonged siege. To gain new territory modern argument requires more time than modern arms, though both of them move with greater rapidity than of yore. But however the convictions of individuals here and there may be influenced, the process must be slow and secular which commends the rival hypothesis of Natural Evolution to the public mind. For what are the core and essence of this hypothesis? Strip it naked and you stand face to face with the notion that not alone the more ignoble forms of animalcular or animal life, not alone the nobler forms of the horse and lion, not alone the exquisite and wonderful mechanism of the human body, but that the human mind itself—emotion, intellect, will, and all their phenomena— were once latent in a fiery cloud. Surely the mere statement of such a notion is more than a refutation. But the hypothesis would probably go even further than this. Many who hold it would probably assent to the position that at the present moment all our philosophy, all our poetry, all our science, and all our art—Plato, Shakspeare, Newton, and Raphael—are potential in the fires of the sun. We long to learn something of our origin. If the Evolution hypothesis be correct, even this unsatisfied yearning must have come to us across the ages which separate the unconscious primeval mist from the consciousness of to-day. I do not think that any holder of the Evolution hypothesis would say that I overstate it or overstrain it in any way. I merely strip it of all vagueness, and bring before you unclothed and unvarnished the notions by which it must stand or fall.

Surely these notions represent an absurdity too monstrous to be entertained by any sane mind. Let us, however, give them fair play. Let us steady ourselves in front of the hypothesis, and, dismissing all terror and excitement from our minds, let us look firmly into it with the hard sharp eye of intellect alone. Why are these notions absurd, and why should sanity reject them? The law of Relativity, of which we have previously spoken, may find its application here. These Evolution notions are absurd, monstrous, and fit only for the intellectual gibbet, in relation to the ideas concerning matter which were drilled into us when young, Spirit and matter have ever been presented to us in the rudest contrast, the one as all-noble, the other as all-vile. But is this correct? Does it represent what our mightiest spiritual teacher would call the Eternal Fact of the Universe? Upon the answer to this question all depends. Supposing, instead of having the foregoing antithesis of spirit and matter presented to our youthful minds, we had been taught to regard them as equally worthy and equally wonderful; to consider them in fact as two opposite faces of the self-same mystery. Supposing that in youth we had been impregnated with the notion of the poet

Goethe, instead of the notion of the poet Young, looking at matter, not as brute matter, but as 'the living garment of God ;' do you not think that under these altered circumstances the law of Relativity might have had an outcome different from its present one ? Is it not probable that our repugnance to the idea of primeval union between spirit and matter might be considerably abated ? Without this total revolution of the notions now prevalent, the Evolution hypothesis must stand condemned ; but in many profoundly thoughtful minds such a revolution has already taken place. They degrade neither member of the mysterious duality referred to ; but they exalt one of them from its abasement, and repeal the divorce hitherto existing between both. In substance, if not in words, their position as regards the relation of spirit and matter is : 'What God hath joined together let not man put asunder.' And with regard to the ages of forgetfulness which lie between the unconscious life of the nebula and the conscious life of the earth, it is, they would urge, but an extension of that forgetfulness which preceded the birth of us all.

I have thus led you to the outer rim of speculative science, for beyond the nebulæ scientific thought has never ventured hitherto, and have tried to state that which I considered ought, in fairness, to be outspoken. I do not think this Evolution hypothesis is to. be flouted away contemptuously ; I do not think it is to be denounced as wicked. It is to be brought before the bar of disciplined reason, and there justified or condemned. Let us hearken to those who wisely support it, and to those who wisely oppose it ; and let us tolerate those, and they are many, who foolishly try to do either of these things. The only thing out of place in the discussion is dogmatism on either side. Fear not the Evolution hypothesis. Steady yourselves in its presence upon that faith in the ultimate triumph of truth which was expressed by old Gamaliel when he said : ' If it be of God, ye cannot overthrow it ; if it be of man, it will come to nought.' Under the fierce light of scientific enquiry, this hypothesis is sure to be dissipated if it possess not a core of truth. Trust me, its existence as a hypothesis in the mind is quite compatible with the simultaneous existence of all those virtues to which the term Christian has been applied. It does not solve—it does not profess to solve—the ultimate mystery of this universe. It leaves in fact that mystery untouched. For granting the nebula and its potential life, the question, whence came they? would still remain to baffle and bewilder us. At bottom, the hypothesis does nothing more than ' transport the conception of life's origin to an indefinitely distant past.'

Those who hold the doctrine of Evolution are by no means ignorant of the uncertainty of their data, and they yield no more to it than a provisional assent. They regard the nebular hypothesis as probable, and in the utter absence of any evidence to prove the act illegal, they extend the method of nature from the present into the past. Here the observed uniformity of nature is their only guide. Within the long range of physical enquiry, they have never discerned in nature the insertion of caprice. Throughout this range the laws of physical and intellectual continuity have run side by side. Having thus determined the elements of their curve in a world of observation and experiment, they prolong that curve into an antecedent world, and accept as probable the unbroken sequence of development from the nebula to the present time. You never hear the really philosophical defenders of the doctrine of Uniformity speaking of *impossibilities* in nature. They never say, what they are constantly charged with saying, that it is impossible for the Builder of the universe to alter His work. Their business is not with the possible, but the actual—not with a world which *might* be, but with a world that *is*. This they explore with a courage not unmixed with reverence, and according to methods which, like the quality of a tree, are tested by their fruits. They have but one desire—to know the truth. They have but one fear—to believe a lie. And if they know the strength of science, and rely upon it with unswerving trust, they also know the limits beyond which science ceases to be strong. They best know that questions offer themselves to thought which science, as now prosecuted, has not even the tendency to solve. They keep such questions open, and will not tolerate any unnecessary limitation of the horizon of their souls. They have as little fellowship with the atheist who says there is no God, as with the theist who professes to know the mind of God. 'Two things,' said Immanuel Kant, ' fill me with awe: the starry heavens and the sense of moral responsibility in man.' And in his hours of health and strength and sanity, when the stroke of action has ceased and the pause of reflection has set in, the scientific investigator finds himself overshadowed by the same awe. Breaking contact with the hampering details of earth, it associates him with a power which gives fulness and tone to his existence, but which he can neither analyse nor comprehend.

A TRANSLATION

OF

GOETHE'S PROEMIUM TO *GOTT UND WELT.*

To Him who from eternity, self-stirred,
Himself hath made by His creative word!
To Him, Supreme, who causeth faith to be,
Trust, hope, love, power, and endless energy!
To Him, who, seek to name Him as we will,
UNKNOWN within Himself abideth still!

Strain ear and eye, till sight and sense be dim;
Thou'lt find but faint similitudes of Him:
Yea, and thy spirit in her flight of flame
Still strives to gauge the symbol and the name:
Charmed and compelled thou climb'st from height to height,
And round thy path the world shines wondrous bright;
Time, space, and size, and distance cease to be,
And every step is fresh infinity.

What were the God who sat outside to scan
The spheres that 'neath His finger circling ran?
God dwells within, and moves the world and moulds,
Himself and Nature in one form enfolds:
Thus all that lives in Him, and breathes, and is,
Shall ne'er His puissance, ne'er His spirit miss.

The soul of man, too, is an universe;
Whence follows it that race with race concurs
In framing all it knows of good and true
God?—yea, its own God; and, with homage due,
Surrenders to His sway both earth and heaven;
Fears Him, and loves, where place for love is given.

J. A. S.

Spectator, September 24, 1870.

VIII.

ON RADIATION.

THE 'REDE' LECTURE.

DELIVERED IN THE SENATE-HOUSE BEFORE THE UNIVERSITY OF
CAMBRIDGE.

On Tuesday, May 16, 1865.

'Forsitan et rosea Sol alte lampade lucens,
  Possideat multum caecis fervoribus ignem
  Circum se, qui sit fulgore notatus,
  Æstifer ut tantum radiorum exaugeat ictum.'

Lucretius, v. 610.

'Perhaps too the sun as he shines aloft with rosy lamp has round about him much fire with heats that are not visible, and thus the fire may be marked by no radiance, so that fraught with heat it increases to such a degree the stroke of the rays.'—*Monro's Translation*.

My attention was drawn to this remarkable passage by the late excellent and accomplished Sir Edmund Head, Bart.

VIII.

RADIATION.

1. *Visible and Invisible Radiation.*

BETWEEN the mind of man and the outer world are interposed the nerves of the human body, which translate, or enable the mind to translate, the impressions of that world into facts of consciousness and thought.

Different nerves are suited to the perception of different impressions. We do not see with the ear, nor hear with the eye, nor are we rendered sensible of sound by the nerves of the tongue. Out of the general assemblage of physical actions, each nerve, or group of nerves, selects and responds to those for the perception of which it is specially organised.

The optic nerve passes from the brain to the back of the eyeball and there spreads out, to form the retina, a web of nerve filaments, on which the images of external objects are projected by the optical portion of the eye. This nerve is limited to the apprehension of the phenomena of radiation, and notwithstanding its marvellous sensibility to certain impressions of this class, it is singularly obtuse to other impressions.

Nor does the optic nerve embrace the entire range even of radiation. Some rays, when they reach it, are incompetent to evoke its power, while others never reach r x all being absorbed by the humours of the eye. To all rays which, whether they reach the retina or not, fail to excite vision, we give the name of invisible or obscure rays. All non-luminous bodies emit such rays. There is no body in nature absolutely cold, and every body not absolutely cold emits rays of heat. But to render radiant heat fit to affect the optic nerve a certain temperature is necessary. A cold poker thrust into a fire remains dark for a time, but when its temperature has become equal to that of the surrounding coals it glows like them. In like manner, if a current of electricity of gradually increasing strength be sent through a wire of the refractory metal platinum, the wire first becomes sensibly warm to the touch; for a time its heat augments, still however remaining obscure; at length we can no longer touch the metal with impunity; and at a certain definite temperature it emits a feeble red light. As the current augments in power the light augments in brilliancy, until finally the wire appears of a dazzling white. The light which it now emits is similar to that of the sun.

By means of a prism Sir Isaac Newton unravelled the texture of solar light, and by the same simple instrument we can investigate the luminous changes of our platinum wire. In passing through the prism all its rays (and they are infinite in variety) are bent or refracted from their straight course; and as different rays are differently refracted by the prism, we are by it enabled to separate one class of rays from another. By such prismatic analysis Dr. Draper has shown, that when the platinum wire first begins to glow, the light emitted is a pure red. As the glow augments the red becomes more brilliant, but at the same time orange rays are added to the emission. Augmenting the temperature still further, yellow rays appear beside the orange, after the yellow green rays are emitted, and after the green come, in succession, blue, indigo and violet rays. To display all these colours at the same time the platinum wire must be *white-hot*: the impression of whiteness being in fact produced by the simultaneous action of all these colours on the optic nerve.

In the experiment just described we began with a platinum wire at an ordinary temperature, and gradually raised it to a white heat. At the beginning, and even before the electric current had acted at all upon the wire, it emitted invisible rays. For some time after the action of the current had commenced, and even for a time after the wire had become intolerable to the touch, its radiation was still invisible. The question now arises,—what becomes of these invisible rays when the visible ones make their appearance? It will be proved in the seqnel that they maintain themselves in the radiation; that a ray once emitted continues to be emitted when the temperature is increased, and hence the emission from our platinum wire, even when it has attained its maximum brilliancy, consists of a mixture of visible and invisible rays. If, instead of the platinum wire, the earth itself were raised to incandescence, the obscure radiation which it now emits would continue to be emitted. To reach incandescence the planet would have to pass through all the stages of non-luminous radiation, and the final emission would embrace the rays of all these stages. There can hardly be a doubt that from the sun itself, rays proceed similar in kind to those which the dark earth pours nightly into space. In fact, the various kinds of obscure rays emitted by all the planets of our system are included in the present radiation of the sun.

The great pioneer in this domain of science was Sir William Herschel. Causing a beam of solar light to pass through a prism he resolved it into its coloured constituents ; he formed what is technically called the solar spectrum. Exposing thermometers to the successive colours he determined their heating power, and found it to augment from the violet or most refracted end, to the red or least refracted end of the spectrum. But he did not stop here. Pushing his thermometers into the dark space beyond the red he found that, though the light had disappeared, the radiant heat falling on the instruments was more intense than that at any visible part of the spectrum. In fact, Sir William Herschel showed, and his results have been verified by various philosophers since his time, that besides its luminous rays, the sun pours forth a multitude of other rays more powerfully calorific than the luminous ones, but entirely unsuited to the purposes of vision.

At the less refrangible end of the solar spectrum, then, the range of the sun's radiation is not limited by that of the eye. The same statement applies to the more refrangible end. Ritter discovered the extension of the spectrum into the invisible region beyond the violet ; and, in recent times, this ultra-violet emission has had peculiar interest conferred upon it by the admirable researches of Professor Stokes. The complete spectrum of the sun consists, therefore, of three distinct parts :—first, of ultra-red rays of high heating power, but unsuited to the purposes of vision ; secondly, of luminous rays which display the succession of colours, red, orange, yellow, green, blue, indigo, violet; thirdly, of ultra-violet rays which, like the ultra-red ones, are incompetent to excite vision, but which, unlike the ultra-red rays, possess a very feeble heating power. In consequence, however, of their chemical energy these ultra-violet rays are of the utmost importance to the organic world.

2. *Origin and Character of Radiation. The Æther.*

When we see a platinum wire raised gradually to a white heat, and emitting in succession all the colours of the spectrum, we are simply conscious of a series of changes in the condition of our own eyes. We do not see the actions in which these successive colours originate, but the mind irresistibly infers that the appearance of the colours corresponds to certain contemporaneous changes in the wire. What is the nature of these changes? In virtue of what condition does the wire radiate at all? We must now look from the wire as a whole to its constituent atoms. Could we see those atoms, even before the electric current has begun to act upon them, we should find them in a state of vibration. In this vibration, indeed, consists such warmth as the wire then possesses. Locke enunciated this idea with great precision, and it seems placed beyond the pale of doubt by the excellent quantitative researches of Mr. Joule. 'Heat,' says Locke, 'is a very brisk agitation of the insensible parts of the object, which produce in us that sensation from which we denominate the object hot: so what in our sensation is *heat* in the object is nothing but *motion*.' When the electric current, still feeble, begins to pass through the wire, its first act is to intensify the vibrations already existing, by causing the atoms to swing through wider ranges.   Technically speaking, the *amplitudes* of the oscillations are increased.   The current does this, however, without altering the *periods* of the old vibrations, or the times in which they were executed.   But besides intensifying the old vibrations the current generates new and more rapid ones, and when a certain definite rapidity has been attained the wire begins to glow. The colour first exhibited is red, which corresponds to the lowest rate of vibration of which the eye is able to take cognisance.   By augmenting the strength of the electric current more rapid vibrations are introduced, and orange rays appear.   A quicker rate of vibration produces yellow, a still quicker green ; and by further augmenting the rapidity, we pass through blue, indigo, and violet, to the extreme ultra-violet rays.

Such are the changes which science recognises in the wire itself, as concurrent with the visual changes taking place in the eye.   But what connects the wire with this organ?   By what means does it send such intelligence of its varying condition to the optic nerve?   Heat being, as defined by Locke, 'a very brisk agitation of the insensible parts of an object,' it is readily conceivable that on *touching* a heated body the agitation may communicate itself to the adjacent nerves, and announce itself to them as light or heat.   But the optic nerve does not touch the hot platinum, and hence the pertinence of the question, By what agency are the vibrations of the wire transmitted to the eye?

The answer to this question involves perhaps the most important physical conception that the mind of man has yet achieved : the conception of a medium filling space and fitted mechanically for the transmission of the vibrations of light and heat, as air is fitted for the transmission of sound. This medium is called the *luminiferous æther.* Every vibration of every atom of our platinum wire raises in this æther a wave, which speeds through it at the rate of 186,000 miles a second. The æther suffers no rupture of continuity at the surface of the eye, the inter-molecular spaces of the various humours are filled with it; hence the waves generated by the glowing platinum can cross these humours and impinge on the optic nerve at the back of the eye. Thus the sensation of light reduces itself to the communication of motion. Up to this point we deal with pure mechanics; but the subsequent translation of the shock of the æthereal waves into consciousness eludes the analysis of science. As an oar dipping into the Cam generates systems of waves, which, speeding from the centre of disturbance, finally stir the sedges on the river's bank, so do the vibrating atoms generate in the surrounding æther undulations, which finally stir the filaments of the retina. The motion thus imparted is transmitted with measurable, and not very great velocity to the brain, where, by a process which science does not even tend to unravel, the tremor of the nervous matter is converted into the conscious impression of light.

Darkness might then be defined as æther at rest; light as æther in motion. But in reality the æther is never at rest, for in the absence of light-waves we have heat-waves always speeding through it. In the spaces of the universe both classes of undulations incessantly commingle. Here the waves issuing from uncounted centres cross, coincide, oppose, and pass through each other, without confusion

N

or ultimate extinction. The waves from the zenith do not jostle out of existence those from the horizon, and every star is seen across the entanglement of wave motions produced by all other stars. It is the ceaseless thrill which those distant orbs collectively create in the æther, which constitutes what we call *the temperature of space.* As the air of a room accommodates itself to the requirements of an orchestra, transmitting each vibration of every pipe and string, so does the inter-stellar æther accommodate itself to the requirements of light and heat. Its waves mingle in space without disorder, each being endowed with an individuality as indestructible as if it alone had disturbed the universal repose.

All vagueness with regard to the use of the terms *radiation* and *absorption* will now disappear. Radiation is the communication of vibratory motion to the æther, and when a body is said to be chilled by radiation, as for example the grass of a meadow on a starlight night, the meaning is, that the molecules of the grass have lost a portion of their motion, by imparting it to the medium in which they vibrate. On the other hand, the waves of æther once generated may so strike against the molecules of a body exposed to their action as to yield up their motion to the latter; and in this transfer of the motion from the æther to the molecules consists the absorption of radiant heat. All the phenomena of heat are in this way reducible to interchanges of motion; and it is purely as the recipients or the donors of this motion, that we ourselves become conscious of the action of heat and cold.

3. *The Atomic Theory in reference to the Æther.*

The word 'atoms' has been more than once employed in this discourse. Chemists have taught us that all matter is reducible to certain elementary forms to which they give this name. These atoms are endowed with powers of mutual attraction, and under suitable circumstances they coalesce to form compounds. Thus oxygen and hydrogen are elements when separate, or merely *mixed*, but they may be made to *combine* so as to form molecules, each consisting of two atoms of hydrogen and one of oxygen. In this condition they constitute water. So also chlorine and sodium are elements, the former a pungent gas, the latter a soft metal; and they unite together to form chloride of sodium or common salt. In the same way the element nitrogen combines with hydrogen, in the proportion of one atom of the former to three of the latter, to form ammonia, or spirit of hartshorn. Picturing in imagination the atoms of elementary bodies as little spheres, the molecules of compound bodies must be pictured as groups of such spheres. This is the atomic theory as Dalton conceived it. Now if this theory have any foundation in fact, and if the theory of an æther pervading space, and constituting the vehicle of atomic motion be founded in fact, we may assuredly expect the vibrations of elementary bodies to be profoundly modified by the act of combination. It is on the face of it almost certain that both as regards radiation and absorption, that is to say, both as regards the communication of motion to the æther and the acceptance of motion from it, the deportment of the uncombined will be different from that of the combined atoms.

4. *Absorption of Radiant Heat by Gases.*

We have now to submit these considerations to the only test by which they can be tried, namely, that of experiment. An experiment is well defined as a question put to Nature; but to avoid the risk of asking amiss we ought to purify the question from all adjuncts which do not necessarily belong to it. Matter has been shown to be composed of elementary constituents, by the compounding of which all its varieties are produced. But besides the chemical unions which they form, both elementary and compound bodies can unite in another and less intimate way. By the attraction of cohesion gases and vapours aggregate to liquids and solids, without any change of their chemical nature. We do not yet know how the transmission of radiant heat may be affected by the entanglement due to cohesion, and as our object now is to examine the influence of chemical union alone, we shall render our experiments more pure by liberating the atoms and molecules entirely from the bonds of cohesion, and employing them in the gaseous or vaporous form.

Let us endeavour to obtain a perfectly clear mental image of the problem now before us. Limiting in the first place our enquiries to the phenomena of absorption, we have to picture a succession of waves issuing from a radiant source and passing through a gas; some of them striking against the gaseous molecules and yielding up their motion to the latter; others gliding round the molecules, or passing through the inter-molecular spaces without apparent hindrance. The problem before us is to determine whether such free molecules have any power whatever to stop the waves of heat, and if so, whether different molecules possess this power in different degrees.

The source of waves which I shall choose for these experiments is a plate of copper, against the back of which a steady sheet of flame is permitted to play. On emerging from the copper, the waves, in the first instance, pass through a space devoid of air, and then enter a hollow glass cylinder, three feet long and three inches wide. The two ends of this cylinder are stopped by two plates of rocksalt, this being the only solid substance which offers a scarcely sensible obstacle to the passage of the calorific waves. After passing through the tube, the radiant heat falls upon the anterior face of a thermo-electric pile,[1] which instantly applies the heat to the generation of an electric current. This current conducted round a magnetic needle deflects it, and the magnitude of the deflection is a measure of the heat falling upon the pile. This famous instrument, and not an ordinary thermometer, is what we shall use in these enquiries, but we shall use it in a somewhat novel way. As long as the two opposite faces of the thermo-electric pile are kept at the same temperature, no matter how high that may be, there is no current generated. The current is a consequence of a *difference* of temperature between the two opposite faces of the pile. Hence, if after the anterior face has received the heat from our radiating source, a second source, which we may call the compensating source, be permitted to radiate against the posterior face, this latter radiation will tend to neutralise the former. When the neutralisation is perfect, the

---

[1] In the Appendix to the first chapter of 'Heat as a Mode of Motion,' the construction of the thermo-electric pile is fully explained.

magnetic needle connected with the pile is no longer deflected, but points to the zero of the graduated circle over which it hangs.

And now let us suppose the glass tube, through which pass the waves from the heated plate of copper, to be exhausted by an air-pump, the two sources of heat acting at the same time on the two opposite faces of the pile. Perfectly equal quantities of heat being imparted to the two faces, the needle points to zero. Let any gas be now permitted to enter the exhausted tube ; if the molecules possess any power of intercepting the calorific waves, the equilibrium previously existing will be destroyed, the compensating source will triumph, and a deflection of the magnetic needle will be the immediate consequence. From the deflections thus produced by different gases, we can readily deduce the relative amounts of wave motion which their molecules intercept.

In this way the substances mentioned in the following table were examined, a small portion only of each being admitted into the glass tube. The quantity admitted was just sufficient to depress a column of mercury associated with the tube one inch : in other words, the gases were examined at a pressure of one-thirtieth of an atmosphere. The numbers in the table express the relative amounts of wave motion absorbed by the respective gases, the quantity intercepted by atmospheric air being taken as unity.

Radiation through Gases.

Name of gas.	Relative absorption.
Air .	1
Oxygen .	1
Nitrogen .	1

Name of gas.							Relative absorption.
Hydrogen	.	.	.	.	.	.	1
Carbonic oxide	.	.	.	.	.	.	750
Carbonic acid .	.	.	.	.	.	.	972
Hydrochloric acid	.	.	.	.	.	.	1,005
Nitric oxide	.	.	.	.	.	.	1,590
Nitrous oxide .	.	.	.	.	.	.	1,860
Sulphide of hydrogen	.	.	.	.	.	.	2,100
Ammonia	.	.	.	.	.	.	5,460
Olefiant gas	.	.	.	.	.	.	6,030
Sulphurous acid	.	.·	.	.	.	.	6,480

Every gas in this table is perfectly transparent to light, that is to say, all waves within the limits of the visible spectrum pass through it without obstruction; but for the waves of slower period, emanating from our heated plate of copper, enormous differences of absorptive power are manifested. These differences illustrate in the most unexpected manner the influence of chemical combination. Thus the elementary gases, oxygen, hydrogen and nitrogen, and the mixture atmospheric air, prove to be practical vacua to the rays of heat; for every ray, or more strictly speaking, for every unit of wave motion, which any one of them is competent to intercept, perfectly transparent ammonia intercepts 5,460 units, olefiant gas 6,030 units, while sulphurous acid gas absorbs 6,480 units. What becomes of the wave motion thus intercepted? It is applied to the heating of the absorbing gas. Through air, oxygen, hydrogen, and nitrogen, on the contrary, the waves of æther pass without absorption, and these gases are not sensibly changed in temperature by the most powerful calorific rays. The position of nitrous oxide in the foregoing table is worthy of particular notice. In this gas we have the same atoms in a state of chemical union, that exist uncombined in the atmosphere; but the absorption of the compound is 1800 times that of air.

5. *Formation of Invisible Foci.*

This extraordinary deportment of the elementary gases naturally directed attention to elementary bodies in another state of aggregation. Some of Melloni's results now attained a new significance; for this celebrated experimenter had found crystals of the element sulphur to be highly pervious to radiant heat; he had also proved that lamp-black and black glass (which owes its blackness to the element carbon) were to considerable extent transparent to calorific rays of low refrangibility. These facts, harmonising so strikingly with the deportment of the simple gases, suggested further enquiry. Sulphur dissolved in bisulphide of carbon was found almost perfectly transparent. The dense and deeply coloured element bromine was examined, and found competent to cut off the light of our most brilliant flames, while it transmitted the invisible calorific rays with extreme freedom. Iodine, the companion element of bromine, was next thought of, but it was found impracticable to examine the substance in its usual solid condition. It however dissolves freely in bisulphide of carbon. There is no chemical union between the liquid and the iodine, it is simply a case of solution, in which the uncombined atoms of the element can act upon the radiant heat. When permitted to do so, it was found that a layer of dissolved iodine, sufficiently opaque to cut off the light of the midday sun, was almost absolutely transparent to the invisible calorific rays.

By prismatic analysis Sir William Herschel separated the luminous from the non-luminous rays of the sun, and he also sought to render the obscure rays visible by concentration. Intercepting the luminous portion of his spectrum he brought, by a converging lens, the ultra-red rays to a focus, but by this condensation he obtained no light. The solution of iodine offers a means of filtering the solar beam, or failing it, the beam of the electric lamp, which renders attainable far more powerful foci of invisible rays than could possibly be obtained by the method of Sir William Herschel. For to form his spectrum he was obliged to operate upon solar light which had passed through a narrow slit or through a small aperture, the amount of the obscure heat being limited by this circumstance. But with our opaque solution we may employ the entire surface of the largest lens, and having thus converged the rays, luminous and non-luminous, we can intercept the former by the iodine, and do what we please with the latter. Experiments of this character, not only with the iodine solution but also with black glass and layers of lamp-black, were publicly performed at the Royal Institution in the early part of 1862, and the effects at the foci of invisible rays then obtained were such as had never been witnessed previously.

In the experiments here referred to, glass lenses were employed to concentrate the rays. But glass, though highly transparent to the luminous, is in a high degree opaque to the invisible heat-rays of the electric lamp, and hence a large portion of those rays was intercepted by the glass. The obvious remedy here is to employ rocksalt lenses instead of glass ones, or to abandon the use of lenses wholly, and to concentrate the rays by a metallic mirror. Both of these improvements have been introduced, and, as anticipated, the invisible foci have been thereby rendered more intense. The mode of operating remains however the same, in principle, as that made known in 1862. It was then found that an instant's exposure of the face of the thermo-electric pile to the focus of invisible rays, dashed the needles of a coarse galvanometer violently aside. It is now found that on substituting for the face of the thermo-electric pile a combustible body, the invisible rays are competent to set that body on fire.

6. *Visible and Invisible Rays of the Electric Light.*

We have next to examine what proportion the non-luminous rays of the electric light bear to the luminous ones. This the opaque solution of iodine enables us to do with an extremely close approximation to the truth. The pure bisulphide of carbon, which is the solvent of the iodine, is perfectly transparent to the luminous, and almost perfectly transparent to the dark rays of the electric lamp. Through the transparent bisulphide the total radiation of the lamp may be considered to pass, while through the solution of iodine only the dark rays are transmitted. Determining, then, by means of a thermo-electric pile, the total radiation, and deducting from it the purely obscure, we obtain the amount of the purely luminous emission. Experiments, performed in this way, prove that if all the visible rays of the electric light were converged to a focus of dazzling brilliancy, its heat would only be one ninth of that produced at the unseen focus of the invisible rays.

.ˑ Exposing his thermometers to the successive colours of the solar spectrum, Sir William Herschel determined the heating power of each, and also that of the region beyond the extreme red. Then drawing a straight line to represent the length of the spectrum, he erected, at various points, perpendiculars to represent the calorific intensity existing at those points. Uniting the ends of all his perpendiculars, he obtained a curve which showed at a glance the manner in which the heat was distributed in the solar spectrum. Professor Müller of Freiburg, with improved instruments, afterwards made similar experiments, and constructed a more accurate diagram of the same kind. We have now to examine the distribution of heat in the spectrum of the electric light; and for this purpose we shall employ a particular form of the thermo-electric pile, devised by Melloni. Its face is a rectangle, which by means of movable side-pieces can be rendered as narrow as desired. We can, for example, have the face of the pile the tenth, the hundredth, or even the thousandth of an inch in breadth. By means of an endless screw, this *linear* thermo-electric pile may be moved through the entire spectrum, from the violet to the red, the amount of heat falling upon the pile at every point of its march, being declared by a magnetic needle associated with the pile.

When this instrument is brought up to the violet end of the spectrum of the electric light, the heat is found to be insensible. As the pile gradually moves from the violet end towards the red, heat soon manifests itself, augmenting as we approach the red. Of all the colours of the visible spectrum the red possesses the highest heating power. On pushing the pile into the dark region beyond the red, the heat, instead of vanishing, rises suddenly and enormously in intensity, until at some distance beyond the red it attains a maximum. Moving the pile still forward, the thermal power falls, somewhat more rapidly than it rose. It then gradually shades away, but for a distance beyond the red greater than the length of the whole visible spectrum, signs of heat may be detected. Drawing a datum line, and erecting along it perpendiculars, proportional in length to the thermal intensity at the respective points, we obtain the extraordinary curve, shown on the adjacent page, which exhibits the distribution of heat in the spectrum of the electric light. In the region of dark rays, beyond the red, the curve shoots up to B, in a steep and massive peak—a kind of Matterhorn of heat, which dwarfs the portion of the diagram C D E, representing the luminous radiation. Indeed, the idea forced upon the mind by this diagram is that the light rays are a mere insignificant appendage to the heat rays represented by the area A B C D, thrown in as it were by nature for the purposes of vision.

The diagram drawn by Professor Müller to represent the distribution of heat in the solar spectrum is not by any means so striking as that just described, and the reason, doubtless, is that prior to reaching the earth the solar rays have to traverse our atmosphere. By the aqueous vapour there diffused, the summit of the peak representing the sun's invisible radiation is cut off. A similar lowering of the mountain of invisible heat is observed when the rays from the electric light are permitted to pass through a film of water, which acts upon them as the atmospheric vapour acts upon the rays of the sun.

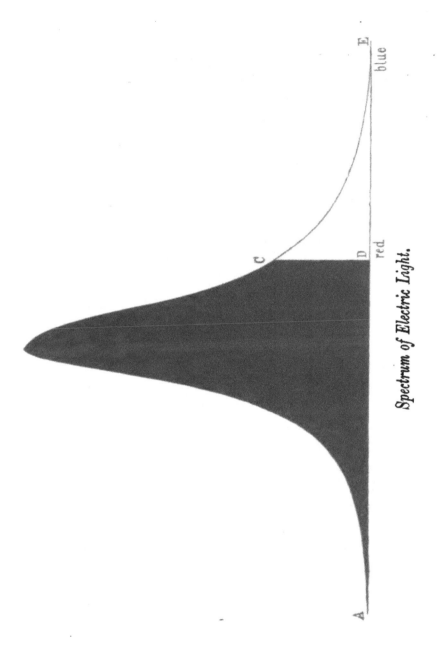

Spectrum of Electric Light.

7. *Combustion by Invisible Rays.*

The sun's invisible rays far transcend the visible ones in heating power, so that if the alleged performances of Archimedes during the siege of Syracuse had any foundation in fact, the dark solar rays would have been the philosopher's chief agents of combustion. On a small scale we can readily produce with the purely invisible rays of the electric light all that Archimedes is said to have performed with the sun's total radiation. Placing behind the electric light a small concave mirror, the rays are converged, the cone of reflected rays and their point of convergence being rendered clearly visible by the dust always floating in the air. Placing, between the luminous focus and the source of rays, our solution of iodine, the light of the cone is entirely cut away, but the intolerable heat experienced when the hand is placed, even for a moment, at the dark focus, shows that the calorific rays pass unimpeded through the opaque solution.

Almost anything that ordinary fire can effect may be accomplished at the focus of invisible rays; the *air* at the focus remaining at the same time perfectly cold, on account of its transparency to the heat-rays. An air thermometer, with a hollow rocksalt bulb, would be unaffected by the heat of the focus : there would be no expansion, and in the open air there is no convection. The *æther* at the focus, and not the air, is the substance in which the heat is embodied. A block of wood, placed at the focus, absorbs the heat, and dense volumes of smoke rise swiftly upwards, showing the manner in which the air itself would rise, if the invisible rays were competent to heat it. At the perfectly dark focus dry paper is instantly inflamed : chips of wood are speedily burnt up : lead, tin, and zinc are fused : and discs of charred paper are raised to vivid incandescence. It might be supposed that the obscure rays would show no preference for black over white ; but they do show a preference, and to obtain rapid combustion, the body, if not already black, ought to be blackened. When metals are to be burned, it is necessary to blacken or otherwise tarnish them, so as to diminish their reflective power. Blackened zinc foil, when brought into the focus of invisible rays, is instantly caused to blaze, and burns with its peculiar purple flame. Magnesium wire flattened, or tarnished magnesium ribbon, also bursts into splendid combustion. Pieces of charcoal suspended in a receiver full of oxygen are also set on fire : the dark rays after having passed through the receiver still possessing sufficient power to ignite the charcoal, and thus initiate the attack of the oxygen. If, instead of being plunged in oxygen, the charcoal be suspended in vacuo, it immediately glows at the place where the focus falls.

8. *Transmutation of Rays:*[1] *Calorescence.*

Eminent experimenters were long occupied in demonstrating the substantial identity of light and radiant heat, and we have now the means of offering a new and striking proof of this identity. A concave mirror produces beyond the object which it reflects an inverted and magnified image of the object ; withdrawing, for example, our iodine solution, an intensely luminous inverted

[1] I borrow this term from Professor Challis, ' Philosophical Magazine,' vol. xii. p. 521.

image of the carbon points of the electric light is formed at the focus of the mirror employed in the foregoing experiments. When the solution is interposed, and the light is cut away, what becomes of this image? It disappears from sight, but an invisible thermograph remains, and it is only the peculiar constitution of our eyes that disqualifies us from seeing the picture formed by the calorific rays. Falling on white paper, the image chars itself out : falling on black paper, two holes are pierced in it, corresponding to the images of the two coal points : but falling on a thin plate of carbon in vacuo, or upon a thin sheet of platinised platinum, either in vacuo or in air, radiant heat is converted into light, and the image stamps itself in vivid incandescence upon both the carbon and the metal. Results similar to those obtained with the electric light have also been obtained with the invisible rays of the lime-light and of the sun.

Before a Cambridge audience it is hardly necessary to refer to the excellent researches of Professor Stokes at the opposite end of the spectrum. The above results constitute a kind of complement to his discoveries. Professor Stokes named the phenomena which he has discovered and investigated *Fluorescence* ; for the new phenomena here described I have proposed the term *Calorescence.* He, by the interposition of a proper medium, so lowered the refrangibility of the ultra-violet rays of the spectrum as to render them visible ; and here, by the interposition of the platinum foil, the refrangibility of the ultra-red rays is so exalted as to render them visible. Looking through a prism at the incandescent image of the carbon points, the light of the image is decomposed, and a complete spectrum obtained. The invisible rays of the electric light, remoulded by the atoms of the platinum, shine thus visibly forth ; ultra-red rays being converted into red, orange, yellow, green, blue, indigo, and ultra-violet ones. Could we, moreover, raise the original source of rays to a sufficiently high temperature, we might not only obtain from the dark rays of such a source a single incandescent image, but from the dark rays of this image we might obtain a second one, from the dark rays of the second a third, and so on—a series of complete images and spectra being thus extracted from the invisible emission of the primitive source.[1]

9. *Deadness of the Optic Nerve to the Calorific Rays.*

The layer of iodine used in the foregoing experiments intercepted the light of the noonday sun. No trace of light from the electric lamp was visible in the darkest room, even when a white screen was placed at the focus of the

[1] On investigating the calorescence produced by rays transmitted through glasses of various colours, it was found that in the case of certain specimens of blue glass, the platinum foil glowed with a *pink* or *purplish* light. The effect was not subjective, and considerations of obvious interest are suggested by it. Different kinds of black glass differ notably as to their power of transmitting radiant heat. In thin plates some descriptions tint the sun with a greenish hue : others make it appear a glowing red without any trace of green. The latter are far more diathermic than the former. In fact, carbon when perfectly dissolved, and incorporated with a good white glass, is highly transparent to the calorific rays, and by employing it as an absorbent, the phenomena of 'calorescence' may be obtained, though in a less striking form than with the iodine. The black glass chosen for thermometers, and intended to absorb completely the solar heat, may entirely fail in this object, if the glass in which the carbon is incorporated be colourless. To render the bulb of a thermometer a perfect absorbent, the glass ought in the first instance to be green. Soon after the discovery of fluorescence the late Dr. William Allen Miller pointed to the lime-light as an illustration of exalted refrangibility. Direct experiments have since entirely confirmed the view expressed at page 210 of his work on 'Chemistry,' published in 1855.

mirror employed to concentrate the light. It was thought, however, that if the retina itself were brought into the focus the sensation of light might be experienced. The danger of this experiment was twofold. If the dark rays were absorbed in a high degree by the humours of the eye, the albumen of the humours might coagulate along the line of the rays. If, on the contrary, no such high absorption took place, the rays might reach the retina with a force sufficient to destroy it. To test the likelihood of these results, experiments were made on water and on a solution of alum, and they showed it to be very improbable that in the brief time requisite for an experiment any serious damage could be done. The eye was therefore caused to approach the dark focus, no defence, in the first instance, being provided; but the heat, acting upon the parts surrounding the pupil, could not be borne. An aperture was therefore pierced in a plate of metal, and the eye, placed behind the aperture, was caused to approach the point of convergence of invisible rays. The focus was attained, first by the pupil and afterwards by the retina. Removing the eye, but permitting the plate of metal to remain, a sheet of platinum foil was placed in the position occupied by the retina a moment before. The platinum became red hot. No sensible damage was done to the eye by this experiment; no impression of light was produced; the optic nerve was not even conscious of heat.

But the humours of the eye are known to be highly impervious to the invisible calorific rays, and the question therefore arises, ' did the radiation in the foregoing experiment reach the retina at all?' The answer is that the rays were in part transmitted to the retina, and in part absorbed by the humours. Experiments on the eye of an ox showed that the proportion of obscure rays which reached the retina amounted to 18 per cent. of the total radiation ; while the luminous emission from the electric light amounts to no more than 10 per cent. of the same total. Were the purely luminous rays of the electric lamp converged by our mirror to a focus, there can be no doubt as to the fate of a retina placed there. Its ruin would be inevitable ; and yet this would be accomplished by an amount of wave motion but little more than half of that which the retina bears, without exciting consciousness, at the focus of invisible rays.

This subject will repay a moment's further attention. At a common distance of a foot the visible radiation of the electric light is 800 times the light of a candle. At the same distance, the portion of the radiation of the electric light which reaches the retina but fails to excite vision, is about 1500 times the luminous radiation of the candle.[1] But a candle on a clear night can readily be seen at a distance of a mile, its light at this distance being less than one 20,000,000th of its light at the distance of a foot. Hence, to make the candle-light a mile off equal in power to the non-luminous radiation received from the electric light at a foot distance, its intensity would have to be multiplied by 1500 × 20,000,000, or by thirty thousand millions. Thus the thirty thousand millionth part of the invisible radiation from the electric light, received by the retina at the distance of a foot, would, if slightly changed in character, be amply sufficient to provoke

---

[1] It will be borne in mind that the heat which any ray, luminous or non-luminous, is competent to generate is the true measure of the energy of the ray.

vision. Nothing could more forcibly illustrate that special relationship supposed by Melloni and others to subsist between the optic nerve and .the oscillating periods of luminous bodies. The optic nerve responds, as it were, to the waves with which it is in consonance, while it refuses to be excited by others of almost infinitely greater energy, whose periods of recurrence are not in unison with its own.

10. *Persistence of Rays.*

At an early part of this lecture it was affirmed that when a platinum wire was gradually raised to a state of high incandescence, new rays were constantly added, while the intensity of the old ones was increased. Thus in Dr. Draper's experiments the rise of temperature that *generated* the orange, yellow, green, and blue rays *augmented* the intensity of the red ones. What is true of the red is true of every other ray of the spectrum, visible and invisible. We cannot indeed *see* the augmentation of intensity in the region beyond the red, but we can measure it and express it numerically. With this view the following experiment was performed. A spiral of platinum wire was surrounded by a small glass globe to protect it from currents of air ; through an orifice in the globe the rays could pass from the spiral and fall afterwards upon a thermo-electric pile. Placing in front of the orifice an opaque solution of iodine, the platinum was gradually raised from a low dark heat to the fullest incandescence, with the following results :—

Appearance of spiral.							Energy of obscure radiation.
Dark	.	.	.	.	.	.	1
Dark, but hotter	.	.	.	.	.	.	3
Dark, but still hotter	.	.	.	.	.	.	5
Dark, but still hotter	.	.	.	.	.	.	10
Feeble red	.	.	.	.	.	. .	19
Dull red	. .	.	.	.	.	.	25
Red	.	.	.	.	.	.	37
Full red	.	.	.	.	.	.	62
Orange	.	.	.	.	.	.	89
Bright orange	.	.	.	.	.	.	144
Yellow	.	.	.	.	.	.	202
White	.	.	.	.	.	.	276
Intense white	.	.	.	.	.	.	440

Thus the augmentation of the electric current, which raises the wire from its primitive dark condition to an intense white heat, exalts at the same time the energy of the obscure radiation, until at the end it is fully 440 times what it was at the beginning.

What has been here proved true of the totality of the ultra-red rays is true for each of them singly. Placing our linear thermo-electric pile in any part of the ultra-red spectrum, it may be proved that a ray once emitted continues to be emitted with increased energy as the temperature is augmented. The platinum spiral so often referred to being raised to whiteness by an electric current, a brilliant spectrum was formed from its light. A linear thermo-electric pile was placed in the region of obscure rays beyond the red, and by diminishing the current the spiral was reduced to a low temperature. It was then caused to pass through various degrees of darkness and incandescence, with the following results :—

Appearance of spiral.								Energy of obscure rays.
Dark	.	.	.	.	.	.	.	1
Dark	.	.	.	.	.	.	.	6
Faint red	.	.	.	.	.	.	.	10
Dull red	.	.	.	.	.	.	.	13
Red	.	.	.	.	.	.	.	18
Full red	.	.	.	.	.	.	.	27
Orange	.	.	.	.	.	.	.	60
Yellow	.	.	.	.	.	.	.	93
White	.	.	.	.	.	.	.	122

Here, as in the former case, the dark and bright radiations reached their maximum together; as the one augmented, the other augmented, until at last the energy of the obscure rays of the particular refrangibility here chosen, became 122 times what it was at first. To reach a white heat the wire has to pass through all the stages of invisible radiation, and in its most brilliant condition it embraces, in an intensified form, the rays of all those stages.

And thus it is with all other kinds of matter, as far as they have hitherto been examined. Coke, whether brought to a white heat by the electric current, or by the oxyhydrogen jet, pours out invisible rays with augmented energy, as its light is increased. The same is true of lime, bricks, and other substances. It is true of all metals which are capable of being heated to incandescence. It also holds good for phosphorus burning in oxygen. Every gush of dazzling light has associated with it a gush of invisible radiant heat, which far transcends the light in energy. This condition of things applies to all bodies capable of being raised to a white heat, either in the solid or the molten condition. It would doubtless also apply to the luminous fogs formed by the condensation of incandescent vapours. In such cases when the curve representing the radiant energy of the body is constructed, the obscure radiation towers upwards like a mountain, the luminous radiation resembling a mere spur at its base. From the very brightness of the light of some of the fixed stars we may infer the intensity of the dark radiation, which is the precursor and inseparable associate of their luminous rays.

We thus find the luminous radiation appearing when the radiant body has attained a certain temperature ; or, in other words, when the vibrating atoms of the body have attained a certain width of swing. In solid and molten bodies a certain amplitude cannot be surpassed without the introduction of periods of vibration, which provoke the sense of vision. How are we to figure this ? If permitted to speculate, we might ask, are not these more rapid vibrations the progeny of the slower ? Is it not really the mutual action of the atoms, when they swing through very wide spaces, and thus encroach upon each other, that causes them to tremble in quicker periods ? If so, whatever be the agency by which the large swinging space is obtained, we shall have light-giving vibrations associated with it. It matters not whether the large amplitudes be produced by the strokes of a hammer, or by the blows of the molecules of a non-luminous gas, such as the air at some height above a gas-flame ; or by the shock of the æther particles when transmitting radiant heat. The result in all cases will be incandescence. Thus, the invisible waves of our filtered electric beam may be regarded as generating synchronous vibrations among the atoms of the platinum on which they impinge ; but once these vibrations have attained a certain amplitude, the mutual jostling of the atoms produces quicker tremors, and the light-giving waves follow as the necessary product of the heat-giving ones.

11. *Absorption of Radiant Heat by Vapours and Odours.*

We commenced the demonstrations brought forward in this lecture by experiments on permanent gases, and we have now to turn our attention to the vapours of volatile liquids. Here, as in the case of the gases, vast differences have been proved to exist between various kinds of molecules, as regards their power of intercepting the calorific waves. While some vapours allow the waves a comparatively free passage, the minutest bubble of other vapours, introduced into the tube already employed for gases, causes a deflection of the magnetic needle. Assuming the absorption effected by air at a pressure of one atmosphere to be unity, the following are the absorptions effected by a series of vapours at a pressure of $\frac{1}{60}$th of an atmosphere :—

Name of vapour.						Absorption.
Bisulphide of carbon .	.	.	.	.	.	47
Iodide of methyl	.	.	.	.	.	115
Benzol .	.	.	.	.	.	136
Amylene .	.	.	.	.	.	321
Sulphuric ether .	.	.	.	.	.	440
Formic ether	.	.	.	.	.	548
Acetic ether	.	.	.	.	.	612

Bisulphide of carbon is the most transparent vapour in this list ; and acetic ether the most opaque ; $\frac{1}{60}$th of an atmosphere of the former, however, produces 47 times the effect of a whole atmosphere of air, while $\frac{1}{60}$th of an atmosphere of the latter produces 612 times the effect of a whole atmosphere of air. Reducing dry air to the pressure of the acetic ether here employed, and comparing them then together, the quantity of wave-motion intercepted by the ether would be many thousand times that intercepted by the air.

Any one of these vapours discharged into the free atmosphere, in front of a body emitting obscure rays, intercepts more or less of the radiation. A similar effect is produced by perfumes diffused in the air, though their attenuation is known to be almost infinite. Carrying, for example, a current of dry air over bibulous paper moistened by patchouli, the scent taken up by the current absorbs 30 times the quantity of heat intercepted by the air which carries it ; and yet patchouli acts more feebly on radiant heat than any other perfume yet examined. Here follow the results obtained with various essential oils, the odour, in each case, being carried by a current of dry air into the tube already employed for gases and vapours :—

Name of perfume.	Absorption.
Patchouli	30
Sandal wood	32
Geranium	33
Oil of Cloves	34
Otto of Roses	37
Bergamot	44
Neroli	47
Lavender	60
Lemon	65
Portugal	67
Thyme	68
Rosemary	74
Oil of Laurel	80
Camomile Flowers	87
Cassia	109
Spikenard	355
Aniseed	372

Thus the absorption by a tube full of dry air being 1, that of the odour of patchouli diffused in it is 30, that of lavender 60, that of rosemary 74, while that of aniseed amounts to 372. It would be idle to speculate on the quantities of matter concerned in these actions.

12. *Aqueous Vapour in relation to the Terrestrial Temperatures.*[1]

We are now fully prepared for a result which, without such preparation, might appear incredible. Water is, to some extent, a volatile body, and our atmosphere, resting as it does upon the surface of the ocean, receives from it a continual supply of aqueous vapour. It would be an error to confound clouds or fog or any visible mist with the vapour of water: this vapour is a perfectly impalpable gas, diffused, even on the clearest days, throughout the atmosphere. Compared with the great body of the air, the aqueous vapour it contains is of almost infinitesimal amount, $99\frac{1}{2}$ out of every 100 parts of the atmosphere being composed of oxygen and nitrogen. In the absence of experiment, we should never think of ascribing to this scant and varying constituent any important influence on terrestrial radiation; and yet its influence is far more potent than that of the great body of the air. To say that on a day of average humidity in England, the atmospheric vapour exerts 100 times the action of the air itself, would certainly be an understatement of the fact. The peculiar qualities of this vapour, and the circumstance that at ordinary temperatures it is very near its point of condensation, render the results which it yields

[1] See Note at the end of this Lecture.

in the apparatus already described, less than the truth ; and I am not prepared to say that the absorption by this substance is not 200 times that of the air in which it is diffused. Comparing a single molecule of aqueous vapour with an atom of either of the main constituents of our atmosphere, I am not prepared to say how many thousand times the action of the former exceeds that of the latter.

But it must be borne in mind that these large numbers depend in part upon the extreme feebleness of the air ; the power of aqueous vapour seems vast, because that of the air with which it is compared is infinitesimal. Absolutely considered, however, this substance, notwithstanding its small specific gravity, exercises a very potent action. Probably from 10 to 15 per cent. of the heat radiated 'from the earth is absorbed within 10 feet of the earth's surface. This must evidently be of the utmost consequence to the life of the world. Imagine the superficial molecules of the earth trembling with the motion of heat, and imparting it to the surrounding æther ; this motion would be carried rapidly away, and lost for ever to our planet, if the waves of æther had nothing but the air to contend with in their outward course. But the aqueous vapour takes up the motion of the æthereal waves, and becomes thereby heated, thus wrapping the earth like a warm garment, and protecting its surface from the deadly chill which it would otherwise sustain. Various philosophers have speculated on the influence of an atmospheric envelope. De Saussure, Fourier, M. Pouillet and Mr. Hopkins have, one and all, enriched scientific literature with contributions on this subject, but the considerations which these eminent men have applied to atmospheric air, have, if my experiments be correct, to be transferred to the aqueous vapour.

The observations of meteorologists furnish important, though hitherto unconscious evidence of the influence of this agent. Wherever the air is dry we are liable to daily extremes of temperature. By day, in such places, the sun's heat reaches the earth unimpeded, and renders the maximum high; by night, on the other hand, the earth's heat escapes unhindered into space, and renders the minimum low. Hence the difference between the maximum and minimum is greatest where the air is driest. In the plains of India, on the heights of the Himalaya, in central Asia, in Australia,—wherever drought reigns, we have the heat of day forcibly contrasted with the chill of night. In the Sahara itself, when the sun's rays cease to impinge on the burning soil, the temperature runs rapidly down to freezing, because there is no vapour overhead to check the calorific drain. And here another instance might be added to the numbers already known, in which nature tends as it were to check her own excess. By nocturnal refrigeration, the aqueous vapour of the air is condensed to water on the surface of the earth, and as only the superficial portions radiate, the act of condensation makes water the radiating body. Now experiment proves that to the rays emitted by water, aqueous vapour is especially opaque. Hence the very act of condensation, consequent on terrestrial cooling, becomes a safeguard to the earth, imparting to its radiation that particular character which renders it most liable to be prevented from escaping into space.

It might however be urged that, inasmuch as we derive all our heat from the sun, the selfsame covering which protects the earth from chill must also shut out the solar radiation. This is partially true, but only partially; the sun's rays are different in quality from the earth's rays, and it does not at all follow that the substance which absorbs the one must necessarily absorb the other. Through a layer of water, for example, one tenth of an inch in thickness, the sun's rays are transmitted with comparative freedom; but through a layer half this thickness, as Melloni has proved, no single ray from the warmed earth could pass. In like manner, the sun's rays pass with comparative freedom through the aqueous vapour of the air : the absorbing power of this substance being mainly exerted upon the heat that endeavours to escape from the earth. In consequence of this differential action upon solar and terrestrial heat, the mean temperature of our planet is higher than is due to its distance from the sun.

13. *Liquids and their Vapours in relation to Radiant Heat.*

The deportment here assigned to atmospheric vapour has been established by direct experiments on air taken from the streets and parks of London, from the downs of Epsom, from the hills and sea-beach of the Isle of Wight, and also by experiments on air in the first instance dried, and afterwards rendered artificially humid by pure distilled water. It has also been established in the following way. Ten volatile liquids were taken at random and the power of these liquids, at a common thickness, to intercept the waves of heat was carefully determined. The vapours of the liquids were next taken, in quantities proportional to the quantities of liquid, and the power of the vapours to intercept the waves of heat was also determined. Commencing with the substance which exerted the least absorptive power, and proceeding upward to the most energetic, the following order of absorption was observed : —

Liquids.	Vapours.
Bisulphide of carbon.	Bisulphide of carbon.
Chloroform.	Chloroform.
Iodide of methyl.	Iodide of methyl.
Iodide of ethyl.	Iodide of ethyl.
Benzol.	Benzol.
Amylene.	Amylene.
Sulphuric ether.	Sulphuric ether.
Acetic ether.	Acetic ether.
Formic ether.	Formic ether.
Alcohol.	Alcohol.
Water.	

We here find the order of absorption in both cases to be the same. We have liberated the molecules from the bonds which trammel them more or less in a liquid condition ; but this change in their state of aggregation does not change their relative powers of absorption. Nothing could more clearly prove that the act of absorption depends upon the individual molecule, which equally asserts its power in the liquid and the gaseous state. We may assuredly conclude from the above table that the position of a vapour is determined by that of its liquid. Now at the very foot of the list of liquids stands *water*, signalising itself above all others by its enormous power of absorption. And from this fact, even if no direct experiment on the vapour of water had ever been made, we should be entitled to rank that vapour as the most powerful absorber of radiant heat hitherto discovered.

It has been proved by experiment that a shell of air two inches in thickness surrounding our planet, and saturated with the vapour of sulphuric ether, would intercept 35 per cent. of the earth's radiation. And though the quantity of aqueous vapour necessary to saturate air is much less than the amount of sulphuric ether vapour which it can sustain, it is still extremely probable that the estimate already made of the action of atmospheric vapour within 10 feet of the earth's surface, is altogether under the mark; and that we are indebted to this wonderful substance, to an extent not accurately determined, but certainly far beyond what has hitherto been imagined, for the temperature now existing at the surface of the globe.

14. *Reciprocity of Radiation and Absorption.*

Throughout the reflections which have hitherto occupied us, the image before the mind has been that of a radiant source generating calorific waves, which on passing among the scattered molecules of a gas or vapour were intercepted by those molecules in various degrees. In all cases it was the transference of motion from the æther to the comparatively quiescent molecules of the gas or vapour. We have now to change the form of our conception, and to figure these molecules not as absorbers but as radiators, not as the recipients but as the originators of wave motion. That is to say, we must figure them vibrating and generating in the surrounding æther undulations which speed through it with the velocity of light. Our object now is to enquire whether the act of chemical combination, which proves so potent as regards the phenomena of absorption, does not also manifest its power in the phenomena of radiation. For the examination of this question it is necessary, in the first place, to heat our gases and vapours to the same temperature, and then examine their power of discharging the motion thus imparted to them upon the æther in which they swing.

A heated copper ball was placed above a ring gas-burner, possessing a great number of small apertures, the burner being connected by a tube with vessels containing the various gases to be examined. By gentle pressure the gases were forced through the orifices of the burner against the copper ball, where each of them, being heated, rose in an ascending column. A thermo-electric pile, entirely screened off from the hot ball, was exposed to the radiation of the warm gas, and the deflection of a magnetic needle connected with the pile, declared the energy of the radiation.

By this mode of experiment it was proved that the selfsame molecular arrangement which renders a gas a powerful absorber, renders it in the same degree a powerful radiator—that the atom or molecule which is competent to intercept the calorific waves is, in the same degree, competent to generate them. Thus, while the atoms of elementary gases proved themselves unable to emit any sensible amount of radiant heat, the molecules of compound gases were shown to be capable of powerfully disturbing the surrounding æther. By special modes of experiment the same was proved to hold good for the vapours of volatile liquids, the radiative power of every vapour being found proportional to its absorptive power.

The method of experiment here pursued, though not of the simplest character, is still within your grasp.

When air is permitted to rush into an exhausted tube, the temperature of the air is raised to a degree equivalent to the *vis viva* extinguished.[1] Such air is said to be dynamically heated, and, if pure, it shows itself incompetent to radiate, even when a rocksalt window is provided for the passage of its rays. But if instead of being empty the tube contain a small quantity of vapour, then the warmed air will communicate heat by contact to the vapour, which will be thus enabled to radiate. Thus the molecules of the vapour convert into the radiant form the heat imparted dynamically to the atoms of the air. By this process, which has been called Dynamic Radiation, the radiative power of both vapours and gases has been determined, and the reciprocity of their radiation and absorption proved.[2]

In the excellent researches of Leslie, De la Provostaye and Desains, and Balfour Stewart, the reciprocity of radiation and absorption as regards solid bodies has been variously illustrated; while the labours, theoretical and experimental, of Kirchhoff have given this subject a wonderful expansion, and enriched it by applications of the highest kind. To their results are now to be added the foregoing, whereby gases and vapours which have been hitherto thought inaccessible to experiments of this kind are proved to exhibit the duality of radiation and absorption, the influence on both of chemical combination being exhibited in the most decisive and extraordinary way.

---

[1] See page 15 for a definition of *vis viva*.

[2] When heated air imparts its motion to another gas or vapour, the transference of heat is accompanied by a change of vibrating period. The Dynamic Radiation of vapours is rendered possible by the transmutation of vibrations.

P

15. *Influence of Vibrating Period and Molecular Form. Physical Analysis of the Human Breath.*

In the foregoing experiments with gases and vapours we have employed throughout invisible rays; some of these bodies are so impervious that in lengths of a few feet only, they intercept every ray as effectually as a layer of pitch would do. The substances however which show themselves thus opaque to radiant heat are perfectly transparent to light. Now the rays of light differ from those of invisible heat only in point of period, the former failing to affect the retina because their periods of recurrence are too slow. Hence, in some way or other the transparency of our gases and vapours depends upon the periods of the waves which impinge upon them. What is the nature of this dependence? The admirable researches of Kirchhoff help us to an answer. The atoms and molecules of every gas have certain definite rates of oscillation, and those waves of æther are most copiously absorbed whose periods of recurrence synchronise with the periods of the molecules amongst which they pass. Thus, when we find the invisible rays absorbed and the visible ones transmitted by a layer of gas, we conclude that the oscillating periods of the gaseous molecules coincide with those of the invisible, and not with those of the visible spectrum.

It requires some discipline of the imagination to form a clear picture of this process. Such a picture is, however, possible, and ought to be obtained. When the waves of æther impinge upon molecules whose periods of vibration coincide with the recurrence of the undulations, the timed strokes of the waves, the vibration of the molecules augments, as a heavy pendulum is set in motion by well-timed puffs of breath. Millions of millions of shocks are received every second from the calorific waves, and it is not difficult to see that as every wave arrives just in time to repeat the action of its predecessor, the molecules must finally be caused to swing through wider spaces than if the arrivals were not so timed. In fact, it is not difficult to see that an assemblage of molecules, operated upon by contending waves, might remain practically quiescent, and this is actually the case when the waves of the visible spectrum pass through a transparent gas or vapour. There is here no sensible transference of motion from the æther to the molecules; in other words, there is no sensible absorption of heat.

One striking example of the influence of period may here be recorded. Carbonic acid gas is one of the feeblest of absorbers of the radiant heat emitted by solid sources. It is, for example, to a great extent transparent to the rays emitted by the heated copper plate already referred to. There are, however, certain rays, comparatively few in number, emitted by the copper, to which the carbonic acid is impervious; and could we obtain a source of heat emitting such rays only, we should find carbonic acid more opaque to the radiation from that source than any other gas. Such a source is actually found in the flame of carbonic oxide, where hot carbonic acid constitutes the main radiating body. Of the rays emitted by our heated plate of copper, olefiant gas absorbs ten times the quantity absorbed by carbonic acid. Of the rays emitted by a carbonic oxide flame, carbonic acid absorbs twice as much as olefiant gas. This wonderful change in the power of the former as an absorber is simply due to the fact, that the periods of the hot and cold carbonic acid are identical, and that the waves from the flame freely transfer their motion to the molecules which synchronise with them.   Thus it is that the tenth of an atmosphere of carbonic acid, enclosed in a tube four feet long, absorbs 60 per cent. of the radiation from a carbonic oxide flame, while one-thirtieth of an atmosphere absorbs 48 per cent. of the heat from the same origin.

In fact, the presence of the minutest quantity of carbonic acid may be detected by its action on the rays from the carbonic oxide flame.   Carrying, for example, the dried human breath into a tube four feet long, the absorption there effected by the carbonic acid of the breath amounts to 50 per cent. of the entire radiation. Radiant heat may indeed be employed as a means of determining practically the amount of carbonic acid expired from the lungs.   My late assistant, Mr. Barrett, has made this determination.   The absorption produced by the breath freed from its moisture, but retaining its carbonic acid, was first determined.   Carbonic acid artificially prepared was then mixed with dry air in such proportions that the action of the mixture upon the rays of heat was the same as that of the dried breath. The percentage of the former being known immediately gave that of the latter.   The same breath analysed chemically by Dr. Frankland, and physically by Mr. Barrett, gave the following results :——

Percentage of Carbonic Acid in the Human Breath.

Chemical analysis.	Physical analysis.
4·66	4·56
5·33	5·22

It is thus proved that in the quantity of æthereal motion which it is competent to take up, we have a practical measure of the carbonic acid of the breath, and hence of the combustion going on in the human lungs.

Still this question of period, though of the utmost importance, is not competent to account for the whole of the observed facts. The æther, as far as we know, accepts vibrations of all periods with the same readiness. To it the oscillations of an atom of oxygen are just as acceptable as those of a molecule of olefiant gas; that the vibrating oxygen then stands so far below the olefiant gas in radiant power must be referred not to period, but to some other peculiarity of the elementary gas. The atomic group which constitutes the molecule of olefiant gas, produces many thousand times the disturbance caused by the oxygen, because the group is able to lay a vastly more powerful hold upon the æther than the single atoms can. The cavities and indentations of a molecule composed of spherical atoms may be one cause of this augmented hold. Another, and probably very potent one may be, that the æther itself, condensed and entangled among the constituent atoms of a compound, virtually increases the magnitude of the group, and hence augments the disturbance. But whatever may be the fate of these attempts to visualise the physics of the process, it will still remain true, that to account for the phenomena of radiation and absorption we must take into consideration the shape, size, and complexity of the molecules by which the æther is disturbed.

16. *Summary and Conclusion.*

Let us now cast a momentary glance over the ground that we have left behind. The general nature of light and heat was first briefly described : the compounding of matter from elementary atoms and the influence of the act of combination on radiation and absorption were considered and experimentally illustrated. Through the transparent elementary gases radiant heat was found to pass as through a vacuum, while many of the compound gases presented almost impassable obstacles to the calorific waves. This deportment of the simple gases directed our attention to other elementary bodies, the examination of which led to the discovery that the element iodine, dissolved in bisulphide of carbon, possesses the power of detaching, with extraordinary sharpness, the light of the spectrum from its heat, intercepting all luminous rays up to the extreme red, and permitting the calorific rays beyond the red to pass freely through it. This substance was then employed to filter the beams of the electric light, and to form foci of invisible rays so intense as to produce almost all the effects obtainable in an ordinary fire. Combustible bodies were burnt and refractory ones were raised to a white heat by the concentrated invisible rays. Thus, by exalting their refrangibility, the invisible rays of the electric light were rendered visible, and all the colours of the solar spectrum were extracted from utter darkness. The extreme richness of the electric light in invisible rays of low refrangibility was demonstrated, one-ninth only of its radiation consisting of luminous rays. The deadness of the optic nerve to those invisible rays was proved, and experiments were then added, to show that the bright and the dark rays of a solid body raised gradually to intense incandescence, are strengthened together; intense dark heat being an invariable accompaniment of intense white heat. A sun could not be formed, or a meteorite rendered luminous, on any other condition. The light-giving rays, constituting only a small fraction of the total radiation, their unspeakable importance to us is due to the fact that their periods are attuned to the special requirements of the eye.

Among the vapours of volatile liquids vast differences were also found to exist, as regards their powers of absorption. We followed various molecules from a state of liquid to a state of gas, and found, in both states of aggregation, the power of the individual molecules equally asserted. The position of a vapour as an absorber of radiant heat was shown to be determined by that of the liquid from which it is derived. Reversing our conceptions, and regarding the molecules of gases and vapours not as the recipients, but as the originators of wave motion; not as absorbers but as radiators; it was proved that the powers of absorption and radiation went hand in hand, the selfsame chemical act which rendered a body competent to intercept the waves of æther, rendering it competent in the same degree to generate them. Perfumes were next subjected to examination, and notwithstanding their extraordinary tenuity, they were found vastly superior, in point of absorptive power, to the body of the air in which they were diffused. We were led thus slowly up to the examination of the most widely diffused and most important of all vapours—the aqueous vapour of our atmosphere, and we found in it a potent absorber of the purely calorific rays. The power of this substance to influence climate, and its general influence on the temperature of the earth, were then briefly dwelt upon. A cobweb spread above a blossom is sufficient to protect it from nightly chill; and thus the aqueous vapour of our air, attenuated as it is, checks the drain of terrestrial heat, and saves the surface of our planet from the refrigeration which would assuredly accrue, were no such substance interposed between it and the voids of space. We considered the influence of vibrating period and molecular form on absorption and radiation, and finally deduced from its action upon radiant heat, the exact amount of carbonic acid expired by the human lungs.

Thus in brief outline were placed before you some of the results of recent enquiries in the domain of Radiation, and my aim throughout has been to raise in your minds distinct physical images of the various processes involved in our researches. It is thought by some that natural science has a deadening influence on the imagination, and a doubt might fairly be raised as to the value of any study which would necessarily have this effect. But the experience of the last hour must, I think, have convinced you, that the study of natural science goes hand in hand with the culture of the imagination. Throughout the greater part of this discourse we have been sustained by this faculty. We have been picturing atoms, and molecules, and vibrations, and waves, which eye has never seen nor ear heard, and which can only be discerned by the exercise of imagination. This, in fact, is the faculty which enables us to transcend the boundaries of sense, and connect the phenomena of our visible world with

those of an invisible one. Without imagination we never
could have risen to the conceptions which have occupied
us here to-day; and in proportion to your power of exer-
cising this faculty aright, and of associating definite mental
images with the terms employed, will be the pleasure
and the profit which you will derive from this lecture.
The outward facts of nature are insufficient to satisfy the
mind. We cannot be content with knowing that the
light and heat of the sun illuminate and warm the world.
We are led irresistibly to enquire what is light, and what
is heat? and this question leads us at once out of the
region of sense into that of imagination.

Thus pondering, and questioning, and striving to supplement that which is felt and seen, but which is incomplete, by something unfelt and unseen which is necessary
to its completeness, men of genius have in part discerned,
not only the nature of light and heat, but also, through
them, the general relationship of natural phenomena. The
working power of nature is the power of actual or potential motion, of which all its phenomena are but special
forms. This motion manifests itself in tangible and in
intangible matter, being incessantly transferred from the
one to the other, and incessantly transformed by the
change. It is as real in the waves of the æther as in the
waves of the sea; the latter, derived as they are from
winds, which in their turn are derived from the sun, being
nothing more than the heaped-up motion of the former.
It is the calorific waves emitted by the sun which heat
our air, produce our winds, and hence agitate our ocean.
And whether they break in foam upon the shore, or rub
silently against the ocean's bed, or subside by the mutual
friction of their own parts, the sea waves, which cannot

subside without producing heat, finally resolve themselves into waves of æther, thus regenerating the motion from which their temporary existence was derived. This connection is typical. Nature is not an aggregate of independent parts, but an organic whole. If you open a piano and sing into it, a certain string will respond. Change the pitch of your voice; the first string ceases to vibrate, but another replies. Change again the pitch; the first two strings are silent, while another resounds. Now in altering the pitch you simply change the form of the motion communicated by your vocal chords to the air, one string responding to one form, and another to another. And thus is sentient man acted on by Nature, the optic, the auditory, and other nerves of the human body being so many strings differently tuned and responsive to different forms of the universal power.

Note.—The statements regarding the action of aqueous vapour, made in sections 12 and 13 of this Lecture, have been controverted by the late Professor Magnus, of Berlin. 1 therefore wish the reader to hold in suspension his judgment of these two sections until new light can be thrown upon the subject. This will soon be done.

IX.

ON RADIANT HEAT

IN RELATION TO THE COLOUR AND CHEMICAL CONSTITUTION OF BODIES.

A DISCOURSE.

DELIVERED IN THE ROYAL INSTITUTION OF GREAT BRITAIN.

19th January, 1866.

'I took a number of little square pieces of broad-cloth from a tailor's pattern-card, of various colours. They were black, deep blue, lighter blue, green, purple, red, yellow, white, and other colours or shades of colour. I laid them all out upon the snow on a bright sunshiny morning. In a few hours (I cannot now be exact as to the time), the black, being warmed most by the sun, was sunk so low as to be below the stroke of the sun's rays; the dark blue almost as low, the lighter blue not quite so much as the dark, the other colours less as they were lighter. The white remained on the surface of the snow, not having entered it at all.

'What signifies philosophy that does not apply to some use? May we not learn from hence that black clothes are not so fit to wear in a hot sunny climate or season as white ones; because in such clothes the body is more heated by the sun when we walk abroad, and are at the same time heated by the exercise, which double heat is apt to bring on putrid dangerous fevers? That soldiers and seamen, who must march and labour in the sun, should, in the East or West Indies, have a uniform of white? That summer hats for men or women should be white, as repelling that heat which gives headaches to so many, and to some the fatal stroke that the French call *coup de soleil?* That the ladies' summer hats, however, should be lined with black, as not reverberating on their faces those rays which are reflected upwards from the earth or water? That the putting of a white cap of paper or linen *within* the crown of a black hat, as most do, will not keep out the heat, though it would if placed *without?* That fruit walls being blacked may receive so much heat from the sun in the daytime as to continue warm in some degree through the night, and thereby preserve the fruit from frosts, or forward its growth?—with sundry other particulars of greater or less importance that will occur from time to time to attentive minds?'

BENJAMIN FRANKLIN.
Letter to Miss Mary Stevenson.

IX.

ON RADIANT HEAT IN RELATION TO THE COLOUR AND CHEMICAL CONSTITUTION OF BODIES.

ONE of the most important functions of physical science, considered as a discipline of the mind, is to enable us by means of the tangible processes of nature to apprehend the intangible. The tangible processes give *direction* to the line of thought ; but this once given, the length of the line is not limited by the boundaries of the senses. Indeed, the domain of the senses in Nature is almost infinitely small in comparison with the vast region accessible to thought which lies beyond them. From a few observations of a comet, when it comes within the range of his telescope, an astronomer can calculate its path in regions which no telescope can reach ; and in like manner, by means of data furnished in the narrow world of the senses, we make ourselves at home in other and wider worlds, which can be traversed by the intellect alone.

From the earliest ages the questions, ' what is light ? ' and ' what is heat ? ' have occurred to the minds of men ; but these questions never would have been answered had they not been preceded by the question, ' what is sound ? ' Amid the grosser phenomena of acoustics the mind was first disciplined, conceptions being thus obtained from direct observation, which were afterwards applied to phenomena of a character far too subtle to be observed directly. Sound we know to be due to vibratory motion. A vibrating tuning-fork, for example, moulds the air around it into undulations or waves, which speed away on all sides with a certain measured velocity, impinge upon the drum of the ear, shake the auditory nerve and awake in the brain the sensation of sound. When sufficiently near a sounding body we can feel the vibrations of the air. A deaf man, for example, plunging his hand into a bell when it is sounded, feels through the common nerves of his body those tremors which, when imparted to the nerves of healthy ears, are translated into sound. There are various ways of rendering those sonorous vibrations not only tangible but visible; and it was not until numberless experiments of this kind had been executed, that the scientific investigator abandoned himself wholly, and without a shadow of uncertainty, to the conviction that what is sound within us is, outside of us, a motion of the air.

But once having established this fact—once having proved beyond all doubt that the sensation of sound is produced by an agitation of the nerve of the ear, the thought soon suggested itself that light might be due to an agitation of the nerve of the eye. This was a great step in advance of that ancient notion which regarded light as something emitted by the eye, and not as anything imparted to it. But if light be produced by an agitation of the optic nerve or retina, what is it that produces the agitation? Newton, you know, supposed minute particles to be shot through the humours of the eye against the retina, which hangs like a target at the back of the eye. The impact of these particles against the target, Newton believed to be the cause of light. But Newton's notion has not held its ground, being entirely driven from the field by the more wonderful and far more philosophical notion that light, like sound, is a product of wave-motion.

The domain in which this motion of light is carried on lies entirely beyond the reach of our senses. The waves of light require a medium for their formation and propagation, but we cannot see, or feel, or taste, or smell this medium. How then has its existence been established? By showing that by the assumption of this wonderful intangible *æther* all the phenomena of optics are accounted for with a fulness and clearness and conclusiveness which leave no desire of the intellect unfulfilled. When the law of gravitation first suggested itself to the mind of Newton, what did he do? He set himself to examine whether it accounted for all the facts. He determined the courses of the planets; he calculated the rapidity of the moon's fall towards the earth; he considered the precession of the equinoxes, the ebb and flow of the tides, and found all explained by the law of gravitation. He therefore regarded this law as established, and the verdict of science subsequently confirmed his conclusion. On similar, and, if possible, on stronger grounds, we found our belief in the existence of the universal æther. It explains facts far more various and complicated than those on which Newton based his law. If a single phenomenon could be pointed out which the æther is proved incompetent to explain, we should have to give it up; but no such phenomenon has ever been pointed out. It is, therefore, at least as certain that space is filled with a medium by means of which suns and stars diffuse their radiant power, as that it is traversed by that force which holds, not only our planetary system, but the immeasurable heavens themselves, in its grasp.

There is no more wonderful instance than this of the production of a line of thought from the world of the senses into the region of pure imagination. I mean by imagination here, not that play of fancy which can give to airy nothing a local habitation and a name, but that power which enables the mind to conceive realities which lie beyond the range of the senses—to present to itself distinct physical images of processes which, though mighty in the aggregate beyond all conception, are so minute individually as to elude all observation. It is the waves of air excited by this tuning-fork which render its vibrations audible. It is the waves of æther sent forth from those lamps overhead which render them luminous to us; but so minute are these waves, that it would take from 30,000 to 60,000 of them placed end to end to cover a single inch. Their number, however, compensates for their minuteness. Trillions of them have entered your eyes and hit the retina at the back of the eye in the time consumed in the utterance of the shortest sentence of this discourse. This is the stedfast result of modern research; but we never could have reached it without previous discipline. We never could have measured the waves of light, nor even imagined them to exist, had we not previously exercised ourselves among the waves of sound. Sound and light are now mutually helpful, the conceptions of each being expanded, strengthened, and defined by the conceptions of the other.

The æther which conveys the pulses of light and heat not only fills the celestial spaces, bathing the sides of suns and planets, but it also encircles the atoms of which these suns and planets are composed. It is the motion of these atoms, and not that of any sensible parts of bodies, that the æther conveys; it is this motion that constitutes the objective cause of what in our sensations are light and heat. An atom, then, sending its pulses through the infinite æther, resembles a tuning-fork sending its pulses through the air. Let us look for a moment at this thrilling æther, and briefly consider its relation to the bodies whose vibrations it conveys. Different bodies, when heated to the same temperature, possess very different powers of agitating the æther : some are good radiators, others are bad radiators; which means that some are so constituted as to communicate their motion freely to the æther, producing therein powerful undulations ; while others are unable thus to communicate their motion, but glide through the æther without materially disturbing its repose. Recent experiments have proved that elementary bodies, except under certain anomalous conditions, belong to the class of bad radiators. An atom vibrating in the æther resembles this naked tuning-fork vibrating in the air. The amount of motion communicated to the air by these thin prongs is too small to evoke at any distance the sensation of sound. But if we permit the atoms to combine chemically and form molecules, the result in many cases is an enormous change in the power of radiation. The amount of æthereal disturbance produced by the combined atoms of a body may be many thousand times that produced by its constituent atoms when uncombined. The effect is roughly typified by this tuning-fork when connected with its resonant case. The fork and its case now swing as a compound system, and the vibrations which were before inaudible, are now the source of a musical sound so powerful that it might be plainly heard by thousands at once. The fork and its case combined may be roughly regarded as a good radiator of sound.

The pitch of a musical note depends upon the rapidity of its vibrations, or, in other words, on the length of its waves. Now, the pitch of a note answers to the colour of light. Taking a slice of white light from the beam of an electric lamp, I cause that light to pass through an arrangement of prisms. It is decomposed, and we have the effect obtained by Newton, who first unrolled the solar beam into the splendours of the solar spectrum. At one end of this spectrum we have red light, at the other violet, and between those extremes lie the other prismatic colours. As we advance along the spectrum from the red to the violet, the pitch of the light—if I may use the expression—heightens, the sensation of violet being produced by a more rapid succession of impulses than that which produces the impression of red. The vibrations of the violet are about twice as rapid as those of the red; in other words, the range of the visible spectrum is about an octave.

There is no solution of continuity in this spectrum; one colour changes into another by insensible gradations. It is as if an infinite number of tuning-forks, of gradually augmenting pitch, were vibrating at the same time. But turning to another spectrum—that, namely, obtained from the incandescent vapour of silver—you observe that it consists of two narrow and intensely luminous green bands. Here it is as if two forks only, of slightly different pitch, were vibrating. The length of the waves which produce this first band is such that 47,460 of them, placed end to end, would fill an inch. The waves which produce the second band are a little shorter; it would take of these 47,920 to fill an inch. In the case of the first band, the number of impulses imparted in one second to every eye which now sees it, is 577 millions of millions; while the number of impulses imparted in the same time by the second band is 600 millions of millions. I now cast upon the screen before you the beautiful stream of green light from which these bands were derived. This luminous stream is the incandescent vapour of silver. The rates of vibration of the atoms of that vapour are as rigidly fixed as those of two tuning-forks; and to whatever height the temperature of the vapour may be raised, the rapidity of its vibrations, and consequently its colour, which wholly depends upon that rapidity, remain unchanged.

The vapour of water, as well as the vapour of silver, has its definite periods of vibration, and these are such as to disqualify the vapour, when acting freely as such, from being raised to a white heat. The oxyhydrogen flame, for example, consists of hot aqueous vapour. It is scarcely visible in the air of this room, and it would be still less visible if we could burn the gas in a clean atmosphere. But the atmosphere, even at the summit of Mont Blanc, is dirty; in London it is more than dirty; and the burning dirt gives to this flame the greater portion of its present light. But the heat of the flame is enormous. Cast iron fuses at a temperature of 2,000° Fahr.; while the temperature of the oxyhydrogen flame is 6,000 Fahr. A piece of platinum is heated to vivid redness at a distance of two inches beyond the visible termination of the flame.

The vapour which produces incandescence is here absolutely dark. In the flame itself the platinum is raised to dazzling whiteness, and is finally pierced by the flame. When this flame impinges on a piece of lime, we have the dazzling Drummond light. But the light is here due to the fact that when it impinges upon the solid body, the vibrations excited in that body by the flame are of periods different from its own.

Thus far we have fixed our attention on atoms and molecules in a state of vibration, and surrounded by a medium which accepts their vibrations, and transmits them through space. But suppose the waves generated by one system of molecules to impinge upon another system, how will the waves be affected? Will they be stopped, or will they be permitted to pass? Will they transfer their motion to the molecules on which they impinge, or will they glide round the molecules, through the intermolecular spaces, and thus escape?

The answer to this question depends upon a condition which may be beautifully exemplified by an experiment on sound. These two tuning-forks are tuned absolutely alike. They vibrate with the same rapidity, and mounted thus upon their resonant stands, you hear them loudly sounding the same musical note. I stop one of the forks, and throw the other into strong vibration. I now bring that other near the silent fork, but not into contact with it. Allowing them to continue in this position for four or five seconds, I stop the vibrating fork; but the sound has not ceased. The second fork has taken up the vibrations of its neighbour, and is now sounding in its turn. I dismount one of the forks, and permit the other to remain upon its stand. I throw the dismounted fork into strong vibration, but you cannot hear it sound. Detached from its stand the amount of motion which it can communicate to the air is too small to make itself sensible to the ear at any distance. I now bring the dismounted fork close to the mounted one, but not into actual contact with it. Out of the silence rises a mellow sound. Whence comes it ? From the vibrations which have been transferred from the dismounted fork to the mounted one.

That motion should thus transfer itself through the air it is necessary that the two forks should be in perfect unison. If I place on one of the forks a morsel of wax not larger than a pea, it is rendered thereby powerless to affect, or to be affected by, the other. It is easy to understand this experiment. The pulses of the one fork can affect the other, because they are *perfectly timed.* A single pulse causes the prong of the silent fork to vibrate through an infinitesimal space. But just as it has completed this small vibration, another pulse is ready to strike it. Thus, the small impulses add themselves together. In the five seconds during which the forks were held near each other, the vibrating fork sent 1,280 waves against its neighbour, and those 1,280 shocks, all delivered at the proper moment, all, as I have said, perfectly timed, have given such strength to the vibrations of the mounted fork as to render them audible to you all.

Let me give you one other curious illustration of the influence of synchronism on musical vibrations. Here are three small gas-flames inserted in three glass tubes of different lengths. Each of these flames can be caused to emit a musical note, the pitch of which is determined by the length of the tube surrounding the flame. The shorter the tube the higher is the pitch. The flames are now silent within their respective tubes, but each of them can be caused to respond to a proper note sounded anywhere in this room. Here is an instrument called a syren, by which a powerful musical note can be produced. Beginning with a note of low pitch, and ascending gradually to a higher one, I finally reach the note of the flame in the longest tube. The moment it is reached, the flame bursts into song. But the other flames are still silent within their tubes. I urge the instrument on to higher notes; the second flame has now started, and the third alone remains. But a still higher note starts it also. Thus, as the sound of the syren rises gradually in pitch, it awakens every flame in passing, by striking it with a series of waves whose periods of recurrence are similar to its own.

Now the wave-motion from the syren is in part taken up by the flame which synchronises with the waves; and had these waves to impinge upon a multitude of flames, instead of upon one flame only, the transference might be so great as to absorb the whole of the original wave-motion. Let us apply these facts to radiant heat. This blue flame is the flame of carbonic oxide; this transparent gas is carbonic acid gas. In the blue flame we have carbonic acid intensely heated, or, in other words, in a state of intense vibration. It thus resembles the sounding tuning-fork, while this cold carbonic acid resembles the silent one. What is the consequence? Through the synchronism of the hot and cold gas *transmission* of motion through the gas is prevented; it is all *transferred.* The cold gas is intensely opaque to the radiation from this particular flame, though highly transparent to heat of every other kind. We are here manifestly dealing with that great principle which lies at the basis of spectrum analysis, and which has enabled scientific men to determine the substances of which the sun, the stars, and even the nebulæ are composed: the principle, namely, that a body which is competent to emit any ray, whether of heat or light, is competent in the same degree to absorb that ray. The absorption depends on the synchronism which exists between the vibrations of the atoms from which the rays, or more correctly the *waves*, issue, and those of the atoms against which they impinge.

To its incompetence to emit white light, aqueous vapour adds incompetence to absorb white light. It cannot, for example, absorb the luminous rays of the sun, though it can absorb the non-luminous rays of the earth. This incompetence of aqueous vapour to absorb luminous rays is shared by water and ice—in fact, by all really transparent substances. Their transparency is due to their inability to absorb luminous rays. The molecules of such substances are in dissonance with the luminous waves, and hence such waves pass through transparent substances without disturbing the molecular rest. A purely luminous beam, however intense may be its heat, is sensibly incompetent to melt the smallest particle of ice. We can, for example, converge a powerful luminous beam upon a surface covered with hoar frost without melting a single spicula of the ice crystals. How then, it may be asked, are the snows of the Alps swept away by the sunshine of summer? I answer they are not swept away by sunshine at all, but by solar rays which have no sunshine whatever in them. The luminous rays of the sun fall upon the snow-fields and are flashed in echoes from crystal to crystal, but they find next to no lodgment within the crystals. They are hardly at all absorbed, and hence they cannot produce fusion. But a body of powerful dark rays is emitted by the sun, and it is these rays that cause the glaciers to shrink and the snows to disappear ; it is they that fill the banks of the Arve and Arveyron, and liberate from their frozen captivity the Rhone and the Rhine.

Placing a concave silvered mirror behind the electric light I converge its rays to a focus of dazzling brilliancy. I place in the path of the rays, between the light and the focus, a vessel of water, and now introduce at the focus a piece of ice. The ice is not melted by the concentrated beam which has passed through the water, though matches are ignited at the focus and wood is set on fire. The powerful heat then of this luminous beam is incompetent to melt the ice. I withdraw the cell of water ; the ice immediately liquefies, and you see the water trickling from it in drops. I re-introduce the cell of water ; the fusion is arrested and the drops cease to fall. The transparent water of the cell exerts no sensible absorption on the luminous rays, still it withdraws something from the beam, which, when permitted to act, is competent to melt the ice. This something is the dark radiation of the electric light. Again, I place a slab of pure ice in front of the electric lamp ; send a luminous beam first through our cell of water and then through the ice. By means of a lens an image of the slab is cast upon a white screen. The beam, sifted by the water, has no power upon the ice. But observe what occurs when the water is removed ; we have here a star and there a star, each star resembling a flower of six petals, and growing visibly larger before our eyes. As the leaves enlarge their edges become serrated, but there is no deviation from the six-rayed type.

We have here, in fact, the crystallisation of the ice inverted by the invisible rays of the electric beam. They take the molecules down in this wonderful way, and reveal to us the exquisite atomic structure of the substance with which nature every winter roofs our ponds and lakes.

Numberless effects, apparently anomalous, might be adduced in illustration of the action of these lightless rays. Here, for example, are two powders, both white, and undistinguishable from each other by the eye. The luminous rays of the lamp are unabsorbed by both powders,—from those rays they acquire no heat; still one of the substances, sugar, is heated so highly by the concentrated beam of the electric lamp that it first smokes violently and then inflames, while the other substance, salt, is barely warmed at the focus. Here, again, are two perfectly transparent liquids placed in a test tube at the focus; one of them boils in a couple of seconds, while the other in a similar position is hardly warmed. The boiling point of the first liquid is 78° C., which is speedily reached; that of the second liquid is only 48° C., which is never reached at all. These anomalies are entirely due to the unseen element which mingles with the luminous rays of the electric beam, and indeed constitutes 90 per cent. of its calorific power.

I have here a substance by which these dark rays may be detached from the total emission of the electric lamp. This ray-filter is a black liquid—that is to say, black as pitch to the luminous, but bright as a diamond to the non-luminous radiation. It mercilessly cuts off the former, but allows the latter free transmission. I bring these invisible rays to a focus at a distance of several feet from the electric lamp; the dark rays form there an invisible image of the source from which they issue. By proper means this invisible image may be transformed into a visible one of dazzling brightness. I could, moreover, show you, if time permitted, how out of those perfectly dark rays we might extract, by a process of transmutation, all the colours of the solar spectrum. I could also prove to you that those rays, powerful as they are, and sufficient to fuse many metals, may be permitted to enter the eye and to break upon the retina without producing the least luminous impression.

The dark rays are now collected before you ; you see nothing at their place of convergence ; with a proper thermometer it could be proved that even the air at the focus is just as cold as the surrounding air. And mark the conclusion to which this leads. It proves the *æther* at the focus to be practically detached from the air,—that the most violent æthereal motion may there exist without the least aërial motion. But though you see it not, there is sufficient heat at that focus to set London on fire. The heat there at the present moment is competent to raise iron to a temperature at which it throws off brilliant scintillations. It can heat platinum to whiteness and almost fuze that refractory metal. It actually can fuse gold, silver, copper, and aluminium. The moment, moreover, that wood is placed at the focus it bursts into a blaze.

It has been already affirmed that whether as regards radiation or absorption the elementary atoms possess but little power. This might be illustrated by a long array of facts ; and one of the most singular of these is furnished by the deportment of that extremely combustible substance, phosphorus, when placed at this dark focus. It is impossible to ignite there a fragment of amorphous phosphorus. But ordinary phosphorus is a far quicker combustible, and its deportment to radiant heat is still more impressive. It may be exposed to the intense radiation of an ordinary fire without bursting into flame. It may also be exposed for twenty or thirty seconds at an obscure focus of sufficient power to raise platinum to a white heat, without ignition. Notwithstanding the energy of the æthereal waves here concentrated, notwithstanding the extremely inflammable character of the elementary body exposed to their action, the atoms of that body refuse to partake of the motion of the waves, and consequently cannot be powerfully affected by their heat.

The knowledge which we now possess will enable us to analyse with profit a practical question. White dresses are worn in summer because they are found to be cooler than dark ones. The celebrated Benjamin Franklin made the following experiment:—He placed bits of cloth of various colours upon snow, exposed them to direct sunshine, and found that they sank to different depths in the snow. The black cloth sank the deepest, the white did not sink at all. Franklin inferred from his experiment that black bodies are the best absorbers, and white ones the worst absorbers, of radiant heat. Let us test the generality of this conclusion. I have here two cards, one of which is coated with a very dark powder, and the other with a perfectly white one. I place the powdered surfaces before the fire, and leave them there until they have acquired as high a temperature as they can attain in this position. Which of the cards is most highly heated? It requires no thermometer to answer this question? Simply pressing the back of the card, on which the white powder is strewn, against my cheek or

forehead, I find it intolerably hot.   Placing the dark card
in the same position I find it cool. The white powder has
absorbed far more heat than the dark one.   This simple
result abolishes a hundred conclusions which have been
hastily drawn from the experiment of Franklin.   Again,
here are suspended two delicate mercurial thermometers
at the same distance from a gas flame.   The bulb of one
of them is covered by a dark substance, the bulb of the
other by a white one. Both bulbs have received the radiation from the flame, but the white bulb has absorbed most,
and its mercury stands much higher than that of the other
thermometer. I might vary this experiment in a hundred
ways, and show you that from the darkness of a body
you can draw no certain conclusion regarding its power
of absorption.

The reason of this simply is, that colour gives us intelligence of only one portion, and that the smallest one, of
the rays impinging on the coloured body.   Were the
rays all luminous we might with certainty infer from the
colour of a body its power of absorption ; but the great
mass of the radiation from our fire, our gas-flame, and
even from the sun itself, consists of invisible calorific rays,
regarding which colour teaches us nothing.   A body
may be highly transparent to one class of rays, and
highly opaque to the other class.   Thus the white
powder, which has shown itself so powerful an absorber,
has been specially selected on account of its extreme
perviousness to the visible rays, and its extreme im-
perviousness to the invisible ones ;   while the dark
powder was chosen on account of its extreme transparency to the invisible, and its extreme opacity to the
visible rays.   In the case of the radiation from our fire,

about 98 per cent. of the whole emission consists of invisible rays; the body, therefore, which was most opaque to these triumphed as an absorber, though that body was a white one.

I would here invite you to consider the manner in which we obtain from natural facts what may be called their intellectual value. Throughout the processes of nature there is interdependence and harmony, and the main value of our science, considered as a mental discipline, consists in the tracing of this interdependence and the demonstration of this harmony. The outward and visible phenomena are with us the counters of the intellect; and our science would not be worthy of its name and fame if it halted at facts, however practically useful, and neglected the laws which accompany and rule phenomena. Let us endeavour, then, to extract from the experiment of Franklin its full intellectual value, calling to our aid the knowledge which our predecessors have already stored. Let us imagine two pieces of cloth of the same texture, the one black and the other white, placed upon sunned snow. Fixing our attention on the white piece, let us enquire whether there is any reason to expect that it will sink into the snow at all. There is knowledge at hand which enables us to reply at once in the negative. There is, on the contrary, reason to expect that after a sufficient exposure the bit of cloth will be found on an eminence instead of in a hollow; that instead of a depression, we shall have a *relative* elevation of the bit of cloth. For, as regards the luminous rays of the sun, the cloth and the snow are alike powerless; the one cannot be warmed, nor the other melted, by such rays. The cloth is white and the snow is white, because their confusedly mingled particles and fibres are incompetent to absorb luminous rays. Whether, then, the cloth will sink or not depends entirely upon the dark rays of the sun. Now the substance which absorbs the dark rays of the sun with the greatest avidity is ice, —or snow, which is merely ice in powder. A less amount of heat will be lodged in the cloth than in the surrounding snow. The cloth must therefore act as a shield to the snow on which it rests; and in consequence of the more rapid fusion of the exposed snow, the cloth must in due time be left behind, perched upon an eminence like a glacier-table.

But though the snow transcends the cloth both as a radiator and absorber it does not *much* transcend it. Cloth is very powerful in both these respects. Let us now turn our attention to the piece of black cloth, the texture and fabric of which I assume to be the same as that of the white. For our object being to compare the effects of colour, we must, in order to study this effect in its purity, preserve all other conditions constant. Let us then suppose the black cloth to be obtained from the dyeing of the white. The cloth itself, without reference to the dye, is nearly as good an absorber of heat as the snow around it. But to the absorption of the dark solar rays by the undyed cloth is now added the absorption of the whole of the luminous rays, and this great additional influx of heat is far more than sufficient to turn the balance in favour of the black cloth. The sum of its actions on the dark and luminous rays exceeds the action of the snow on the dark rays alone. Hence the cloth will sink in the snow, and this is the philosophy of Franklin's experiment.

Throughout this discourse the main stress has been laid on chemical constitution, as influencing most powerfully the phenomena of radiation and absorption. With regard to gases, vapours, and to the liquids from which these vapours are derived, it has been proved by the most varied and conclusive experiments that the acts of radiation and absorption are *molecular*—that they depend upon chemical and not upon mechanical condition. In attempting to extend this principle to solids I was met by a multitude of facts obtained by celebrated experimenters, which seemed flatly to forbid such extension. Melloni, for example, found the same radiant and absorbent power for chalk and lampblack. MM. Masson and Courtépée performed a most elaborate series of experiments on chemical precipitates of various kinds, and found that they one and all manifested the same power of radiation. They concluded from their researches, that where bodies are reduced to an extremely fine state of division the influence of this state is so powerful as entirely to mask and override whatever influence may be due to chemical constitution.

But it appears to me that through the whole of these researches a serious oversight has run, the mere mention of which will show you what caution is essential in the operations of experimental philosophy. Let me state wherein I suppose this oversight to consist. I have here a metal cube with two of its sides brightly polished. I fill the cube with boiling water and determine the quantity of heat emitted by the two bright surfaces. One of them far transcends the other as a radiator of heat. Both surfaces appear to be metallic; what then is the cause of the observed difference in their radiative power?

Simply this: I have coated one of the surfaces with transparent gum, through which, of course, is seen the metallic lustre behind. Now this varnish, though so perfectly transparent to luminous rays, is as opaque as pitch or lampblack to non-luminous ones. It is a powerful emitter of dark rays; it is also a powerful absorber. While, therefore, at the present moment it is copiously pouring forth radiant heat itself, it does not allow a single ray from the metal behind to pass through it. The varnish then, and not the metal, is the real radiator.

Now Melloni, and Masson, and Courtépée experimented thus: they mixed their powders and precipitates with gum-water, and laid them by means of a brush upon the surfaces of a cube like this. True they saw their red powders red, their white ones white, and their black ones black, but they saw these colours *through the coat of varnish which encircled every particle of their powders.* When, therefore, it was concluded that colour had no influence on radiation, no chance had been given to it of asserting its influence; when it was found that all chemical precipitates radiated alike, it was the radiation from a varnish common to them all which showed the observed constancy. Hundreds, perhaps thousands, of experiments on radiant heat have been performed in this way by various enquirers, but I fear the work will have to be done over again. I am not, indeed, acquainted with an instance in which an oversight of so trivial a character has been committed in succession by so many able men, and vitiated so large an amount of otherwise excellent work.

Basing our reasonings then on demonstrated facts, we arrive at the extremely probable conclusion that the envelope of the particles, and not the particles themselves, was the real radiator in the experiments just referred to. To reason thus, and deduce their more or less probable consequences from experimental facts, is an incessant exercise of the student of physical science. But having thus followed for a time the light of reason alone through a series of phenomena, and emerged from them with a purely intellectual conclusion, our duty is to bring that conclusion to an experimental test. In this way we fortify our science, sparing no pains, shirking no toil to secure sound materials for the edifice which it is our privilege to raise.

For the purpose of testing our conclusion regarding the influence of the gum I take two powders of the same physical appearance; one of them is a compound of mercury and the other a compound of lead. On two surfaces of this cube are spread these bright red powders without varnish of any kind. Filling the cube with boiling water, and determining the radiation from the two surfaces, one of them is found to emit thirty-nine rays, while the other emits seventy-four. This, surely, is a great difference. Here, however, is a second cube, having two of its surfaces coated with the same powders, the only difference being that now the powders are laid on by means of a transparent gum. Both surfaces are now absolutely alike in radiative power. Both of them emit somewhat more than was emitted by either of the unvarnished powders, simply because the gum employed is a better radiator than either of them. Excluding all varnish, and comparing white with white, I find vast differences; comparing black with black I find them also different; and when black and white are compared, in some cases the black radiates far more than the white, while in other cases the white radiates far more than the black. Determining, moreover, the absorptive power of those powders, it is found to go hand-in-hand with their radiative power. The good radiator is a good absorber, and the bad radiator is a bad absorber. From all this it is evident that as regards the radiation and absorption of non-luminous heat, colour teaches us nothing; and that even as regards the radiation of the sun, consisting as it does mainly of non-luminous rays, conclusions as to the influence of colour may be altogether delusive. This is the strict scientific upshot of our researches. But it is not the less true that in the case of wearing apparel—and this for reasons which I have given in analysing the experiment of Franklin—black dresses are more potent than white ones as absorbers of solar heat.

Thus, in brief outline, I have brought before you a few of the results of recent enquiry. If you ask me what is the use of them, I can hardly answer you, unless you define the term use. If you meant to ask me whether those dark rays which clear away the Alpine snows will ever be applied to the roasting of turkeys or the driving of steam-engines, while affirming their power to do both, I would frankly confess that they are not at present capable of competing profitably with coal in these particulars. Still they may have great uses unknown to me ; and when our coal-fields are exhausted, it is possible that a more æthereal race than ourselves may cook their victuals and perform their work in this transcendental way. But is it necessary that the student of science should have his labours tested by their possible practical applications ? What is the practical value of Homer's Iliad ? You smile, and possibly think that Homer's Iliad is good as a means of culture. There's the rub. The people who demand of science practical uses, forget, or do not know, that it also is great as a means of culture ; that the knowledge of this wonderful universe is a thing profitable in itself, and requiring no practical application to justify its pursuit. But while the student of nature distinctly refuses to have his labours judged by their practical issues, unless the term practical be made to include mental as well as material good, he knows full well that the greatest practical triumphs have been episodes in the search after pure natural truth. The electric telegraph is the standing wonder of this age, and the men whose scientific knowledge and mechanical skill have made the telegraph what it is are deserving of all honour. In fact, they have their reward, both in reputation and in those more substantial benefits which the direct service of the public always carries in its train. But who, I would ask, put the soul into this telegraphic body? Who snatched from heaven the fire that flashes along the line? This, I am bound to say, was done by two men, the one a dweller in Italy,[1] the other a dweller in England, and therefore not a thousand miles distant from the spot where I now stand,[2] who never in their enquiries consciously set a practical object before them,—whose only stimulus was the fascination which draws the climber to a never-trodden peak, and would have made Cæsar quit his victories to seek the sources of the Nile. That the knowledge brought us by those prophets, priests, and kings of science is what the world calls useful knowledge, the triumphant application of their discoveries proves. But science has another

[1] Volta.   [2] Faraday.

R 2

function to fulfil, in the storing and the training of the human mind ; and I would base my appeal to you on the poor specimen which has been brought before you this evening, whether any system of education at the present day can be deemed even approximately complete in which the knowledge of nature is neglected or ignored.

X.

ON CHEMICAL RAYS AND THE STRUCTURE AND LIGHT OF THE SKY.

A DISCOURSE.

DELIVERED IN THE ROYAL INSTITUTION OF GREAT BRITAIN.

On Friday, 15th January, 1869.

'This is a very mysterious and a very beautiful phenomenon when observed by the aid of a polariscope, consisting of a tourmaline plate, with a slice of Iceland crystal or nitre, cut at right angles to the optic axis, and applied on the side of the tourmaline farthest from the eye. In a cloudless day, if the sky be explored in all parts by looking through this compound plate, the polarised rings will be seen developed with more or less intensity in every region but that nearest the sun and that most distant from it—the maximum of polarisation taking place on a zone of the sky 90° from the sun, or in a great circle, having the sun for one of its poles, so that the cause of polarisation is evidently a reflection of the sun's light *on something*. The question is, on what? Were the angle of maximum polarisation 76°, we should look to water or ice for the reflecting body. But though we were once of this opinion (art. Light, *Encycl. Metropol.* § 1143), careful observation has satisfied us that 90°, or thereabouts, is the correct angle, and that therefore, whatever be the body on which the light has been reflected, *if polarised by a single reflection*, the " polarising angle " must be 45°, and the index of refraction, which is the tangent of that angle, unity; in other words, the reflection would require to be made *in* air *upon* air! The only imaginable way in which this could happen would be at the plane of contact of two portions of air differently heated, such as *might* be supposed to occur at almost every point of the atmosphere in a bright sunny day; but against this there seems to be an insuperable objection. The polarisation is most regular and complete, as we have lately been able to satisfy ourselves under the most favourable possible atmospheric conditions, after sunset, in the bright twilight of a summer night, with the sun some degrees below the horizon, and long after all the tremor and turmoil of the air, due to irregular heating, must have completely subsided. On the other hand, if effected by several successive reflections, what is to secure a large majority of them being in one plane (in which case only their polarising effect would accumulate); and of those which become ultimately effective, what is there to determine an ultimate deviation of 90° as that of the maximum? The more the subject is considered, the more it will be found beset with difficulties; and its explanation, when arrived at, will probably be found to carry with it that of the blue colour of the sky itself.'

SIR JOHN HERSCHEL.

X.

ON CHEMICAL RAYS AND THE STRUCTURE AND LIGHT OF THE SKY.

THE first physical investigation of any importance in which, jointly with my friend Professor Knoblauch, I took part, bore the title, ' The Magneto-optic Properties of Crystals, and the Relation of Magnetism and Diamagnetism to Molecular Arrangement.'[1] This investigation compelled me to reflect upon the structure of crystals, on their optical properties in relation to that structure, and more particularly on the striking phenomena exhibited by many of them in the field of a sufficiently powerful magnet. These were evidently due to the manner in which the molecules of the crystals were built together by the force of crystallisation; and it was natural, if not necessary for me, to employ such strength of imagination as I possessed in obtaining a mental picture of this molecular architecture. The enquiry gave a tinge and bias to my subsequent scientific thought, rendering, as it did, the conceptions and pursuit of molecular physics pleasant to me. Its influence is to be traced in most of my scientific work. The first lecture, for example, which I ever delivered in this theatre, was ' On the Influence of Material Aggregation on the Manifestations of Force ;' by ' material aggregation ' being meant the way in which,

---

[1] *Philosophical Magazine*, July, 1850.

by nature or by art, the molecules of matter are arranged together. In 1853 I also published a paper 'On Molecular Influences,' in which common heat was made the explorer of organic structure. In the 'Bakerian Lecture,' given before the Royal Society in 1855, the same idea and phraseology crop out. The Bakerian Lecture for 1864 bears the title 'Contributions to Molecular Physics.' And all through the investigations which have occupied me during the last ten years, my wish and aim have been to make radiant heat an instrument by which to lay hold of the ultimate particles of matter.

The labours now to be considered lie in the same direction. In the researches just referred to, tubes of glass and brass were employed, called, for the sake of distinction, 'experimental tubes,' in which radiant heat was acted upon by the gases and vapours subjected to examination. Two or three months ago, with a view of seeing what occurred within these tubes on the entrance of the gases or vapours, it was found necessary to intensely illuminate their interiors. The source of illumination chosen was the electric light, the beam of which, converged by a suitable lens, was sent along the axis of the tube. The dirt and filth in which we habitually live were strikingly revealed by this method of illumination. For, wash the tube as we might with water, alcohol, acid, or alkali, until its appearance in ordinary daylight was that of absolute purity, the delusive character of this appearance was in most cases revealed by the electric beam. In fact, in air so charged with suspended matter as that which supplies our lungs in London, it is not possible to be more than approximately cleanly.

Vapours of various kinds were sent into a glass experimental tube a yard in length, and about three inches in diameter. As a general rule, the vapours were perfectly transparent ; the tube when they were present appearing as empty as when they were absent. In two or three cases, however, a faint cloudiness showed itself within the tube. This caused me a momentary anxiety, for I did not know how far, in describing my previous experiments, actions might have been ascribed to pure cloudless vapour which were really due to those newly-observed nebulæ. Intermittent discomfort, however, is the normal feeling of the investigator ; for it drives him to closer scrutiny, to greater accuracy, and often, as a consequence, to new discovery. It was soon found that the nebulæ revealed by the beam were also *generated* by the beam, and the observation opened a new door into that region inaccessible to sense, which embraces so much of the intellectual life of the physical investigator.

What *are* those vapours of which we have been speaking ? They are aggregates of *molecules,* or small masses of matter, and every molecule is itself an aggregate of smaller parts called *atoms.* A molecule of aqueous vapour, for example, consists of two atoms of hydrogen and one of oxygen. A molecule of ammonia consists of three atoms of hydrogen and one of nitrogen, and so of other substances. Thus the molecules, themselves inconceivably small, are made up of distinct parts still smaller. When, therefore, a compound vapour is spoken of, the corresponding mental image is an aggregate of molecules separated from each other, though still exceedingly near, each of these being composed of a group of atoms still nearer to each other. So much for the *matter* which enters into our conception of a vapour.[1] To this must now be added the idea of *motion.* The molecules have motions of their own as *wholes*; their constituent atoms have also motions of their own, which are executed independently of those of the molecules; just as the various movements of the earth's surface are executed independently of the orbital revolution of our planet.

The vapour molecules are kept asunder by forces which, virtually or actually, are forces of repulsion. Between these elastic forces and the atmospheric pressure under which the vapour exists, equilibrium is established as soon as the proper distances between the molecules have been assumed. If, after this, the molecules be urged nearer to each other by a momentary force, they recoil as soon as the force is expended. If they be separated more widely apart, when the separating force ceases to act they again approach each other. The case is different as regards the constituent atoms.

And here let it be remarked, that we are now upon the very outmost verge of molecular physics; and that I am attempting to familiarise your minds with conceptions which have not yet obtained universal currency even among chemists; which many chemists, moreover, might deem untenable. But, tenable or untenable, it is of the highest scientific importance to discuss them. Let us, then, look mentally at our atoms grouped together to form a molecule. Every atom is held apart from its neighbours by a force of repulsion; why, then, do not

---

[1] Newton seemed to consider that the molecules might be rendered visible by microscopes; but of the atoms he appears to have entertained a different opinion. He finely remarks :—' It seems impossible to see the more secret and noble works of nature within the corpuscles, by reason of their transparency.' (Herschel, *On Light*, Art. 1145.)

the mutually repellent members of this group part company? The molecules separate from each other when the external pressure is lessened or removed, but the atoms do not. The reason of this stability is that two forces, the one attractive and the other repulsive, are in operation between every two atoms; and the position of every atom—its distance from its fellows—is determined by the equilibration of these two forces. If the atoms come too near, repulsion predominates and drives them apart; if too distant, attraction predominates and draws them together. The point at which attraction and repulsion are equal to each other is the atom's *position of equilibrium.* If not absolutely cold—and there is no such thing as absolute coldness in our corner of nature—the atoms are always in a state of vibration, their vibrations being executed to and fro across their positions of equilibrium.

Into a vapour thus constituted, we have now to pour a beam of light; which most of you know to be a train of minute waves, excited in, and propagated through, an almost infinitely attenuated and elastic medium, which fills all space, and which we name the Æther. It is hardly necessary to remind you that these waves of light are not all of the same size; that some of them are much longer and higher than others; that the short waves and the long ones move with the same rapidity through space, just as short and long waves of sound travel with the same rapidity through air, and that, therefore, the shorter waves must follow each other in quicker succession than the longer ones; that the different rapidities with which the waves of light impinge upon the retina, or optic nerve, give rise in consciousness to differences of shown that there are, moreover, numberless waves emitted by the sun and other luminous bodies which reach the retina, but which are incompetent to excite the sensation of light; for if the lengths of the waves exceed a certain limit, or if they fall short of a certain other limit, they cannot generate vision. And it is to be particularly borne in mind, that the capacity to excite vision does not depend so much on the strength of the waves as on their periods of recurrence. I have, as many of you know, permitted waves to enter my own eye, which, if their energy were that of light, would have instantly and utterly ruined the optic nerve, but which failed to produce any impression whatever upon consciousness, because their periods were not those competent to excite the retina.

The elements of all the conceptions with which we shall have subsequently to deal are now in your possession. And you will observe that though we are speaking of things which lie entirely beyond the range of the senses, the conceptions are as truly *mechanical* as they would be if we were dealing with ordinary masses of matter, and with waves of sensible magnitude. No really scientific mind at the present day will be disposed to draw a substantial distinction between chemical and mechanical phenomena. They differ from each other as regards the magnitude of the masses involved; but in this sense the phenomena of astronomy differ, also, from those of ordinary mechanics. The main bent of the natural philosophy of a future age will probably be to chasten into order, by subjecting it to mechanical laws, the existing chaos of chemical phenomena.

Whether we see rightly or wrongly—whether our intellection be real or imaginary—it is of the utmost importance in science to aim at perfect clearness in the description of all that comes, or seems to come, within the range of the intellect. For if we are right, clearness of utterance forwards the cause of right ; while, if we are wrong, it ensures the speedy correction of error. In this spirit, and with the determination at all events to speak plainly, let us deal with our conceptions of æther waves and molecules. Supposing a wave, or a train of waves, to impinge upon a molecule so as to urge all its parts with the same motion, the molecule would move bodily as a whole, but because they are animated by a *common motion* there would be no tendency of its constituent atoms to separate from each other. *Differential motions* among the atoms themselves would be necessary to effect a separation, and if such motions be not introduced by the shock of the waves, there is no mechanical ground for the decomposition of the molecule.

Thus the conception of the decomposition of compound molecules by the waves of æther comes to us recommended by *à priori* probability. But a closer examination of the question compels us to supplement, if not materially to qualify, this conception. It is a most remarkable fact, that the waves which have thus far been found most effectual in shaking asunder the atoms of compound molecules are those of least mechanical power. Billows, to use a strong comparison, are incompetent to produce effects which are readily produced by ripples. It is, for example, the violet and ultra-violet rays of the sun that are most effectual in producing these chemical decompositions ; and, compared with the red and ultra-red solar rays, the energy of these ' chemical rays ' is infinitesimal. This energy would probably in some cases have to be multiplied by millions to bring it up to that of the ultra-red rays : and still the latter are powerless where the smaller waves are potent. We here observe a remarkable similarity between the behaviour of chemical molecules and that of the human retina. The energy transmitted to the eye from a candle-flame half a mile distant is more than sufficient to inform consciousness ; while waves of a different period, possessing twenty thousand million times this energy, have been suffered to impinge upon my own retina, with an absolute unconsciousness of any effect whatever—mechanical, physiological, chemical, or thermal.

If, then, the power of these smaller waves to unlock the bonds of chemical union be not a result of their strength, it must be, as in the case of vision, a result of their periods of recurrence. But how are we to figure this action? The shock of a single wave produces no more than an infinitesimal effect upon an atom or a molecule. To produce a larger effect, the motion must accumulate, and for wave-impulses to accumulate, they must arrive in periods identical with the periods of vibration of the atoms on which they impinge. In this case each successive wave finds the atom in a position which enables that wave to add its shock to the sum of the shocks of its predecessors. The effect is mechanically the same as that due to the timed impulses of a boy upon a swing. The single tick of a clock has no appreciable effect upon the unvibrating and equally long pendulum of a distant clock ; but a succession of ticks, each of which adds, at the proper moment, its infinitesimal push to the sum of the pushes preceding it, will, as a matter of fact, set the second clock going. So likewise a single puff of air against the prong of a heavy tuning-fork produces no sensible motion, and, consequently, no audible sound; but a succession of puffs, which follow each other in periods identical with the tuning-fork's period of vibration, will render the fork sonorous. I think the chemical action of light is to be regarded in this way. Fact and reason point to the conclusion that it is the heaping up of motion on the atoms, in consequence of their synchronism with the shorter waves, that causes them to part company. This I take to be the mechanical cause of these decompositions which are effected by the waves of æther.

And now let us return to that faint cloudiness already mentioned, from which, as from a germ, these considerations and speculations have sprung. It has been long known that light effected the decomposition of a certain number of bodies. The transparent iodide of ethyl, or of methyl, for example, becomes brown and opaque on exposure to light, through the discharge of its iodine. The art of photography is founded on the chemical actions of light; so that it is well known that the effects for which the foregoing theoretic considerations would have prepared us, are not only probable, but actual.

But the method employed in the experiments in which the cloudiness above referred to was observed, and which consists simply in offering the vapours of volatile substances to the action of light, enables us not only to give such experiments a beautiful form, but also to give a great extension to the operations of light, or rather of radiant force, as a chemical agent. It also enables us to illustrate in our laboratories actions which have been hitherto performed only in the laboratory of nature. A few of these actions of a representative character will now be brought before you ; and advantage will be taken of the fact that, in a great number of cases, one or more of the substances into which the waves of light break up compound molecules are comparatively *involatile*. The products of decomposition require a greater heat than is required by the vapours from which they are derived to keep them in the gaseous form ; and hence, if the space in which these new bodies are liberated be of the proper temperature, they will not remain in the vaporous condition, but will precipitate themselves as liquid particles, thus forming visible clouds upon the beam, to the action of which they owe their existence.

The little flask, F, in the annexed figure is stopped by a cork, pierced in two places. Through one orifice passes a narrow glass tube, *a*, which terminates immediately under the cork ; through the other orifice passes a similar tube, *b*, descending to the bottom of the little flask, which is filled to a height of about an inch with a transparent liquid. The name of this liquid is *nitrite of amyl,* in every molecule of which we have 5 atoms of carbon, 11 of hydrogen, 1 of nitrogen, and 2 of oxygen. Upon this group the waves of our electric light will be immediately let loose. The large horizontal tube that you see before you is called an ' experimental tube ; ' it is connected with our small flask ; between them, however, a stop-cock intervenes, by means of which the passage between the flask and the experimental tube can be opened or closed at pleasure. The other tube, passing through the cork of the flask and descending into the liquid, is connected with a U-shaped vessel, filled with fragments of clean glass, covered with sulphuric acid. In front of the U-shaped vessel is a narrow tube stuffed with cotton-wool. At one end of the experimental tube is our electric lamp; and here, finally, is an air-pump, by means of which the tube has been exhausted. We are now ready for experiment.

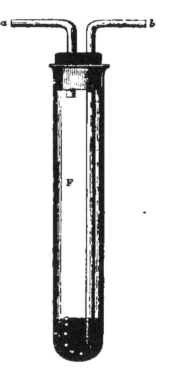

Opening the cock cautiously, the air of the room passes, in the first place, through the cotton-wool, which holds back the numberless organic germs and inorganic dust-particles floating in the atmosphere. The air, thus cleansed, passes into the U-shaped vessel, where it is *dried* by the sulphuric acid. It then descends through the narrow tube to the bottom of the little flask, and escapes there through a small orifice into the liquid. Through this it bubbles, loading itself to some extent with the nitrite of amyl vapour, and then the air and vapour enter the experimental tube together.

The closest scrutiny would now fail to discover anything within this tube; it is, to all appearance, absolutely empty The air and the vapour are both invisible. We will permit the electric beam to play upon this mixture. The lens of the lamp is so situated as to render the beam slightly convergent, the focus being formed in the vapour at about the middle of the tube. You will notice that the tube remains dark for a moment after the turning on of the beam; but the chemical action will be so rapid that attention is requisite to mark this interval of darkness. I ignite the lamp; the tube for a moment seems empty; but suddenly the beam darts through a luminous white cloud, which has banished the preceding darkness. It has, in fact, shaken asunder the molecules of the nitrite of amyl, and brought down upon itself a shower of liquid particles which cause it to flash forth in your presence like a solid luminous spear. It is worth while to mark how this experiment illustrates the fact, that however intense a luminous beam may be, it remains invisible unless it has something to shine upon. *Space*, though traversed by the rays from all suns and all stars, is itself unseen. Not even the æther which fills space, and whose motions are the light of the universe, is itself visible.

You notice that the end of the experimental tube most distant from the lamp is free from cloud. Now the nitrite of amyl vapour is there also, but it is unaffected by the powerful beam passing through it. Let us make the transmitted beam more concentrated by receiving it on a concave silver mirror, and causing it to return by reflection into the tube. It is still powerless. Though a cone of light of extraordinary intensity now traverses the vapour, no precipitation occurs, no trace of cloud is formed. Why? Because the very small portion of the beam competent to decompose the vapour is quite exhausted by its work in the frontal portions of the tube. The great body of the light which remains, after this sifting out of the few effectual rays, has no power over the molecules of nitrite of amyl. We have here, strikingly illustrated, what has been already stated regarding the influence of *period*, as contrasted with that of *strength*. For the portion of the beam which is here ineffectual has probably more than a million times the absolute energy of the effectual portion. It is energy specially related to the atoms that we here need, which specially related energy being possessed by the feeble waves, invests them with their extraordinary power. When the experimental tube is reversed so as to bring the undecomposed vapours under the action of the *unsifted* beam, you have instantly this fine luminous cloud precipitated.

The light of the sun also effects the decomposition of the nitrite of amyl vapour. A small room in the Royal Institution, into which the sun shone, was partially darkened, the light being permitted to enter through an open portion of the window-shutter. In the track of the beam was placed a large plano-convex lens, which formed a fine convergent cone in the dust of the room behind it. The experimental tube was filled in the laboratory, covered with a black cloth, and carried into the partially darkened room. On thrusting one end of the tube into the cone of rays behind the lens, precipitation within the cone was copious and immediate. The vapour at the distant end of the tube was shielded by that in front; but on reversing the tube, a second and similar splendid cone was precipitated.

Now let us pause for a moment and glance at the ground over which we have passed. We have defined a vapour as an aggregate of molecules mutually repellent, but hindered from indefinitely retreating from each other by an external pressure. We have defined a molecule as an aggregate of atoms maintained in positions of equilibrium by the equalised action of two opposing forces, and always oscillating to and fro across those positions. We have defined a beam of light as a train of innumerable waves, and have illustrated their chemical action. We have learned that it is not the magnitude or power of the waves, so much as their periods of recurrence, that renders them effectual as chemical agents. We have also seen how the luminous beam is sifted by the vapour which it decomposes, and deprived of those rays which are competent to effect the decomposition. The effects, moreover, obtained with the electric beam are also produced by the beams of the sun.

And here I would ask you to make familiar to your minds the idea that no chemical action can be produced by a ray that does not involve the destruction of the ray. But the term ' ray ' is unsatisfactory to us at present, when our desire is to abolish all vagueness, and to affix a definite physical significance to each of our terms. Abandoning the term ray as loose and indefinite, we have to fix our thoughts upon the *waves* of light ; and to render clear to our minds that those waves which produce chemical action do so by delivering up their own motion to the molecules which they decompose. We have here forestalled to some extent a question of great importance in molecular physics, which, however, is worthy of being fixed more definitely in your mind ; it is this : When the waves of æther are intercepted by a compound vapour, is the motion of the waves transferred to the molecules of the vapour, or to the atoms of the molecules ? We have thus far leaned to the conclusion that the motion is communicated to the atoms ; for if not to these individually, why should they be shaken asunder ? The question, however, is capable of, and is worthy of, another test, the bearing and significance of which you will immediately appreciate.

As already explained, the molecules are held in their positions of equilibrium by their mutual repulsion on the one side, and by an external pressure on the other. Their rate of vibration, if they vibrate at all, must depend upon the elastic force which they mutually exert. If this force be changed, the rate of vibration must change along with it ; and after the change the molecules could no longer absorb the waves which they absorbed prior to the change. Now the elastic force between molecule and molecule is utterly altered when a vapour passes to the liquid state. Hence, if the liquid absorbs waves of the same period as its vapour, it is a proof that the absorption is not effected by the molecules. Let us be perfectly clear on this important point. Those waves are absorbed whose vibrations synchronise with those of the molecules or atoms on which they impinge ; a principle which is sometimes expressed by saying that bodies radiate and absorb the same rays. This great law, as you know, is the foundation of spectrum-analysis ; it enabled Kirchhoff to explain the lines of Frauenhofer, and to determine the chemical composition of the atmosphere of the sun. If then, after such a change as that involved in the passage of a vapour to the liquid state, the same waves are absorbed as were absorbed prior to the passage, it is a proof that the molecules, which must have utterly changed *their* periods, cannot be the seat of the absorption ; and we are driven to conclude that it is to the *atoms*, whose rates of vibration are unchanged by the change of aggregation, that the wave-motion is transferred. If experiment should prove this identity of action on the part of a vapour and its liquid, it would establish in a new and striking manner the conclusion to which we have previously leaned.

We will now resort to the experimental test. In front of this experimental tube, which contains a quantity of the nitrite of amyl vapour, is placed a glass cell a quarter of an inch in thickness, filled with the liquid nitrite of amyl. I send the electric beam first through the liquid and then through its vapour. The luminous power of this beam is very great, but it can make no impression upon the vapour. The liquid has robbed it completely of its effective waves. When the liquid is removed chemical action immediately commences, and in a moment we have the apparently empty tube filled with this bright cloud, precipitated by one portion of the beam, and illuminated by another. Thus we uncover to some extent the secrets of this world of molecules and atoms.

Instead of employing air as the vehicle by which the vapour is carried into the experimental tube, we may employ oxygen, hydrogen, or nitrogen. With hydrogen curious effects are observed, due to the sinking of the clouds through the extremely light gas in which they float. They illustrate, but do not prove, the untenable notion of those who say that the clouds of our own atmosphere could not float if the cloud particles were not little bladders instead of full spheres. Before you is a tube filled with the nitrite of amyl vapour, which has been carried into the tube by hydrogen gas. On sending the beam through the tube a delicate bluish-white cloud is precipitated. A few strokes of the pump clear the tube of this cloud, but leave a residue of vapour behind. Again turning on the beam we have a second cloud, more delicate than the first. This may be done half-a-dozen times in succession. A residue of vapour will still linger in the tube sufficient to yield a cloud of exquisite delicacy, both as regards colour and texture.

Besides the nitrite of amyl a great number of other substances might be employed, which, like the nitrite, have been hitherto not known to be chemically susceptible to light. This is, in fact, a representative case. One point in addition I wish to illustrate, chiefly because the effect is the same in kind as one of great importance in nature. Our atmosphere contains carbonic acid gas, which furnishes food to the vegetable world. But this food, as many of you know, could not be consumed by plants and vegetables without the intervention of the sun's rays. As far as we know, however, these rays are powerless upon the free carbonic acid of our atmosphere ; the sun can only decompose the gas when it is absorbed by the leaves of plants. In the leaves the carbonic acid is in close proximity with substances ready to take advantage of the loosening of the molecules by the waves of light. Incipient disunion being introduced by the solar rays, the carbon of the gas is seized upon by the leaf and appropriated, while the oxygen is discharged into the atmosphere.

The experimental tube now before you contains a quantity of a different vapour from that which we have hitherto employed. The liquid from which this vapour is derived is called the nitrite of butyl. On sending the electric beam through the vapour, which has been carried in by air, the chemical action is insensible. I add to the vapour a quantity of air which has been permitted to bubble through hydrochloric acid. When the beam is now turned on, so rapid is the action and so dense the clouds precipitated, that you could hardly by an effort of attention observe the dark interval which preceded the precipitation of the cloud. This enormous augmentation of the action is due to the presence of the hydrochloric acid. Like the chlorophyl in the leaves of plants, it takes advantage of the loosening of the molecules of nitrite of butyl by the waves of the electric light.

In these experiments we have employed a luminous beam for two different purposes. A small portion of it has been devoted to the decomposition of our vapours, while the great body of the light has served to render luminous the clouds resulting from the decomposition. It is possible to impart to these clouds any required degree of tenuity, for it is in our power to limit at pleasure the amount of vapour in our experimental tube. When the quantity is duly limited, the precipitated particles are at first inconceivably small, defying the highest microscopic power to bring them within the range of vision. Probably their diameters might then be expressed in millionths of an inch. They grow gradually, and as they augment in size, they scatter a continually increasing quantity of wave-motion, until, finally, the cloud which they form becomes so luminous as to fill this theatre with light. During the growth of the particles the most splendid iridescences are often exhibited. Such I have sometimes seen with delight and wonder in the atmosphere of the Alps, but never anything so gorgeous as those which our laboratory experiments reveal. It is not, however, with the iridescences, however beautiful they may be, that we have now to occupy our thoughts, but with other effects which bear upon the two great standing enigmas of meteorology —the colour of the sky and the polarisation of its light.

It is possible, as stated, by duly regulating the quantity of vapour, to make our precipitated particles grow from an infinitesimal and altogether ultra-microscopic size to masses of sensible magnitude ; and by means of these particles, in a certain stage of their growth, we can produce a blue which shall rival, if it does not transcend, that of the deepest and purest Italian sky. Let this point be in the first place established. Associated with our experimental tube is a barometer, the mercurial column of which now indicates that the tube is exhausted. Into the tube is introduced a quantity of the mixed air and nitrite of butyl vapour sufficient to depress the mercurial column one-twentieth of an inch ; that is to say, the air and vapour together exert a pressure of one six-hundredth of an atmosphere. I now add a quantity of air and hydrochloric acid sufficient to depress the mercury half-an-inch further, and into this compound and highly attenuated atmosphere I discharge the beam of the electric light. The effect is slow ; but gradually within the tube arises this splendid azure, which strengthens for a time, reaches a maximum of depth and purity, and then, as the particles grow larger, passes into whitish blue. This experiment is representative, and it illustrates a general principle. Various other colourless substances of the most diverse properties, optical and chemical, might be employed for this experiment. The *incipient cloud* in every case would exhibit this superb blue ; thus proving to demonstration that particles of infinitesimal size, without any colour of their own, and irrespective of those optical properties exhibited by the substance in a massive state, are competent to produce the colour of the sky.

But there is another subject connected with our firmament, of a more subtle and recondite character than even its colour. I mean that 'mysterious and beautiful phenomenon,'[1] the polarisation of the light of the sky. The polarity of a magnet consists in its *two-endedness*, both ends, or poles, acting in opposite ways. Polar forces, as most of you know, are those in which the duality of attraction and repulsion is manifested. And a kind of *two-sidedness*—noticed by Huygens, commented on by Newton, and discovered by a French philosopher, named Malus, in a beam of light which had been reflected from one of the windows of the Luxembourg Palace in Paris—receives the name of *polarisation*. We must now, however, attach a distinctness to the idea of a polarised beam, which its discoverers were not able to attach to it. For in their day men's thoughts were not sufficiently ripe, nor optical theory sufficiently advanced, to seize upon or express the physical meaning of polarisation. When a gun is fired, the explosion is propagated as a wave through the air. The shells of air, if I may use the term, surrounding the centre of concussion, are successively thrown into motion, each shell yielding up its motion to that in advance of it, and returning to its position of equilibrium. Thus, while the *wave* travels through long distances, each individual particle of air concerned in its transmission performs merely a small excursion to and fro.[2] In the case of sound, the vibration of the air-particles are executed *in* the direction in which the sound

---

[1] Herschel's *Meteorology*, Art. 233.
[2] *Lectures on Sound*, p. 3. (Longmans.)

travels. They are therefore called *longitudinal* vibrations. In the case of light, on the contrary, the vibrations are *transversal*; that is to say, the individual particles of æther move to and fro *across* the direction in which the light is propagated. In this respect waves of light resemble ordinary water-waves, more than waves of sound. In the case of an ordinary beam of light, the vibrations of the æther particles are executed in every direction perpendicular to it; but let the beam impinge obliquely, upon a plane glass surface, as in the case of Malus, the portion reflected will no longer have its particles vibrating in all directions round it. By the act of reflection, *if it occur at the proper angle*, the vibrations are all confined to a single plane, and light thus circumstanced is called *plane polarised light.*

A beam of light passing through ordinary glass executes its vibrations within the substance exactly as it would do in air, or in æther-filled space. Not so when it passes through many transparent crystals. For these have also their two-sidedness, the arrangement of their molecules being such as to tolerate vibrations only in certain definite directions. There is the well-known crystal tourmaline, which shows a marked hostility to all vibrations executed at right angles to the axis of the crystal. It speedily extinguishes such vibrations, while those executed parallel to the axis are freely propagated. The consequence is, that a beam of light, after it has passed through any thickness of this crystal, emerges from it polarised. So also as regards the beautiful crystal known as Iceland spar, or as double refracting spar. In one direction, but in one only, it acts like a piece of glass; in all other directions it splits the beam of light passing through it into two distinct halves, both of which are perfectly polarised, their vibrations being executed in two planes, at right angles to each other.

It is possible by a suitable contrivance to get rid of one of the two polarised beams into which Iceland spar divides an ordinary beam of light. This was done so ingeniously and effectively by a man named Nicol, that the Iceland spar, cut in his fashion, is now universally known as Nicol's prism. Such a prism can polarise a beam of light, and if the beam, before it impinges on the prism, be already polarised, in one position of the prism it is stopped, while in another position it is transmitted. Our way is now, to some extent, cleared towards an examination of the light of the sky. Looking at various points of the blue firmament through a Nicol's prism, and turning the prism round its axis, we immediately notice variations of brightness. In certain positions of the prism, and from certain points of the firmament, the light appears to be freely transmitted ; while it is only necessary to turn the prism round its axis through an angle of ninety degrees to materially diminish the intensity of the light. On close scrutiny it is found that the difference produced by the rotation of the prism is greatest when the sky is regarded in a direction at right angles to that of the solar rays through the air.

Let me describe a few actual observations made some days ago on Primrose Hill. The sun was near setting, and a few scattered neutral-tint clouds, which failed to catch the dying light, were floating in the air. When these were looked at across the track of the solar beams, it was possible, by turning the Nicol round, to see them either as white clouds on a dark ground, or as dark clouds on a bright ground.[1] In certain positions of the prisms the sky-light was in great part quenched, and then the clouds, projected against the darkness of space, appeared white. Turning the Nicol ninety degrees round its axis, the brightness of the sky was restored, the clouds becoming dark through contrast with this brightness. Experiments of this kind prove that the blue light sent to us by the firmament is polarised, and that the direction of most perfect polarisation is perpendicular to the solar rays. Were the heavenly azure like the light scattered from a thick cloud, the turning of the prism would have no effect upon it ; it would be transmitted equally during the entire rotation of the prism. The light of the sky is in great part quenched, because it is in great part polarised.

When a luminous beam impinges at the proper angle on a plane glass surface it is polarised by reflection. It is polarised, in part, by all oblique reflections ; but at one particular angle, the reflected light is perfectly polarised. An exceedingly beautiful and simple law, discovered by Sir David Brewster, enables us readily to find *the polarising angle* of any substance whose refractive index is known. This law was discovered experimentally by Brewster ; but the Wave Theory of light renders a complete reason for the law. A geometrical image of it is thus given. When a beam of light impinges obliquely upon a plate of glass it is in part reflected and in part refracted. At one particular incidence the reflected and the refracted portions of the beam are at right angles to each other. The angle of incidence is *then* the polarising

---

[1] I was not aware when these words were written that this observation was made by the indefatigable Brewster.

angle. It varies with the refractive index of the substance; being for water 52½, for glass 57½, and for diamond 68 degrees.

It has been already stated, that in order to obtain the most perfect polarisation of the firmamental light, the sky must be regarded in a direction at right angles to the solar beams. · This is sometimes expressed by saying that the place of maximum polarisation is at an angular distance of 90° from the sun. This angle, enclosed as it is between the direct and reflected rays, comprises both the angles of incidence and reflection, supposing the polarisation to be due to a single reflection. Hence the angle of incidence is half of 90°, or 45°. This is the atmospheric polarising angle, and the question is, what known substance possesses an index of refraction to correspond with this polarising angle? 'If,' says Sir John Herschel, 'we knew this substance, we might be tempted to conclude that particles of it, scattered in the atmosphere, produce the polarisation of the sky. Were the angle of maximum polarisation 76° (instead of 90°), we should look to *water* or ice, as the reflecting body, however inconceivable the existence in a cloudless atmosphere and a hot summer day, of unevaporated particles of water.' But a polarising angle of 45° corresponds to a refractive index of 1 ; this means that there is no refraction at all, in which case we ought to have no reflection. Brewster, therefore, and others came to the conclusion that the reflection was from the particles of air themselves. Dr. Rubenson, of Upsala, made the angle enclosed between the direct and reflected beams 90° 2'; 'the half of which,' says Mr. Buchan, in his excellent little 'Handy Book of Meteorology,' ' is so near the polarising angle of air as to leave no doubt that the light of the sky, as first stated by Brewster, is polarised by reflection from the particles of air.'

If you doubt the wisdom, acknowledge, at all events, the faith in your capacity which has caused me to bring so entangled a subject before you. I would fain believe, however, that even the intellect which draws its culture from a totally different source, may have its interest excited in subjects like the present, dark and difficult though they seem. I do not expect that you will grasp all the details of this discussion ; but everybody present will, I think, see the extremely important part hitherto played by the law of Brewster in speculations as to the colour and polarisation of the sky. Let me now endeavour to demonstrate in your presence, firstly, and in confirmation of our former experiments, that sky-blue may be produced by exceedingly minute particles of any kind of matter ; secondly, that polarisation identical with that of the sky is produced by such particles ; and thirdly, that matter in this fine state of division, where its particles are small in comparison with the height and span of a wave of light, releases itself completely from the law of Brewster ; the direction of maximum polarisation being absolutely independent of the polarising angle as hitherto defined.

Into this experimental tube, in the manner already described, I introduce a vapour which is decomposable by the waves of light. The mixed air and vapour are sufficient to depress the mercurial column one inch. I add to this mixture air, which has been permitted to bubble through dilute hyrochloric acid, until the column is depressed thirty inches : in other words, until the tube is full. And now I permit the electric beam to play upon the mixture. For some time nothing is seen. The chemical action is doubtless progressing, and condensation is going on ; but the condensing molecules have not yet coalesced to particles sufficiently large to reflect sensibly the waves of light. As before stated—and the statement rests upon an experimental basis—the particles here generated are at first so small that their diameters would probably have to be expressed in millionths of an inch ; while to form each of these *particles* whole crowds of *molecules* are probably aggregated. Helped by such considerations the intellectual vision plunges more profoundly into atomic nature, and shows us, among other things, how far we are from the realisation of Newton's hope that the molecules might one day be seen by microscopes. While I am speaking, you observe this delicate blue colour, forming and strengthening within the experimental tube. No sky-blue could exceed it in richness and purity ; but the particles which produce this colour lie wholly beyond our microscopic range. A uniform colour is here developed, which has as little breach of continuity,— which yields as little evidence of the particles concerned in its production, as that yielded by a body whose colour is due to true molecular absorption. This blue is at first as deep and dark as the sky seen from the highest Alpine peaks, and for the same reason. But it grows gradually brighter, still maintaining its blueness, until at length a whitish tinge mingles with the pure azure ; announcing that the particles are now no longer of that infinitesimal size which mainly scatters the shortest waves.[1]

The liquid here employed is the iodide of allyl, but I

---

[1] Possibly a photographic impression might be taken long before the blue becomes visible, for the ultra-blue rays are first reflected.

might choose any one of a dozen substances here before me to produce the effect. You have seen what may be done with the nitrite of butyl. With nitrite of amyl, bisulphide of carbon, benzol, benzoic ether, &c. the same blue colour may be produced. In all cases, where matter slowly passes from the molecular to the massive state the transition is marked by the production of the blue. More than this :—you have seen me looking at the blue colour (I hardly like to call it a blue ' cloud,' its texture and properties are so different from ordinary clouds) through this bit of spar. This is a Nicol's prism, and it is to be wished that one of them could be placed in the hands of each of you. Now, this blue that I have been regarding turns out to be, if the expression be allowed, a bit of more perfect sky than the sky itself. On looking across the illuminating beam as we look across the solar rays in the atmosphere, we obtain not only partial polarisation, but *perfect* polarisation. In one position of the Nicol the blue light passes freely to the eye ; in the other it is absolutely cut off, the experimental tube being reduced to optical *emptiness.* It is well to place a black surface behind the experimental tube, so as to prevent foreign light from troubling the eye. In one position of the prism this black surface is seen without softening or qualification ; for the particles within the tube are themselves invisible, and the light which they scatter is quenched. If the light of the sky were polarised with the same perfection, on looking properly towards it through a Nicol we should meet, not the mild radiance of the firmament, but the unillumined blackness of space.

The construction of a Nicol's prism is such that it allows the passage of vibrations which are executed in a certain

T

determinate direction, and these only. All vibrations executed at right angles to this direction are completely stopped: while components only of those executed obliquely to it are transmitted. It is easy, therefore, to see that from the position in which the prism must be held to transmit or to quench the light of our incipient cloud, we can infer the direction of the vibrations of that light. You will be able to picture those vibrations without difficulty. Suppose a line drawn from any point of the 'cloud' perpendicular to the illuminating beam. The particles of æther along that line, which carry the light from the cloud to the eye, vibrate in a direction perpendicular both to the line and to the beam. And if any number of lines be drawn in the same way from the cloud, like the spokes of a˘ wheel, the particles of æther along all of them oscillate in the same manner. Wherefore, if a plane surface be imagined cutting the incipient cloud at right angles to its length, the vibrations discharged laterally are all parallel to this surface. This is the plane of vibration of the polarised light.

Our incipient blue cloud is a virtual Nicol's prism, and, between it and the real prism, we can produce all the effects obtainable between the polariser and analyser of a polariscope. When, for example, a thin plate of selenite, which is crystallised sulphate of lime, is placed between the Nicol and the incipient cloud, we obtain the splendid chromatic phenomena of polarised light. The colour of the gypsum-plate, as many of you know, depends upon its thickness. If this be uniform, the colour is uniform. If, on the contrary, the plate be wedge-shaped, thickening gradually and uniformly from edge to back, we have brilliant bands of colour produced parallel to the edge of the wedge. Perhaps the best form of plate for experiments of this character is that now in my hand, which was prepared for me some years ago by a man of genius in his way, the late Mr. Darker of Lambeth. It consists of a plate of selenite thin at the centre, and gradually thickening towards the circumference. Placing this film between the Nicol and the cloud, we obtain, instead of a series of parallel bands, a system of splendidly coloured rings. Precisely the same phenomena are observed when we look at the blue firmament in a direction perpendicular to the solar rays.

We have thus far illuminated our artificial sky with ordinary light. We will now examine the effects produced when the light which illuminates the particles is itself polarised. In front of the electric lamp, and between it and the experimental tube, is placed this fine Nicol's prism, which is sufficiently large to embrace and to polarise the entire beam. The plane of vibration of the light now emergent from the prism, and falling upon the cloud, is vertical; and we find that this formless aggregate of infinitesimal particles, without definite structure, is absolutely incompetent to scatter the light upwards or downwards, while it freely discharges the light horizontally, right and left. I turn the polarising Nicol so as to render the plane of vibration horizontal; the cloud now freely scatters the light vertically upwards and downwards, but it is absolutely incompetent to shed a ray horizontally to the right or left.

Suppose the atmosphere of our planet to be surrounded by an envelope impervious to light, with an aperture on the sunward side, through which a solar beam could enter. Surrounded on all sides by air not directly illuminated, the track of the sunlight would resemble that of the electric beam in a dark space filled with our incipient cloud. The course of the sunbeam would be *blue,* and it would discharge laterally, in all directions round it light in precisely the same polarised condition as that discharged from the incipient cloud. In fact, the azure revealed by the sunbeam would be the azure of such a cloud. And if, instead of permitting the ordinary light of the sun to enter the aperture, a Nicol's prism were placed there, which should polarise the sunlight on its entrance into our atmosphere, the particles producing the colour of the sky would act precisely like those of our incipient cloud. In two directions we should have the solar light reflected ; in two others unreflected. In fact, out of such a solitary beam, traversing the unilluminated air, we should be able to extract every effect shown by our incipient cloud. In the production of such clouds we virtually carry bits of the sky into our laboraties, and obtain with them all the effects obtainable in the open firmament of heaven.

The real sky is, as I have said, less perfect than our artificial one may be made. For mingled with the infinitesimal particles which constitute the true matter of the sky, there are others too coarse, at right angles to the solar beams, to scatter perfectly polarised light. Hence, when the brilliancy of the sky is diminished to the uttermost, there is still a residue of light ; the extinction is partial, and not total, as in the case of our incipient cloud. Let us consider this matter. The perfect polarisation can only be produced by excessively minute particles ; imagine them growing gradually larger as they actually do in our experiments. The extinction by the Nicol is perfect as long as the polarisation is complete. But what would you expect? Manifestly, that after a time the polarisation would cease to be perfect. But here again the relation of the size of the particles to the size of the waves must come into play. In relation to the blue waves the particles are larger than in relation to the red; the blue waves, therefore, will be the first liberated from a condition dependent on the smallness of the particles. They will first escape from the trammels of polarisation; and on their liberation they exhibit an azure far purer and more brilliant than that produced by the first precipitation of the particles. Could we overarch ourselves with a sky of this colour for a single day, it would make us discontented with our present lack-lustre firmament ever afterwards. It will be observed that in all these cases reason and experiment go hand in hand, the one predicting, the other verifying; every such verification lending its weight of proof to the undulatory theory on which the predictions are founded.

The selenite ring-system, already referred to, is a most delicate reagent for the detection of polarised light. When we look *normally*, or perpendicularly, at an incipient cloud, the colours of the rings are most vividly developed, a diminution of the colour being immediately apparent when the incipient cloud is regarded obliquely. But let us continue to look through the Nicol and selenite normally at the cloud: the particles augment in size, the cloud becomes coarser and whiter, the strength of the selenite colours becoming gradually feebler. At length the cloud ceases to discharge polarised light along the normal, and then the selenite colours entirely disappear.

If now the cloud be regarded *obliquely* the colours are restored, very vividly, if not with their first vividness and clearness. Thus the cloud that has ceased to discharge polarised light at right angles to the illuminating beam, pours out such light copiously in oblique directions. The direction of maximum polarisation changes with the texture of the cloud.

But this is not all ; and to understand, even partially, what remains, a word must be said regarding the appearance of the colours of our plate of selenite. If, as before stated, the plate be of uniform thickness, its hue in polarised light is uniform. Suppose, then, that by arranging the Nicol the colour of the plate is raised to its maximum brilliancy, and suppose the colour produced to be *green* ; on turning the Nicol round its axis the green becomes fainter. When the angle of rotation amounts to 45 degrees the colour disappears ; we then pass what may be called a neutral position, where the selenite behaves, not as a crystal, but as a bit of glass. Continuing the rotation, a colour reappears, but it is no longer green, but *red*. This attains its maximum at a distance of 45 degrees from the neutral position, or, in other words, at a distance of 90 degrees from the position which showed the green at its maximum. At a further distance of 45 degrees from the position of maximum red, the colour disappears a second time. We have there a second neutral point, beyond which the green comes again into view, attaining its maximum brilliancy at the end of a rotation of 180 degrees. By the rotation of the Nicol, therefore, through an angle of 90 degrees, we produce a colour *complementary* to that with which we started.

As may be inferred from this result, the selenite ring system changes its character when the Nicol is turned. It is possible to have the centre of the circle dark, the surrounding rings being vividly coloured. The turning of the Nicol through an angle of 90 degrees renders the centre bright, while every point occupied by a certain colour in the first instance is occupied by the complement of that colour in the second. By curious internal actions, not here to be described, the cloud in our experimental tube sometimes divides itself into sections of different textures. Some sections are coarser than others, while it often happens that some are iridescent to the naked eye, and others not. Looking normally at such a cloud through the selenite and Nicol, it often happens that in passing from section to section the whole character of the ring-system is changed. You start with a section producing a *dark* centre and a corresponding system of rings ; you pass through a neutral point to another section and find there the centre *bright*, and each of the first rings displaced by one of the complementary colour. Sometimes as many as four such reversions occur in the cloud of an experimental tube a yard long. Now, the changes here indicated mean that in passing from section to section of the cloud the plane of vibration of the polarised light turns suddenly through an angle of 90 degrees ; this change being entirely due to the different texture of the two parts of the cloud.

You will now be able to understand, as far as it is capable of being understood, a very beautiful effect which, under favourable circumstances, might be observed in our atmosphere. This experimental tube contains an inch of the iodide of allyl vapour, the remaining

29 inches necessary to fill the tube being air, which has bubbled through aqueous hydrochloric acid. Besides, therefore, the vapour of iodide of allyl, we have those of water and of acid within the tube. The light has been acting on the mixture for some time, a beautiful incipient blue cloud being formed. As before stated, the 'incipient cloud' is wholly different in texture and optical properties from an ordinary cloud; but it is possible to precipitate in the midst of the azure the aqueous vapour so as to cause it to form in the tube a cloud similar to the clouds of our atmosphere. An exhausted vessel of about one-third of the capacity of the experimental tube is connected with it, the passage uniting both being closed by a stop-cock. On opening this cock the mixed air and vapour rush from the experimental tube into the empty vessel; and, in consequence of the chilling due to rarefaction, the vapour in the experimental tube is precipitated as a true cloud. What is the result? Instantly the centre of the system of coloured rings becomes bright, and the whole series of colours corresponding to definite radial distances, complementary. While you continue to look at the cloud, it gradually melts away as an atmospheric cloud might do in the azure of heaven. And *there* is our azure also remaining behind. The coarser cloud seems drawn aside like a veil, the blue reappears, the first ring-system, with its dark centre and correspondingly coloured circles, being restored.

Thus patiently you have accompanied me over a piece of exceedingly difficult ground; and I think as a prudent guide, we ought to halt upon the eminence we have now attained. We might go higher, but the boulders begin here to be very rough. At a future day we shall, I doubt not, be able to overcome this difficulty, and to reach together a greater elevation.

THE SKY OF THE ALPS.

THE vision of an object always implies a differential action on the retina of the observer. The object is distinguished from surrounding space by its excess or defect of light in relation to that space. By altering the illumination, either of the object itself or of its environment, we alter the appearance of the object. Take the case of clouds floating in the atmosphere with patches of blue between them. Anything that changes the illumination of either alters the appearance of both, that appearance depending, as stated, upon differential action. Now the light of the sky, being polarised, may, as the reader of the foregoing pages knows, be in great part quenched by a Nicol's prism, while the light of a cloud, being unpolarised, cannot be thus extinguished. Hence the possibility of very remarkable variations, not only in the aspect of the firmament, which is really changed, but also in the aspect of the clouds which have that firmament as a background. It is possible, for example, to choose clouds of such a depth of shade that when the Nicol quenches the light behind them, they shall vanish, being undistinguishable from the residual dull tint which outlives the extinction of the brilliance of the sky. A cloud less deeply shaded, but still deep enough, when viewed with the naked eye, to appear dark on a bright ground, is suddenly changed to a white cloud on a dark ground by the quenching of the sky behind it. When a reddish cloud at sunset chances to float in the region of maximum polarisation, the quenching of the sky behind it causes it to flash with a brighter crimson. Last Easter eve the Dartmoor sky, which had just been cleansed by a snow storm, wore a very wild appearance. Round the horizon it was of steely brilliancy, while reddish cumuli and cirri floated southwards. When the sky was quenched behind them these floating masses seemed like dull embers suddenly blown upon ; they brightened like a fire. In the Alps we have the most magnificent examples of crimson clouds and snows, so that the effects just referred to may be here studied under the best possible conditions. On August 23, 1869, the evening Alpenglow was very fine, though it did not reach its maximum depth and splendour. Towards sunset I walked up the slopes to obtain a better view of the Weisshorn. The side of the peak seen from the Bel Alp, being turned from the sun, was tinted *mauve* ; but I wished to see one of the rose-coloured buttresses of the mountain. Such was visible from a point a few hundred feet above the hotel. The Matterhorn also, though for the most part in shade, had a crimson projection, while a deep ruddy red lingered along its western shoulder. Four distinct peaks and buttresses of the Dom, in addition to its dominant head—all covered with pure snow—were reddened by the light of sunset. The shoulder of the Alphubel was similarly coloured, while the great mass of the Fletschorn was all a-glow, and so was the snowy spine of the Monte Leone.

Looking at the Weisshorn through the Nicol, the glow of its protuberance was strong or weak according to the

position of the prism. The summit also underwent a
change. In one position of the prism it exhibited a pale
white against a dark background; in the rectangular
position, it was a dark mauve against a light background.
The red of the Matterhorn changed in a similar manner;
but the whole mountain also passed through striking
changes of definition. The air at the time was filled
with a silvery haze, in which the Matterhorn almost
disappeared. This could be wholly quenched by the
Nicol, and then the mountain sprang forth with astonishing solidity and detachment from the surrounding air.
The changes of the Dom were still more wonderful.
A vast amount of light could be removed from the
sky behind it, for it occupied the position of maximum
polarisation. By a little practice with the Nicol it was
easy to render the extinction of the light, or its restoration, almost instantaneous. When the sky was quenched,
the four minor peaks and buttresses, and the summit of
the Dom, together with the shoulder of the Alphubel,
glowed as if set suddenly on fire. This was immediately
dimmed by turning the Nicol through an angle of 90°.
It was not the stoppage of the light of the sky behind
the mountains alone which produced this startling effect;
the air between them and me was highly opalescent, and
the quenching of this intermediate glare augmented
remarkably the distinctness of the mountains.

On the morning of August 24 similar effects were finely
shown. At 10 A.M. all three mountains, the Dom, the
Matterhorn, and the Weisshorn, were powerfully affected
by the Nicol. But in this instance also the line drawn to
the Dom being accurately perpendicular to the direction
of the solar shadows, and consequently very nearly per-

pendicular to the solar beams, the effects on this mountain were most striking. The grey summit of the Matterhorn at the same time could scarcely be distinguished from the opalescent haze around it; but when the Nicol quenched the haze, the summit became instantly isolated, and stood out in bold definition. It is to be remembered that in the production of these effects the only things changed are the sky behind and the luminous haze in front of the mountains; that these are changed because the light emitted from the sky and from the haze is plane polarised light,[1] and that the light from the snows and from the mountains being sensibly unpolarised, is not directly affected by the Nicol. It will also be understood that it is not the interposition of the haze *as an opaque body* that renders the mountains indistinct, but that it is the *light* of the haze which dims and bewilders the eye, and thus weakens the definition of objects seen through it.

These results have a direct bearing upon what artists call ' aërial perspective.' As we look from the summit of the Aletschhorn, or from a lower elevation, at the serried crowd of peaks, especially if the mountains be darkly coloured—covered with pines, for example—every peak and ridge is sèparated from the mountains behind it by a thin blue haze which renders the relations of the mountains as to distance unmistakable. When this haze is regarded through the Nicol perpendicular to the sun's rays, it is in many cases wholly quenched, because the light which it emits in this direction is wholly polarised. When this happens, aërial perspective is abolished, and mountains very differently distant appear to rise in the same vertical plane. Close to the Bel Alp, for instance,

---

[1] Defined at page 267.

is the gorge of the Massa, and beyond the gorge is a high ridge darkened by pines. This ridge may be projected upon the dark slopes at the opposite side of the Rhone valley, and between both we have the blue haze referred to, throwing the distant mountains far away. But at certain hours of the day this haze may be quenched, and then the Massa ridge and the mountains beyond the Rhone seem almost equally distant from the eye. The one appears, as it were, a vertical continuation of the other. The haze varies with the temperature and humidity of the atmosphere. At certain times and places it is almost as blue as the sky itself; but to see its colour, the attention must be withdrawn from the mountains and from the trees which cover them. In point of fact, the haze is a piece of more or less perfect sky ; it is produced in the same manner, and is subject to the same laws, as the firmament itself. We live *in* the sky, not *under* it.

These points were further elucidated by the deportment of the selenite plate, with which the readers of the fore-going discourse are already acquainted. On some of the sunny days of August the haze in the valley of the Rhone, as looked at from the Bel Alp, was very remarkable. Towards evening the sky above the mountains opposite to my place of observation yielded a series of the most splendidly-coloured iris-rings ; but on lowering the selenite until it had the darkness of the pines at the opposite side of the Rhone valley, instead of the darkness of space as a background, the colours were not much diminished in brilliancy. I should estimate the distance across the valley, as the crow flies, to the opposite mountains, at nine miles ; so that a body of air nine miles thick can, under favourable circumstances, produce chromatic effects of polarisation almost as vivid as those produced by the sky itself.

Again : the light of a landscape, as of most other things, consists of two parts ; the one part comes purely from superficial reflection, and this light is always of the same colour as that which falls upon the landscape ; the other part comes to us from a certain depth within the objects which compose the landscape, and it is this portion of the total light which gives these objects their distinctive colours.   The white light of the sun enters all substances to a certain depth, and is partially ejected by internal reflection ; each distinct substance absorbing and reflecting the light in accordance with the laws of its own molecular constitution.   Thus the solar light is *sifted* by the landscape, which appears in such colours and variations of colour as, after the sifting process, reach the observer's eye.   Thus the bright green of grass, or the darker colour proper to the pine, never comes to us alone, but is always mingled with an amount of really foreign light derived from superficial reflection.   A certain hard brilliancy is conferred upon the woods and meadows by this superficially-reflected light.   Under certain circumstances, it may be quenched by a Nicol's prism, and we then obtain the true colour of the grass and foliage. Trees and meadows thus regarded exhibit a richness and softness of tint which they never show as long as the superficial light is permitted to mingle with the true interior emission.   The needles of the pines show this effect very well, large-leaved trees still better ; while a glimmering field of maize exhibits the most extraordinary variations when looked at through the rotating Nicol.

Thoughts and questions like those here referred to took

me, in August 1869, to the top of the Aletschhorn. The effects described in the foregoing paragraphs were for the most part reproduced in the summit of the mountain. I scanned the whole of the sky with my Nicol. Both alone and in conjunction with the selenite it pronounced the perpendicular to the solar beams to be the direction of maximum polarisation. But at no portion of the firmament was the polarisation complete. The artificial sky produced in the experiments recorded in the preceding discourse could, in this respect, be rendered more perfect than the natural one; while the gorgeous 'residual blue' which makes its appearance when the polarisation of the artificial sky ceases to be perfect, was strongly contrasted with the lack-lustre hue which, in the case of the firmament, outlived the extinction of the brilliance. With certain substances, however, artificially treated, this dull residue may also be obtained.

All along the arc from the Matterhorn to Mont Blanc the light of the sky immediately above the mountains was powerfully acted upon by the Nicol. In some cases the variations of intensity were astonishing. I have already said that a little practice enables the observer to shift the Nicol from one position to another so rapidly as to render the alternate extinction and restoration of the light immediate. When this was done along the arc to which I have referred, the alternations of light and darkness resembled the play of sheet lightning behind the mountains. My notes state that there was an element of awe connected with the suddenness with which the mighty masses, ranged along the line referred to, changed their aspect and definition under the operation of the prism.

XI.

DUST AND DISEASE.

A DISCOURSE.

DELIVERED IN THE ROYAL INSTITUTION OF GREAT BRITAIN.

21st January, 1870. *With Additions.*

U

'Tout miasme contagieux a les propriétés, 1° de reproduire son analogue dans une maladie qu'il a occasionnée ; 2° de se répandre et de s'entendre à l'infini, en vertu de ce développement secondaire, c'est-à-dire, aussi longtemps qu'il existe une matière propre à recevoir le miasme, et en à produire un nouveau.   Ces deux propriétés lui sont communes avec les germes des animaux et des plantes.'

HILDEBRAND.

XI.

ON DUST AND DISEASE.

Experiments on Dusty Air.

SOLAR light in passing through a dark room reveals its track by illuminating the dust floating in the air. 'The sun,' says Daniel Culverwell, ' discovers atomes, though they be invisible by candle-light, and makes them dance naked in his beams.'[1]

In my researches on the decomposition of vapours by light I was compelled to remove these 'atomes' and this dust. It was essential that the space containing the vapours should embrace no visible thing; that no substance capable of scattering the light in the slightest sensible degree should, at the outset of an experiment, be found in the ' experimental tube ' traversed by the luminous beam.

For a long time I was troubled by the appearance there of floating dust, which though invisible in diffuse daylight was at once revealed by a powerfully condensed beam. Two tubes were placed in succession in the path of the air: the one containing fragments of glass wetted with concentrated sulphuric acid ; the other, fragments of marble wetted with a strong solution of caustic potash. To my astonishment the dust passed through both. The air

[1] On a day of transient shadows there is something almost magical in the rise and dissolution of the luminous beams among the scaffolding poles of the Royal Albert Hall.

of the Royal Institution sent through these tubes at a rate sufficiently slow to dry it, and to remove its carbonic acid, carried into the experimental tube a considerable amount of mechanically suspended matter, which was illuminated when the beam passed through the tube. The effect was substantially the same when the air was permitted to bubble through the liquid acid and through the solution of potash.

Thus, on October 5, 1868, successive charges of air were admitted through the potash and sulphuric acid into the exhausted experimental tube. Prior to the admission of the air the tube was *optically empty*; it contained nothing competent to scatter the light. After the air had entered the tube, the conical track of the electric beam was in all cases clearly revealed. This indeed was a daily observation at the time to which I now refer.

I tried to intercept this floating matter in various ways; and on the day just mentioned, prior to sending the air through the drying apparatus, I carefully permitted it to pass over the tip of a spirit-lamp flame. The floating matter no longer appeared, having been burnt up by the flame. It was therefore of *organic origin*. I was by no means prepared for this result; for I had thought that the dust of our air was, in great part, inorganic and non-combustible.

I had constructed a small gas-furnace, now much employed by chemists, containing a platinum tube, which could be heated to vivid redness.[1] The tube contained a roll of platinum gauze, which, while it permitted the air to pass through it, ensured the practical contact of the

---

[1] Pasteur was, I believe, the first to employ such a tube.

dust with the incandescent metal. The air of the laboratory was permitted to enter the experimental tube, sometimes through the cold, and sometimes through the heated, tube of platinum. The rapidity of admission was also varied. In the first column of the following table the quantity of air operated on is expressed by the number of inches which the mercury gauge of the air-pump sank when the air entered. In the second column the condition of the platinum tube is mentioned, and in the third the state of the air which had entered the experimental tube.

Quantity of Air.		State of Platinum Tube.		State of Experimental Tube.
15 inches	. .	Cold	. .	Full of particles.
15 ,,	. .	Red-hot .	.	Optically empty.

The phrase ' optically empty ' shows that when the conditions of perfect combustion were present, the floating matter totally disappeared. It was wholly burnt up, leaving no sensible residue. The experiment was repeated many times with the same invariable result.

The whole of the visible particles floating in the air of London rooms being thus proved to be of organic origin,[1] I sought to burn them up at the focus of a concave reflector. One of the powerfully convergent mirrors employed in my experiments on combustion by dark rays was here made

---

[1] According to an analysis kindly furnished to me by Dr. Percy, the dust collected *from the walls* of the British Museum contains fully 50 per cent. of inorganic matter. I have every confidence in the results of this distinguished chemist; they show that the *floating* dust of our rooms is, as it were, winnowed from the heavier matter. As bearing directly upon this point I may quote the following passage from Pasteur:—' Mais ici se présente une remarque : la poussière que l'on trouve à la surface de tous les corps est soumise constamment à des courants d'air, qui doivent soulever ses particules les plus légères, au nombre desquelles se trouvent, sans doute, de préférence les corpuscules organisés, œufs ou spores, moins lourds généralement que les particules minérales.'

use of, but I failed in the attempt. Doubtless the floating particles are in part transparent to radiant heat, and are so far incombustible by such heat. Their rapid motion through the focus also aids their escape. They do not linger there sufficiently long to be consumed. A flame it was evident would burn them up, but I at first thought the presence of the flame would mask its own action among the particles.

In a cylindrical beam, which strongly illuminated the dust of the laboratory, was placed an ignited spirit-lamp. Mingling with the flame, and round its rim, were seen curious wreaths of darkness resembling an intensely black smoke. On lowering the flame below the beam the same dark masses stormed upwards. They were at times blacker than the blackest smoke that I have ever seen issuing from the funnel of a steamer; and their resemblance to smoke was so perfect as to lead the most practised observer to conclude that the apparently pure flame of the alcohol lamp required but a beam of sufficient intensity to reveal its clouds of liberated carbon.

But is the blackness smoke? This question presented itself in a moment. A red-hot poker was placed underneath the beam, and from it the black wreaths also ascended. A large hydrogen flame was next employed, and it produced those whirling masses of darkness far more copiously than either the spirit-flame or poker. Smoke was therefore out of the question.

What then was the blackness? It was simply that of stellar space; that is to say, blackness resulting from the absence from the track of the beam of all matter competent to scatter its light. When the flame was placed below the beam the floating matter was destroyed *in situ*;

and the air, freed from this matter, rose into the beam, jostled aside the illuminated particles, and substituted for their light the darkness due to its own perfect transparency. Nothing could more forcibly illustrate the invisibility of the agent which renders all things visible. The beam crossed, unseen, the black chasm formed by the transparent air, while at both sides of the gap the thick-strewn particles shone out like a luminous solid under the powerful illumination.

But here a rather perplexing difficulty meets us. It is not necessary to burn the particles to produce a stream of darkness. Without actual combustion, currents may be generated which shall exclude the floating matter, and therefore appear dark amid the surrounding brightness. I noticed this effect first on placing a red-hot copper ball below the beam and permitting it to remain there until its temperature had fallen below that of boiling water. The dark currents, though much enfeebled, were still produced. They may also be produced by a flask filled with hot water.

To study this effect a platinum wire was stretched across the beam, the two ends of the wire being connected with the two poles of a voltaic battery. To regulate the strength of the current a rheostat was placed in the circuit. Beginning with a feeble current the temperature of the wire was gradually augmented; but before it reached the heat of ignition, a flat stream of air rose from it, which when looked at edgeways appeared darker and sharper than one of the blackest lines of Fraunhofer in the solar spectrum. Right and left of this dark vertical band the floating matter rose upwards, bounding definitely the non-luminous stream of air. What is the explanation? Simply this. The hot wire rarefied the air in contact

with it, but it did not equally lighten the floating matter.
The convection current of pure air therefore passed upwards *among the inert particles*, dragging them after it
right and left, but forming between them an impassable
black partition. This elementary experiment enables us
to render an account of the dark currents produced by
bodies at a temperature below that of combustion.

When the wire is white hot, it sends up a band of
intense darkness. This, I say, is due to the *destruction* of
the floating matter. But even when its temperature does
not exceed that of boiling water the wire produces a dark
ascending current. This, I say, is due to the *distribution*
of the floating matter. Imagine the wire clasped by the
mote-filled air. My idea is that it heats the air and
lightens it, without in the same degree lightening the
floating matter. The tendency, therefore, is to start a
current of clean air through the mote-filled air. Figure
the motion of the air all round the wire. Looking at its
transverse section we should see the air at the bottom of
the wire bending round it right and left in two branch
currents, ascending its sides and turning to fill the partial
vacuum created above the wire. Now as each new supply
of air filled with its motes comes in contact with the hot
wire, the clean air, as just stated, is first started through
the inert motes. They are dragged after it, but there is a
fringe of cleansed air in advance of the motes. The two
purified fringes of the two branch currents unite above
the wire, and, keeping the motes that once belonged to
them right and left, they form by their union the dark
band observed in the experiment. This process is incessant. Always the moment the mote-filled air touches the
wire this distribution is effected, a permanent dark band

being thus produced. Could the air and the particles
under the wire pass *through* its mass we should have a
vertical current of particles, but no dark band. For here,
though the motes would be left behind at starting, they
would hotly follow the ascending current and thus abolish
the darkness.

It has been said that when the platinum wire is intensely
heated, the floating matter is not only distributed, but
destroyed. Let this be proved. I stretched a wire about
4 inches long through the air of an ordinary glass shade
resting on its stand. Its lower rim rested on cotton
wool, which also surrounded the rim. The wire was
raised to a white heat by an electric current. The air
expanded, and some of it was forced through the cotton
wool, while when the current was interrupted and the
air within the shade cooled, the expelled air in its return
did not carry motes along with it. At the beginning of
this experiment the shade was charged with floating
matter; at the end of half an hour it was optically empty.

A second experiment was thus arranged: on the
wooden base of a cubical glass shade measuring 11¼ inches
a side, upright supports were fixed, and from one support to the other 38 inches of platinum wire were
stretched in four parallel lines. The ends of the platinum
wire were soldered to two stout copper wires which
passed through the base of the shade and could be connected with a battery. As in the last experiment the
shade rested upon cotton wool. A beam sent through
the shade revealed the suspended matter. The platinum
wire was then raised to whiteness. In five minutes there
was a sensible diminution of the matter, and in ten
minutes it was totally consumed. This proves that when

the platinum wire is sufficiently heated, the floating matter, instead of being distributed, is destroyed.

But is not the matter really of a character which permits of its destruction by the moderately heated platinum wire? Here is the reply :—

1. A platinum tube with its plug of platinum gauze was connected with an experimental tube, through which a powerful beam could be sent from an electric lamp placed at its end. The platinum tube was heated till it glowed feebly but distinctly in the dark. The experimental tube was exhausted and then filled with air which had passed through the red-hot tube. A considerable amount of floating matter which had escaped combustion was revealed by the electric beam.

2. The tube was raised to brighter redness and the air permitted to pass slowly through it. Though diminished in quantity, a certain amount of floating matter passed into the exhausted experimental tube.

3. The platinum tube was rendered still hotter; a barely perceptible trace of the floating matter now passed through it.

4. The experiment was repeated, with the difference that the air was sent more slowly through the red-hot tube. The floating matter was totally destroyed.

5. The platinum tube was now lowered until it bordered upon a visible red heat. The air sent through it still more slowly than in the last experiment carried with it a cloud of floating matter.

If then the suspended matter is destroyed by a bright red heat, much more is it destroyed by a flame, whose temperature is vastly higher than any here employed. So that the blackness introduced into a luminous beam where a flame is placed beneath it is due, as stated, to the destruction of the suspended matter. At a dull red heat, however, and still more when only on the verge of redness, the platinum tube permitted the motes to pass freely. In the latter case the temperature was 800° or 900° Fahr. This was unable to destroy the suspended matter; much less, therefore, would a platinum wire heated to 212° be competent to do so. Such a wire can only distribute the matter, not destroy it.

The floating dust is revealed by intense local illumination. It is seen by contrast with the adjacent illuminated space; the brighter the illumination the more sensible is the difference. Now the beam employed in the foregoing experiments is not of the same brightness throughout its entire transverse section. Pass a white switch, or an ivory paper-cutter, rapidly across the beam, the impression of its section will linger on the retina. The section seems to float for a moment in the air as a luminous circle with a rim much brighter than its central portion. The core of the beam is thus seen to be enclosed by an intensely luminous sheath. An effect complementary to this is observed when the beam is intersected by the dark band from the platinum wire. The brighter the illumination the greater must be the relative darkness consequent on the withdrawal of the light. Hence the cross section of the sheath surrounds the dark band as a darker ring.

Oxygen, hydrogen, nitrogen, carbonic acid, so prepared as to exclude all floating particles, produce the darkness when poured or blown into the beam. Coal-gas does the same. An ordinary glass shade placed in the air with its mouth downwards permits the track of the beam to be seen crossing it. Let coal gas or hydrogen enter the shade by a tube reaching to its top, the gas gradually fills the shade from the top downwards. As soon as it occupies the space crossed by the beam, the luminous track is instantly abolished. Lifting the shade so as to bring the common boundary of gas and air above the beam, the track flashes forth. After the shade is full, if it be inverted, the gas passes upwards like a black smoke among the illuminated particles.

The air of our London rooms is loaded with this organic dust, nor is the country air free from its presence. However ordinary daylight may permit it to disguise itself, a sufficiently powerful beam causes dust suspended in air to appear almost as a semi-solid. Nobody could, in the first instance, without repugnance, place the mouth at the illuminated focus of the electric beam and inhale the thickly-massed dust revealed there. Nor is the repugnance abolished by the reflection that, although we do not see the floating particles, we are taking them into our lungs every hour and minute of our lives.

The Germ Theory of Contagious Disease.

There is no respite to this contact with the floating matter of the air ; and the wonder is, not that we should suffer occasionally from its presence, but that so small a portion of it, and even that but rarely diffused over large areas, should appear to be deadly to man. And what is this portion? It was some time ago the current belief that epidemic diseases generally were propagated by a kind of malaria, which consisted of organic matter in a state of *motor-decay*; that when such matter was taken into the body through the lungs, skin, or stomach, it had the power of spreading there the destroying process which had attacked itself. Such a power was visibly exerted in the case of yeast. A little leaven was seen to leaven the whole lump, a mere speck of matter in this supposed state of decomposition being apparently competent to propagate indefinitely its own decay. Why should not a bit of rotten malaria work in a similar manner within the human frame? In 1836 a very wonderful reply was given to this question. In that year Cagniard de la Tour discovered the *yeast plant*, a living organism, which when placed in a proper medium feeds, grows, and reproduces itself, and in this way carries on the process which we name fermentation. By this striking discovery fermentation was connected with organic growth.

Schwann, of Berlin, discovered the yeast plant independently about the same time; and in February, 1837, he also announced the important result, that when a decoction of meat is effectually screened from ordinary air, and supplied solely with calcined air, putrefaction never sets in. Putrefaction, therefore, he affirmed to be caused by something derived from the air, which something could be destroyed by a sufficiently high temperature. The results of Schwann were confirmed by the independent experiments of Helmholtz, Ure, and Pasteur, while other methods, pursued by Schultze and by Schroeder and Dusch, led to the same result. But as regards fermentation, the minds of chemists, influenced probably by the great authority of Gay-Lussac, fell back upon the old notion of matter in a state of decay. It was not the living yeast plant, but the dead or dying parts of it, which, assailed by oxygen, produced the fermentation. This notion was finally exploded by Pasteur. He proved

that the so-called ' ferments' are not such ; that the true
ferments are organised beings which find in the reputed
ferments their necessary food.

Side by side with these researches and discoveries, and
fortified by them and others, has run the *germ theory* of
epidemic disease.   The notion was expressed by Kircher,
and favoured by Linnæus, that epidemic diseases are due
to germs which float in the atmosphere, enter the body,
and produce disturbance by the development within the
body of parasitic life.   While it was still struggling against
great odds, this theory found an expounder and a de-
fender in the President of this Institution.   At a time
when most of his medical brethren considered it a wild
dream, Sir Henry Holland contended that some form of
the germ theory was probably true.   The strength of this
theory consists in the perfect parallelism of the phenomena
of contagious disease with those of life.   As a planted
acorn gives birth to an oak competent to produce a whole
crop of acorns, each gifted with the power of reproducing
its parent tree ; and as thus from a single seedling a whole
forest may spring ; so, it is contended, these epidemic
diseases literally plant their seeds, grow, and shake abroad
new germs, which, meeting in the human body their
proper food and temperature, finally take possession of
whole populations.   There is nothing to my knowledge in
pure chemistry which resembles the power of self-multiplication possessed by the matter which produces epidemic
disease.   If you sow wheat you do not get barley ; if you
sow small-pox you do not get scarlet-fever, but small-pox
indefinitely multiplied, and nothing else.   The matter of
each contagious disease reproduces itself as rigidly as if it
were (as Miss Nightingale puts it.) dog or cat.

Parasitic Diseases of Silkworms.   Pasteur's Researches.

It is admitted on all hands that some diseases are the product of parasitic growth.   Both in man and lower creatures, the existence of such diseases has been demon-strated.   I am enabled to lay before you an account of an epidemic of this kind, thoroughly investigated and successfully combated by M. Pasteur.   For fifteen years a plague had raged among the silkworms of France. They had sickened and died in multitudes, while those that succeeded in spinning their cocoons furnished only a fraction of the normal quantity of silk.   In 1853 the silk culture of France produced a revenue of one hun-dred and thirty millions of francs.   During the twenty previous years the revenue had doubled itself, and no doubt was entertained as to its future augmentation. The weight of the cocoons produced in 1853 was twenty-six millions of kilogrammes ; in 1865 it had fallen to four millions, the fall entailing in the single year last mentioned a loss of one hundred millions of francs.

The country chiefly smitten by this calamity happened to be that of the celebrated chemist Dumas, now perpetual secretary of the French Academy of Sciences.   He turned to his friend, colleague, and pupil, Pasteur, and besought him with an earnestness which the circumstances ren-dered almost personal, to undertake the investigation of the malady.   Pasteur at this time had never seen a silk-worm, and he urged his inexperience in reply to his friend.   But Dumas knew too well the qualities needed for such an enquiry to accept Pasteur's reason for de-clining it.   'Je mets,' said he, 'un prix extrême à voir votre attention fixée sur la question qui intéresse mon pauvre pays; la misère surpasse tout ce que vous pouvez imaginer.' Pamphlets about the plague had been showered upon the public, the monotony of waste paper being broken at rare intervals by a more or less useful publication. 'The Pharmacopœia of the Silkworm,' wrote M. Cornalia in 1860, 'is now as complicated as that of man. Gases, liquids, and solids have been laid under contribution. From chlorine to sulphurous acid, from nitric acid to rum, from sugar to sulphate of quinine,—all has been invoked in behalf of this unhappy insect.' The helpless cultivators, moreover, welcomed with ready trustfulness every new remedy, if only pressed upon them with sufficient hardihood. It seemed impossible to diminish their blind confidence in their blind guides. In 1863 the French Minister of Agriculture himself signed an agreement to pay 500,000 francs for the use of a remedy which its promoter declared to be infallible. It was tried in twelve different departments of France, and found perfectly useless. In no single instance was it successful. It was under these circumstances that M. Pasteur, yielding to the entreaties of his friend, betook himself to Alais in the beginning of June, 1865. As regards silk husbandry, this was the most important department in France, and it was also that which had been most sorely smitten by the epidemic.

The silkworm had been previously attacked by *muscardine*, a disease proved by Bassi to be caused by a vegetable parasite. Though not hereditary, this malady was propagated annually by the parasitic spores, which, wafted by winds, often sowed the disease in places far removed from the centre of infection. Muscardine is now said to be very rare; but for the last fifteen or twenty years a deadlier malady has taken its place. A frequent outward sign of this new disease are the black spots which cover the silkworms, hence the name *pébrine*, first applied to the plague by M. de Quatrefages, and adopted by Pasteur. Pébrine declares itself in the stunted and unequal growth of the worms, in the languor of their movements, in their fastidiousness as regards food, and in their premature death. The track of discovery as regards the epidemic is this. In 1849 Guerin Méneville noticed in the blood of silkworms vibratory corpuscles which he supposed to be endowed with independent life. Filippi proved him wrong, and showed that the motion of the corpuscles was the well-known Brownian motion. But Filippi himself committed the error of supposing the corpuscles to be normal to the life of the insect. They are really the cause of its mortality—the form and substance of its disease. This was well described by Cornalia; while Lebert and Frey subsequently found the corpuscles not only in the blood, but in all the tissues of the insect. Osimo, in 1857, discovered them in the eggs, and on this observation Vittadiani founded, in 1859, a practical method of distinguishing healthy from diseased eggs. The test often proved fallacious, and it was never extensively applied.

These corpuscles take possession of the intestinal canal, and spread thence throughout the body of the worm. They fill the silk cavities, the stricken insect often going through the motions of spinning without any material to answer to the act. Its organs, instead of being filled with the clear viscous liquid of the silk, are packed to distension by the corpuscles. On this feature of the plague Pasteur fixed his entire attention. The cycle of the silkworm's life is briefly this:—From the fertile egg comes

x

the little worm, which grows, and casts its skin. This process of moulting is repeated two or three times at subsequent intervals during the life of the insect. After the last moulting the worm climbs the brambles placed to receive it, and spins among them its cocoon. It passes thus into a chrysalis; the chrysalis becomes a moth, and the moth when liberated lays the eggs which form the starting-point of a new cycle. Now Pasteur proved that the plague-corpuscles might be incipient in the egg, and escape detection; they might also be germinal in the worm, and still baffle the microscope. But as the worm grows, the corpuscles grow also, becoming larger and more defined. In the aged chrysalis they are more pronounced than in the worm; while in the moth, if either the egg or the worm from which it comes should have been at all stricken, the corpuscles infallibly appear, offering no difficulty of detection. This was the first great point made out in 1865 by Pasteur. The Italian naturalists, as aforesaid, recommended the examination of the eggs before risking their incubation. Pasteur showed that both eggs and worms might be smitten and still pass muster, the culture of such eggs or such worms being sure to entail disaster. He made the moth his starting-point in seeking to regenerate the race.

Pasteur made his first communication on this subject to the Academy of Sciences in September, 1865. It raised a cloud of criticism. Here forsooth was a chemist rashly quitting his proper *métier* and presuming to lay down the law for the physician and biologist on a subject which was eminently theirs. 'On trouva étrange que je fusse si peu au courant de la question; on m'opposa des travaux qui avaient paru depuis longtemps en Italie, dont les résultats montraient l'inutilité de mes efforts, et l'impossibilité d'arriver à un résultat pratique dans la direction que je m'étais engagé. Que mon ignorance fut grande au sujet des recherches sans nombre qui avaient paru depuis quinze années.' Pasteur heard the buzz, but he continued his work. In choosing the eggs intended for incubation, the cultivators selected those produced in the successful 'educations' of the year. But they could not understand the frequent and often disastrous failures of their selected eggs; for they did not know, and nobody prior to Pasteur was competent to tell them, that the finest cocoons may envelope doomed corpusculous moths. It was not, however, easy to make the cultivators accept new guidance. To strike their imagination, and if possible determine their practice, Pasteur hit upon the expedient of prophecy. In 1866 he inspected at St. Hippolyte-du-Fort fourteen different parcels of eggs intended for incubation. Having examined a sufficient number of the moths which produced these eggs, he wrote out the prediction of what would occur in 1867, and placed the prophecy as a sealed letter in the hands of the Mayor of St. Hippolyte.

In 1867 the cultivators communicated to the mayor their results. The letter of Pasteur was then opened and read, and it was found that in twelve out of fourteen cases there was absolute conformity between his prediction and the observed facts. Many of the groups had perished totally; the others had perished almost totally; and this was the prediction of Pasteur. In two out of the fourteen cases, instead of the prophesied destruction, half an average crop was obtained. Now, the parcels of eggs here referred to were considered healthy by their owners. They had been hatched and tended in the firm hope that the labour expended on them would prove remunerative. The application of the moth-test for a few minutes in 1866 would have saved the labour and averted the disappointment. Two additional parcels of eggs were at the same time submitted to Pasteur. He pronounced them healthy; and his words were verified by the production of an excellent crop. Other cases of prophecy still more remarkable, because more circumstantial, are recorded in Pasteur's work.

Pasteur subjected the development of the corpuscles to a searching investigation. With admirable skill and completeness he examined the various modes by which the plague is propagated. He obtained perfectly healthy worms from moths perfectly free from corpuscles, and selecting from them 10, 20, 30, 50, as the case might be, he introduced into the worms the corpusculous matter. It was first permitted to accompany the food. Let us take a single example out of many. Rubbing up a small corpusculous worm in water, he smeared the mixture over the mulberry leaves. Assuring himself that the leaves had been eaten, he watched the consequences from day to day. Side by side with the infected worms he reared their fellows, keeping them as much as possible out of the way of infection. These constituted his 'lot temoign,' his standard of comparison. On the 16th of April, 1868, he thus infected thirty worms. Up to the 23rd they remained quite well. On the 25th they seemed well, but on that day corpuscles were found in the intestines of two of them. They first form in the tunic of the intestine. On the 27th, or eleven days after the infected repast, two fresh worms were examined, and not only was the intestinal canal found in each case invaded, but the silk organ itself was found charged with corpuscles. On the 28th the twenty-six remaining worms were covered by the black spots of pébrine. On the 30th the difference of size between the infected and non-infected worms was very striking, the sick worms being not more than two-thirds of the size of the healthy ones. On the 2nd of May a worm which had just finished its fourth moulting was examined. Its whole body was so filled with corpuscles as to excite astonishment that it could live. The disease advanced, the worms died and were examined, and on the 11th of May only six out of the thirty remained. They were the strongest of the lot, but on being searched they also were found charged with corpuscles. Not one of the thirty worms had escaped ; a single corpusculous meal had poisoned them all. The standard lot, on the contrary, spun their fine cocoons, and two only of their moths were found to contain any trace of corpuscles, which had doubtless been introduced during the rearing of the worms.

As his acquaintance with the subject increased, Pasteur's desire for precision augmented, and he finally gives the growing number of corpuscles seen in the field of his microscope from day to day. After a contagious repast the number of worms containing the parasite gradually augmented until finally it became cent. per cent. The number of corpuscles would at the same time rise from 0 to 1, to 10, to 100, and sometimes even to 1,000 or 1,500 for a single field of his microscope. He then varied the mode of infection. He inoculated healthy worms with the corpusculous matter, and watched the consequent growth of the disease. He showed how the worms inoculate each other by the infliction of visible wounds with their claws. In various cases he washed the claws, and found corpuscles in the water. He demonstrated the spread of infection by the simple association of healthy and diseased worms. The diseased worms sullied the leaves by their dejections, they also used their claws, and spread infection in both ways. It was no hypothetical injected medium that killed the worms, but a definitely organised and isolated thing. He examined the question of contagion at a distance, and demonstrated its existence. In fact, as might be expected from Pasteur's antecedents, the investigation was exhaustive, the skill and beauty of his manipulation finding fitting correlatives in the strength and clearness of his thought.

The following quotation from Pasteur's work clearly shows the relation in which his researches stand to this great question:

'Place,' he says, 'the most skilful educator, even the most expert microscopist, in presence of large educations which present the symptoms described in our experiments; his judgment will necessarily be erroneous if he confines himself to the knowledge which preceded my researches. The worms will not present to him the slightest spot of pébrine; the microscope will not reveal the existence of corpuscles; the mortality of the worms will be null or insignificant; and the cocoons leave nothing to be desired. Our observer would, therefore, conclude without hesitation that the eggs produced will be good for incubation. The truth is, on the contrary, that all the worms of these fine crops have been poisoned; that from the beginning they carried in them the germ of the malady; ready to multiply itself beyond measure in the chrysalides and the moths, thence to pass into the eggs and smite with sterility the next generation. And what is the first cause of the evil concealed under so deceitful an exterior? In our experiments we can, so to speak, touch it with our fingers. It is entirely the effect of a single corpusculous repast; an effect more or less prompt according to the epoch of life of the worm that has eaten the poisoned food.'

Pasteur describes in detail his method of securing healthy eggs. which is nothing less than a mode of restoring to France her ancient prosperity in silk husbandry. And the justification of his work is to be found in the reports which reached him of the application, and the unparalleled success of his method, at the time he was putting his researches together for final publication. In France and Italy his method has been pursued with the most surprising results. It was an up-hill fight which led to this triumph, but opposition stimulated Pasteur, and thus, without meaning it, did good service. ' Ever,' he says, ' since the commencement of these researches, I have been exposed to the most obstinate and unjust contradictions ; but I have made it a duty to leave no trace of these contests in this book.' And in reference to parasitic diseases he uses the following weighty words : ' Il est au pouvoir de l'homme de faire disparaître de la surface du globe les maladies parasitaires, si, comme c'est ma conviction, la doctrine des générations spontanées est une chimère.'

Pasteur dwells upon the ease with which an island like Corsica might be absolutely isolated from the silkworm epidemic. And with regard to other epidemics, Mr. Simon describes the extraordinary exemption of the Scilly Isles for the ten years extending from 1851 to 1860. Of the 627 registration districts of England, one only had an entire escape from diseases which, in whole or in part, were prevalent in all the others : ' In all the ten years it had not a single death by measles, nor a single death by small-pox, nor a single death by scarlet fever. And why? Not because of its general sanitary merits, for it had an average amount of other evidence of unhealthiness.

Doubtless, the reason of its escape was that it was insular. It was the *district of the Scilly Isles*; to which it was most improbable that any febrile 'contagion should come from without. And its escape is an approximative proof that, at least for those ten years, no contagium of measles, nor any contagium of scarlet-fever, nor any contagium of small-pox had arisen spontaneously within its limits.' It may be added that there were only seven districts in England in which no death from diphtheria occurred, and that, of those seven districts, the district of the Scilly Isles was one.

A second parasitic disease of silkworms, called in France *la flacherie*, co-existent with pébrine, but quite distinct from it, has also been investigated. Enough, however, has been said to send such of you as are interested in these questions to the original volumes for further information. To one important practical point M. Pasteur, in a letter written to me, directs attention :

'Permettez-moi de terminer ces quelques lignes que je dois dicter, vaincu que je suis par la maladie, en vous faisant observer que vous rendriez service aux Colonies de la Grande-Bretagne en répandant la connaissance de ce livre, et des principes que j'établis touchant la maladie des vers à soie. Beaucoup de ces colonies pourraient cultiver le mûrier avec succès, et en jetant les yeux sur mon ouvrage vous vous convaincrez aisément qu'il est facile aujourd'hui, non-seulement d'éloigner la maladie régnante, mais en outre de donner aux récoltes de la soie une prospérité qu'elles n'ont jamais eue.'

Origin and Propagation of Contagious Matter.

Prior to Pasteur, the most diverse and contradictory opinions were entertained as to the contagious character of pébrine ; some stoutly affirmed it, others as stoutly denied it. But on one point all were agreed.

' They believed in the existence of a deleterious medium, rendered epidemic by some occult and mysterious influence, to which was attributed the cause of the disease.' Those acquainted with medical literature will not fail to observe an instructive analogy here. We have on the one side accomplished writers ascribing epidemic diseases to 'deleterious media' which arise spontaneously in crowded hospitals and over ill-smelling drains. According to them the *matter* of epidemic disease is formed *de novo* in a putrescent atmosphere. On the other side we have writers, clear, vigorous, with well-defined ideas and methods of research, contending that the matter which produces epidemic disease comes always from a parent stock. It behaves as germinal matter, and they do not hesitate to regard it as such. They no more believe in the spontaneous generation of such diseases than they do in the spontaneous generation of mice. Pasteur, for example, found that pébrine had been known for an indefinite time as a disease among silkworms. The development of it which he combated was merely the expansion of an already existing power, the bursting into open conflagration of a previously smouldering fire. There is nothing surprising in this. For though epidemic disease requires a special contagium to produce it, surrounding conditions must have a potent influence on its development. Common seeds may be duly sown, but the conditions of temperature and moisture may be such as to restrict, or altogether prevent, the subsequent growth. Looked at, therefore, from the point of view of the germ theory, the exceptional energy which epidemic disease from time to time exhibits is not out of harmony with the method of Nature. You sometimes hear diphtheria spoken of as if

it were a new disease of the last twenty years; but Mr.
Simon tells me that from about three centuries ago, when
tremendous epidemics of it began to rage in Spain (where
it was named *Garrotillo*), and soon afterwards in Italy,
the disease has been well known to all successive generations of doctors; and that, for instance, in or about 1758,
Dr. Starr, of Liskeard, in a communication to the Royal
Society, particularly described the disease, with all the
characters which have recently again become familiar,
but under the name of *morbus strangulatorius*, as then
severely epidemic in Cornwall; a fact the more interesting, as diphtheria, in its more modern re-appearance,
again showed predilection for that remote county. Many
also believe that the Black Death of five centuries ago
has disappeared as mysteriously as it came; but Mr.
Simon finds that it is believed to be prevalent at this
hour in some of the north-western parts of India.

Let me here state an item of my own experience.
When I was at the Bel Alp last year the clergyman
appointed to that station received letters informing him
of the breaking out of scarlet fever among his children.
He lived, if I remember rightly, on the healthful eminence
of Dartmoor, and it was difficult to imagine how scarlet
fever could have been wafted to the place. A drain ran
close to his house, and on it his suspicions were manifestly fixed. Some of our medical writers would fortify
him in this notion, while those of another school would
deny to a drain, however foul, the power of producing
a specific disease. After close enquiry he recollected that
a hobby-horse had been used both by his boy and another
that a short time previously had passed through scarlet
fever. Drains and cesspools are by no means in such
evil odour as they used to be. A fetid Thames and a low

death-rate occur from time to time together in London. For, if the special matter or germs of epidemic disorder be not present, a corrupt atmosphere, however obnoxious otherwise, will not produce the disorder. Corrupted air may promote an epidemic, but cannot originate it. On the other hand, through the transport of the special germ or virus, disease may develope itself in regions where the drainage is good and the atmosphere pure.

If you see a new thistle growing in your field you feel sure that its seed has been wafted thither. Just as sure does it seem that the contagious matter of scarlatina, or any other contagious fever, has been transplanted to the place where it newly appears. With a clearness and conclusiveness not to be surpassed Dr. William Budd has traced such diseases from place to place ; showing how they plant themselves at distinct foci among populations subjected to the same atmospheric influences, just as grains of corn might be carried in the pocket and sown. Hildebrand, to whose remarkable work, *Du Typhus contagieux*, Dr. de Mussy has directed my attention, gives the following striking case, both of the durability and the transport of the virus of scarlatina : ' Un habit noir que j'avais en visitant une malade attaquée de scarlatine, et que je portai de Vienne en Podolie, sans l'avoir mis depuis plus d'un an et demi, me communiqua, dès que je fus arrivé, cette maladie contagieuse, que je répandis ensuite dans cette province, où elle était jusqu'alors presque inconnue.' Some years ago Dr. de Mussy himself was summoned to a country house in Surrey to see a young lady who was suffering from a dropsy, evidently the consequence of scarlatina. The original disease being of a very mild character had been quite overlooked, but circumstances were recorded which could

... upon the mind as to the nature and cause of the ... But then the question arose, how did the young lady catch the scarlatina? She had come there a few minutes previously, and it was only after she had been a minute in the house that she was taken ill. The housekeeper at once cleared up the mystery. The young lady on her arrival had expressed a particular wish to occupy a nice room in an isolated ... and in that room six months previously a visitor had been seized with an attack of scarlatina. The room had been swept and whitewashed, but the carpets had been permitted to remain.

Thousands of cases could probably be cited in which the disease has shown itself in this mysterious way, but where a strict examination has revealed its true parentage and extraction. Is it then philosophical to take refuge in the fortuitous occurrence of atoms as a cause of specific disease, merely because in special cases the parentage may be indistinct? Those best acquainted with atomic nature, and who are most ready to admit, as regards even higher things than this, the potentialities of matter, will be the last to accept these rash hypotheses.

The Germ Theory applied to Surgery.

Not only medical but surgical science is now seeking light and guidance from this germ theory. Upon it the antiseptic system of Professor Lister of Edinburgh is founded; and if the facts be correctly given, the results are extraordinary. As already stated, the germ theory of putrefaction was started by Schwann, but the illustrations of this theory adduced by Professor Lister are of such public moment as not only to justify, but to render imperative, their introduction here.

Schwann's observations, says Professor Lister, did not receive the attention which they appear to me to have deserved. The fermentation of sugar was generally allowed to be occasioned by the *torula cerevisiœ*; but it was not admitted that putrefaction was due to an analogous agency. And yet the two cases present a very striking parallel. In each a stable chemical compound, sugar in the one case, albumen in the other, undergoes extraordinary chemical changes under the influence of an excessively minute quantity of a substance which, regarded chemically, we should suppose inert. As an example of this in the case of putrefaction, let us take a circumstance often witnessed in the treatment of large chronic abscesses. In order to guard against the access of atmospheric air, we used to draw off the matter by means of a canula and trocar, such as you see here, consisting of a silver tube with a sharp-pointed steel rod fitted into it, and projecting beyond it. The instrument, dipped in oil, was thrust into the cavity of the abscess, the trocar was withdrawn, and the pus flowed out through the canula, care being taken by gentle pressure over the part to prevent the possibility of regurgitation. The canula was then drawn out with due precaution against the reflux of air. This method was frequently successful as to its immediate object, the patient being relieved from the mass of the accumulated fluid, and experiencing no inconvenience from the operation. But the pus was pretty certain to reaccumulate in course of time, and it became necessary again and again to repeat the process. And unhappily there was no absolute security of immunity from bad consequences. However carefully the procedure was conducted, it sometimes happened, even though the puncture seemed healing by first intention, that feverish symptoms declared themselves in the course of the first or second day, and, on inspecting the seat of the abscess, the skin was perhaps seen to be red, implying the presence of some cause of irritation, while a rapid reaccumulation of the fluid was found to have occurred. Under these circumstances, it became necessary to open the abscess by free incision, when a quantity, large in proportion to the size of the abscess, say, for example, a quart, of pus escaped, fetid from putrefaction. Now, how had this change been brought about ? Without the germ theory, I venture to say, no rational explanation of it could have been given. It must have been caused by the introduction of something from without. Inflammation of the punctured wound, even supposing it to have occurred, would not explain the phenomenon. For mere inflammation, whether acute or chronic, though it occasions the formation of pus, does not induce putrefaction. The pus originally evacuated was perfectly sweet, and we know of nothing to account for the alteration in its quality but the influence of something derived from the external world. And what could that something be? The dipping of the instrument in oil, and the subsequent precautions, prevented the entrance of oxygen. Or even if you allowed that a few atoms of the gas did enter, it would be an extraordinary assumption to make that these could in so short a time effect such changes in so large a mass of albuminous material. Besides, the pyogenic membrane is abundantly supplied with capillary vessels, through which arterial blood, rich in oxygen, is perpetually flowing; and there can be little doubt that the pus, before it was evacuated at all, was liable to any action which the element might be disposed to exert upon it.

On the oxygen theory, then, the occurrence of putrefaction under these circumstances is quite inexplicable. But if you admit the germ theory, the difficulty vanishes at once. The canula and trocar having been lying exposed to the air, dust will have been deposited upon them, and will be present in the angle between the trocar and the silver tube, and in that protected situation will fail to be wiped off when the instrument is thrust through the tissues. Then when the trocar is withdrawn, some portions of this dust will naturally remain upon the margin of the canula, which is left projecting into the abscess, and nothing is more likely than that some particles may fail to be washed off by the stream of out-flowing pus, but may be dislodged when the tube is taken out, and left behind in the cavity. The germ theory tells us that these particles of dust will be pretty sure to contain the germs of putrefactive organisms, and if one such is left in the albuminous liquid, it will rapidly develope at the high temperature of the body, and account for all the phenomena.

But striking as is the parallel between putrefaction in this instance and the vinous fermentation, as regards the greatness of the effect produced, compared with the minuteness and the inertness, chemically speaking, of the cause, you will naturally desire further evidence of the similarity of the two processes. You can see with the microscope the torula of fermenting must or beer. Is there, you may ask, any organism to be detected in the putrefying pus? Yes, gentlemen, there is. If any drop of the putrid matter is examined with a good glass, it is found to be teeming with myriads of minute jointed bodies, called vibrios, which indubitably proclaim their vitality by the energy of their movements. It is not an affair of probability, but a fact, that the entire mass of that quart of pus has become peopled with living organisms as the result of the introduction of the canula and trocar ; for the matter first let out was as free from vibrios as it was from putrefaction. If this be so, the greatness of the chemical changes that have taken place in the pus ceases to be surprising. We know that it is one of the chief peculiarities of living structures that they possess extraordinary powers of effecting chemical changes in materials in their vicinity, out of all proportion to their energy as mere chemical compounds. And we can hardly doubt that the animalcules which have been developed in the albuminous liquid, and have grown at its expense, must have altered its constitution, just as we ourselves alter that of the materials on which we feed.[1]

Secured from the danger of putrefaction, it is amazing how, under the hands of a really able surgeon, the human flesh and bones may be cut, torn, and crunched with impunity. The accounts of the operations of our eminent surgeons read like romance. On this, however, I must not dwell further than to recommend to your attention a case described in the ' British Medical Journal ' for the 14th of January last. In the operations of Professor Lister care is taken that every portion of tissue laid bare by the knife shall be defended from germs ; that

[1] Introductory Lecture before the University of Edinburgh.

if they fall upon the wound they shall be killed as they fall. With this in view he showers upon his exposed surfaces the spray of diluted carbolic acid, which is particularly deadly to the germs, and he surrounds the wound in the most careful manner with antiseptic bandages. To those accustomed to strict experiment it is manifest that we have a strict experimenter here—a man with a perfectly distinct object in view, which he pursues with never-tiring patience and unwavering faith. And the result, in his hospital practice, as described by himself, has been, that even in the midst of abominations too shocking to be mentioned here, and in the neighbourhood of wards where death was rampant from pyæmia, erysipelas, and hospital gangrene, he was able to keep his patients absolutely free from these terrible scourges. Let me here recommend to your attention Professor Lister's ' Introductory Lecture before the University of Edinburgh,' which I have already quoted ; his paper on ' The Effect of the Antiseptic System of Treatment on the Salubrity of a Surgical Hospital ;' and the article in the ' British Medical Journal,' to which I have just referred.

If, instead of using carbolic acid spray, he could surround his wounds with properly filtered air, the result would, he contends, be the same. In a room where the germs not only float but cling to clothes and walls, this would be difficult, if not impossible. But surgery is acquainted with a class of wounds in which the blood is freely mixed with air that has passed through the lungs, and it is a most remarkable fact that such air does not produce putrefaction. Professor Lister, as far as I know, was the first to give a philosophical interpretation of this fact, which he describes and comments upon thus :—

I have explained to my own mind the remarkable fact that

in simple fracture of the ribs, if the lung be punctured by a fragment, the blood effused into the pleural cavity, though freely mixed with air, undergoes no decomposition. The air is sometimes pumped into the pleural cavity in such abundance that, making its way through the wound in the pleura costalis, it inflates the cellular tissue of the whole body. Yet this occasions no alarm to the surgeon (although if the blood in the pleura were to putrefy, it would infallibly occasion dangerous suppurative pleurisy). Why air introduced into the pleural cavity through a wounded lung should have such wholly different effects from that entering directly through a wound in the chest was to me a complete mystery until I heard of the germ theory of putrefaction, when it at once occurred to me that it was only natural that air should be filtered of germs by the air-passages, one of whose offices is to arrest inhaled particles of dust, and prevent them from entering the air-cells.

I shall have occasion to refer to this remarkable hypothesis further on.

The advocates of the germ theory, both of putrefaction and epidemic disease, hold that both arise, not from the air, but from something contained in the air. They hold, moreover, that 'something' to be not a vapour nor a foreign gas, nor indeed a molecule of any kind, but a *particle*.[1] The term 'particulate' has been used in the Reports of the Medical Department of the Privy Council to describe this supposed constitution of contagious matter; and Dr. Sanderson's experiments render it in the highest degree probable, if they do not actually demonstrate, that the virus of small-pox is 'particulate.' Definite knowledge upon this point is of exceeding importance, because in

---

[1] As regards size, there is probably no sharp line of division between molecules and particles; the one gradually shades into the other. But the distinction that I would draw is this :—the atom or the molecule, if free, is always part of a gas, the particle is never so. A particle is a bit of liquid or solid matter formed by the aggregation of atoms or molecules.

the treatment of *particles* methods are available which it would be futile to apply to *molecules.*

Application of Luminous Beams to researches of this nature.

My own interference with this great question, while sanctioned by eminent names, has been also an object of varied and ingenious attack. On this point I will only say that when feeling escapes from behind the intellect, where it is a useful urging force, and places itself in front of the intellect, it is liable to produce glamour and all manner of delusions. Thus my censors, for the most part, have levelled their remarks against positions which I never assumed, and against claims which I never made. The simple history of the matter is this :—During the autumn of 1868 I was much occupied with the observations referred to at the beginning of this discourse. For fifteen years I had habitually employed the electric light, making use of the floating dust to reveal the paths of luminous beams ; but until 1868, when I was driven to it, I did not intentionally reverse the process and employ a luminous beam to reveal and examine the dust. In a paper presented to the Royal Society in December, 1869, I thus described the observations which induced me to give more special attention to the question of spontaneous generation and the germ theory of epidemic disease.

The Floating Matter of the Air.

Prior to the discovery of the foregoing action (the chemical action of light upon vapours), and also during the experiments just referred to, the nature of my work compelled me to aim at obtaining experimental tubes absolutely clean upon the surface, and absolutely empty within. Neither condition is, however, easily attained.

For however well the tubes might be washed and polished, and however bright and pure they might appear in ordinary daylight, the electric beam infallibly revealed signs and tokens of dirt. The air was always present, and it was sure to deposit some impurity. All chemical processes, not conducted in a vacuum, are open to this disturbance. When the experimental tube was exhausted it exhibited no trace of floating matter, but on admitting the air through the U-tubes containing caustic potash and sulphuric acid, a *dust-cone* more or less distinct was always revealed by the powerfully condensed electric beam.

The floating motes resembled minute particles of liquid which had been carried mechanically into the experimental tube. Precautions were therefore taken to prevent any such transfer. They produced little or no mitigation. I did not imagine at the time that the dust of the external air could find such free passage through the caustic potash and the sulphuric-acid tubes. But the motes really came from without. They also passed with freedom through a variety of ethers and alcohols. In fact, it requires long-continued action on the part of an acid first to *wet* the motes and afterwards to destroy them. By carefully passing the air through the flame of a spirit-lamp or through a platinum tube heated to bright redness, the floating matter was sensibly destroyed. It was therefore combustible, in other words, *organic* matter. I tried to intercept it by a large respirator of cotton-wool. Close pressure was necessary to render the wool effective. A plug of the wool rammed pretty tightly into the tube through which the air passed was finally found competent to hold back the motes. They appeared from time to time afterwards and gave me much trouble; but they were invariably traced in the end to some defect in the purifying apparatus—to some crack or flaw in the sealing-wax employed to render the tubes air-tight. Thus through proper care, but not without a great deal of searching out of disturbances, the experimental tube, even when filled with pure air or vapour, contains nothing competent to scatter the light. The space within it has the aspect of an absolute vacuum.

An experimental tube in this condition I call *optically empty*.

The simple apparatus employed in these experiments will be at once understood by reference to the figure on page 325. S S' is the glass experimental tube which has varied in length from

1 to 5 feet, and which may be from 2 to 3 inches in diameter. From the end S the pipe $p\,p'$ passes to an air-pump. Connected with the other end S' we have the flask F, containing the liquid whose vapour is to be examined; then follows a U-tube, T, filled with fragments of clean glass wetted with sulphuric acid; then a second U-tube, T', containing fragments of marble wetted with caustic potash; and finally a narrow straight tube $t\,t'$, containing a tolerably tightly fitting plug of cotton-wool. To save the air-pump gauge from the attack of such vapours as act on mercury, as also to facilitate observation, a separate barometer tube was employed.

Through the cork which stops the flask F two glass tubes, a and b, pass air-tight. The tube a ends immediately under the cork; the tube b, on the contrary, descends to the bottom of the flask and dips into the liquid. The end of the tube b is drawn out so as to render very small the orifice through which the air escapes into the liquid.

The experimental tube S S' being exhausted, a cock at the end S' is carefully turned on. The air passes slowly through the cotton-wool, the caustic potash, and the sulphuric acid in succession. Thus purified it enters the flask F and bubbles through the liquid. Charged with vapour it finally passes into the experimental tube, where it is submitted to examination. The electric lamp L placed at the end of the experimental tube furnished the necessary beam.

Wanting the cotton-wool the floating matter of the air ran the gauntlet of this system. The fact thus forced upon my attention had a bearing too obvious to be overlooked. It rendered at once evident to the senses why air filtered through cotton-wool is incompetent to generate animalcular life. The air is rendered by this treatment optically pure; in other words, freed from all floating matter, germs included. But the observations also revealed the great liability to error in experiments of this nature. They showed that without an amount of care which was hardly to be expected in all cases, error

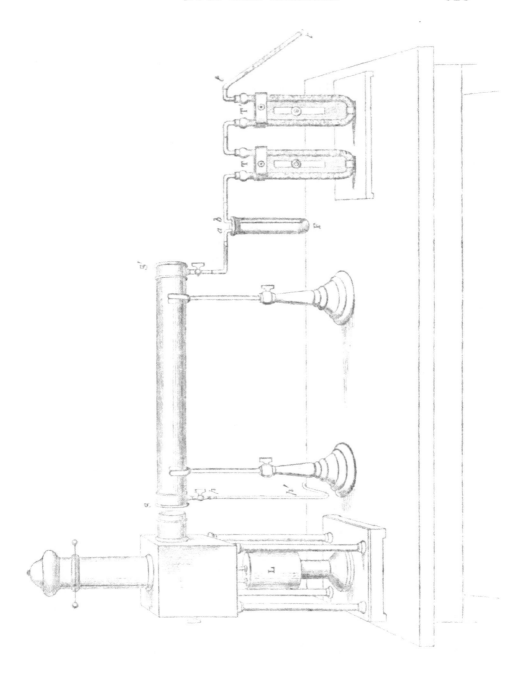

would be inevitable. It was especially manifest that the chemical method of Schultze might lead to the most erroneous consequences ; that neither acids nor alkalies had the power of rapid destruction which they had been supposed to possess. In short, the ̄ employment of the luminous beam rendered evident the cause of success in experiments rigidly conducted like those of Pasteur ; while it made equally evident the certainty of failure in experiments less severely and less skilfully carried out.

Dr. Bennett's Experiments.

Take, for example, the well-conceived experiments of Dr. Hughes Bennett, described before the Royal Society of Surgeons in Edinburgh on January 17, 1868.[1] Into flasks containing decoctions of liquorice root, hay, or tea, Dr. Bennett, by an ingenious method, forced air. The air was driven through two U-tubes, the one containing a solution of caustic potash, the other sulphuric acid. ' All the bent tubes,' says Dr. Bennett, ' were filled with fragments of pumice stone to break up the air, so as to prevent the possibility of any germs passing through in the centre of bubbles.' The air also passed through a Liebig's bulb containing sulphuric acid, and also through a bulb containing gun-cotton.

It was only natural for Dr. Bennett to believe that his bent tubes entirely cut off the germs. Previous to the observations just referred to I also believed in their competence to do this. But these observations destroy any such notion. The gun-cotton, moreover, will fail to arrest the whole of the floating matter unless it is tightly packed, and there is no indication in Dr. Bennett's memoir that it was so packed. On the whole, I should

[1] *British Medical Journal,* 13, pt. ii. 1868.

infer from the mere inspection of the apparatus the very results which Dr. Bennett has described—a retardation of the development of life, a total absence of it in some cases, and its presence in others.

In his first series of experiments eight flasks were fed with his sifted air, and five with common air. In ten or twelve days all the five had fungi in them; whilst it required from four to nine months to develope fungi in the others. In one case, moreover, even after this interval, no fungi appeared. In a second series of experiments there was a similar exception. In a third series of experiments he abandoned the cork stoppers used in the first and second series, and employed glass stoppers. Flasks containing decoctions of tea, beef, and hay were filled with common air, and other flasks with sifted air. In every one of the former fungi appeared, and in not one of the latter. These experiments simply ruin the doctrine that Dr. Bennett finally espouses.

In all these cases the prepared air was forced into the infusion when it was boiling hot. Dr. Bennett made a fourth series of experiments, in which, previous to forcing in the air, he permitted the flasks to cool. Into four bottles thus treated he forced prepared air, and after a time found fungi in all of them. What is his conclusion? Not that the boiling hot liquid employed in his first experiments had destroyed such germs as had run the gauntlet of his apparatus; but that air which, previous to being sealed up, had been exposed to a temperature of 212° is too rare to support life. This conclusion is so remarkable that it ought to be stated in Dr. Bennett's own words. 'It may be easily conceived that air subjected to a boiling temperature is so expanded as scarcely to merit the name of air, and that it is more or less unfit for the purpose of sustaining animal or vegetable life.'

Now numerical data are attainable here, and, as a matter of fact, I live and flourish for a considerable portion of each year in air of less density than that which Dr. Bennett describes as scarcely meriting the name of air ; the Swiss men, women, children, flocks, herds, tadpoles, grasshoppers, flowers, and grasses, do the same, while the chamois rears its kids in air rarer still.

In a fifth series of experiments sixteen bottles were filled with infusions. Into four of them, while cold, ordinary unheated and unsifted air was pumped. In these four bottles fungi were developed. Into four other bottles, containing a boiling infusion, ordinary air was also pumped—no fungi were here developed. Into four other bottles containing an infusion which had been boiled and permitted to cool, sifted air was pumped—no fungi were developed. Finally, into four bottles containing a boiling infusion sifted air was pumped—no fungi were developed. Only, therefore, in the four cases where the infusions were cold infusions, and the air ordinary air, did fungi appear.

Dr. Bennett does not draw from these experiments the conclusion to which they so obviously point. On them, on the contrary, their author founds a defence of the doctrine of spontanous generation, and a general theory of spontaneous development. So strongly was he impressed with the idea that the germs could not possibly pass through his potash and sulphuric acid tubes, that the appearance of fungi, even in a small minority of cases, where the air had been sent through these tubes, was to him conclusive evidence of the spontaneous origin of such fungi. And he accounts for the absence of life in many of his experiments by resorting to an hypothesis which will not bear a moment's consideration. But now that we know that organic particles may pass unscathed through alkalies and acids, the experiments of Dr. Bennett are precisely what ought under the circumstances to be expected. Indeed, their harmony with the conditions now revealed, is a proof of the honesty and accuracy with which they were executed.

On another point also the luminous beam will cast a light. Pasteur opened flasks upon the Mer de Glace, and, being careful not to come between the wind and his flasks, found the air incompetent, in the great majority of cases, to generate life. M. Pouchet repeated Pasteur's experiment in the Pyrenees, adding the precaution of holding the flasks, when they were opened, above his head. The luminous beam at once shows us the effect of this additional precaution. Let smoking brown paper be placed at the open mouth of a glass shade so that the smoke shall ascend and fill the shade. A beam sent through the shade forms a bright track through the smoke. When the closed fist is placed underneath the shade, a vertical wind of surprising violence, considering the small elevation of temperature, rises from the hand, displacing by comparatively dark air the illuminated smoke. Such a wind infallibly rose from M. Pouchet's body as he held his flasks above his head, and thus the precaution of Pasteur of not coming between the wind and the flask was annulled.

Again, in order to utterly destroy all germs M. Pouchet produced water from the combustion of hydrogen in air; but even in this water he found organisms. Had he seen, however, as you have, the manner in which the air is clouded with floating matter, would he have concluded that the deportment of water which had been permitted to trickle through such air could have the least influence in deciding this great question? I think not. Here is a quantity of water produced and collected by allowing a hydrogen flame to play upon the polished bottom of a silver basin, in which ice had been placed. This water is clear in the common light; but in the condensed electric beam it is seen to be laden with particles, so thick-strewn and minute, as to produce a continuous cone of light. In passing through the air the water loaded itself with this matter, and doubtless became charged with incipient life.

Let me now draw your attention to another experiment of Pasteur. He prepared twenty-one flasks, each containing a decoction of yeast, filtered and clear. He boiled the decoction, so as to destroy whatever germs it might contain, and while the space above the liquid was filled with pure steam he sealed his flasks with a blow-pipe. He opened ten of them in the deep, damp caves of the Paris Observatory, and eleven of them in the courtyard of the establishment. Of the former, one only showed signs of life subsequently. In nine out of the ten flasks no organisms of any kind were developed. In all the others organisms speedily appeared.

Now here is an experiment conducted in Paris; let us see whether we cannot throw light upon it in London. I place this large flask in the beam, and you see the luminous track crossing it from side to side. The flask is filled with the air of this room, charged with its germs and its dust, and hence capable of illumination. But here is another similar flask, which cuts a clear gap out of the beam. It is filled with unfiltered air, and still no trace of the beam is visible. Why? By pure accident I stumbled on this flask in our apparatus room, where it had remained quiet for some time. Here are three other flasks which have also been kept quiet for a couple of days; they are all optically empty. The still air of the flasks has deposited its dust, germs and all, and is itself practically free from suspended matter. Hence, manifestly, the result of Pasteur.

I have had a chamber erected, the lower half of which is of wood, its upper half being enclosed by four glazed window-frames. The chamber tapers to a truncated cone at the top. It measures in plan 3 ft. by 2 ft. 6 in., and its height is 5 ft. 10 in. On the 6th of February this chamber was closed, every crevice that could admit dust, or cause displacement of the air, being carefully pasted over with paper. The electric beam at first revealed the dust within the chamber as it did in the air of the laboratory. The chamber was examined almost daily; a perceptible diminution of the floating matter being noticed as time advanced. At the end of a week the chamber was optically empty, exhibiting no trace of matter competent to scatter the light. But where the beam entered, and where it quitted the chamber, the white circles stamped upon the interior surfaces of the glass showed what had become of the dust. It clung to those surfaces, and from them instead of from the air, the light was scattered. If the electric beam were sent through the air of the Paris caves, the cause of its impotence as generator of life would, I venture to predict, be revealed.

These experiments illustrate the application of a luminous beam to researches of this kind. They prove that the germs which produce infusorial and fungoid life share the fate of the ordinary visible dust with which they are intermixed ; that such germs attach themselves to the sides of vessels, and fall gradually to the bottom of spaces filled with perfectly still air. But I will now turn to a far more interesting application of the luminous beam than any hitherto described. My reference to Professor Lister's interpretation of the fact that air which has passed through the lungs cannot produce putrefaction is fresh in your memories. He there assumed that the air was rendered innocuous by the filtering action of the lungs. Can this filtering process be taken out of the region of assumption and placed in that of demonstration ? It can.

Here is the concentrated beam with which we operated at the commencement of this discourse. Its track through the dust is luminous, and you have seen the blackness introduced when the dust is burnt, or otherwise removed. I fill my lungs with ordinary air and breathe through a glass tube across the beam. The condensation of the aqueous vapour of the breath is here shown by the formation of a luminous white cloud of delicate texture. It is necessary to abolish this cloud, and this may be done by drying the breath previous to its entering the beam ; or, still more simply, by warming the glass tube. When this is done the luminous track of the beam is for a time uninterrupted, because the dust returning from the lungs makes good, in great part, the particles displaced. After some time, however, an obscure disc appears in the beam, the darkness of which increases, until finally, towards the end of the expiration, the beam is, as it were, pierced by an intensely black hole, in which no particles whatever can be discerned. The deeper air of the lungs is absolutely free from suspended matter. It is therefore in the precise condition required by Professor Lister's explanation. This experiment may be repeated any number of times with the same result. I think it must be regarded as a crowning piece of evidence both of the correctness of Professor Lister's views and of the impotence, as regards vital development, of optically pure air.

Cotton-wool Respirator.

I now empty my lungs as perfectly as possible, and placing a handful of cotton-wool against my mouth and nostrils, inhale through it. There is no difficulty in thus filling the lungs with air. On expiring this air through a glass tube, its freedom from floating matter is at once manifest. From the very beginning of the act of expiration the beam is pierced by a black aperture. The first puff from the lungs abolishes the illuminated dust, and puts a patch of darkness in its place; and the darkness continues throughout the entire course of the expiration. When the tube is placed below the beam and moved to and fro, the same smoke-like appearance as that obtained with a flame is observed. In short, the cotton-wool, when used in sufficient quantity, and with due care, completely intercepts the floating matter on its way to the lungs.[1]

[1] Since the first publication of these results Professor Lister has availed himself of the filtering power of cotton-wool in the treatment of wounds. He first destroys the germs adhering to the wool, and by a proper lotion kills those that may be scattered on the flesh. The cleansed wool placed upon the wound permits of a free diffusion of the air, but entirely intercepts the germs, and thus keeps the blood perfectly sweet. It is here essential that no matter from the wound should reach the outside air, for such matter would open a highway to the organisms.

The application of these experiments is obvious. If a physician wishes to hold back from the lungs of his patient, or from his own, the germs or virus by which contagious disease is propagated, he will employ a cotton-wool respirator. If perfectly filtered, attendants may breathe the air unharmed. In all probability the protection of the lungs and mouth will be the protection of the entire system. For it is exceedingly probable that the germs which lodge in the air-passages, or find their way with the saliva into the stomach with its absorbent system, are those which sow in the body epidemic disease. If this be so, then disease can be warded off by carefully prepared filters of cotton-wool. I should be most willing to test their efficacy in my own person. But apart from all doubtful applications, it is perfectly certain that various noxious trades in England may be rendered harmless by the use of such filters. I have had conclusive evidence of this from people engaged in such trades. A form of respirator devised by Mr. Garrick, a hotel proprietor in Glasgow, in which inhalation and exhalation occur through two different valves, the one permitting the air to enter through the cotton-wool, and the other permitting the exit of the air direct into the atmosphere, is well adapted for this purpose. But other forms might readily be devised.

Fireman's Respirator.

Smoke is often the fireman's greatest obstacle in his efforts to save life ; I thought therefore of inventing a respirator for the use of firemen. Schroeder was the first to use cotton-wool as a filter. To catch the atmospheric germs, M. Pouchet employed a film of adhesive glycerine spread upon glass; while Dr. Stenhouse turned charcoal to important account in respirators. By a combination of all three a respirator of peculiar efficacy is obtained. For the smoke of dried leaves the cotton-wool alone was found an adequate protection; but for the far more pungent smoke of resinous deal it was found totally inadequate. At the suggestion of a friend I moistened the wool with glycerine, and found it a great improvement. It was the notion of Pouchet in another form. Still about five minutes in dense smoke was all that could be endured. I then associated fragments of charcoal with the moistened cotton; the effect was excellent.[1] Armed with a respirator of this kind, one can breathe without annoyance in a space so crammed with smoke that a single inhalation without the respirator would be intolerable. I wrote to the chief officer of the Metropolitan Fire Brigade, asking him whether such a respirator would be useful. He replied to me that it would, but added that he was aware of every invention of the kind in all the countries of Europe, and that none had been found of any use. At my invitation he was kind enough to come to the Royal Institution with two firemen and an assistant. The three latter, wearing such respirators, went in succession into the smoke-filled space, and on returning stated that they had not experienced the slightest discomfort, that they could have remained there all day long. Captain Shaw himself repeated the experiment with the same result. I am confident that sooner or later this respirator will be employed, to the great benefit of a class of men whose actions in critical circumstances I have often had occasion to admire.

[1] Mr. Ladd, of Beak Street, makes these respirators.

Application of Luminous Beams to Water.

The method of examination here pursued is also applicable to water. It is in some sense complementary to that of the microscope, and may I think materially aid enquiries conducted with that instrument. In microscopic examination attention is directed to a small portion of the liquid, and the aim is to detect the individual suspended particles. By the present method a large portion of the liquid is illuminated, its general condition being revealed, through the light scattered by suspended particles. Care is taken to defend the eye from the access of all other light, and thus defended, it becomes an organ of inconceivable delicacy. Indeed, an amount of impurity so infinitesimal as to be scarcely expressible in numbers, and the individual particles of which are so small as wholly to elude the microscope, may, when examined by the method alluded to, produce not only sensible but striking effects upon the eye.

I take, for instance, this bottle of water intended to quench your lecturer's thirst. In the track of the beam it simply reveals itself as dirty water. So you see that we are invaded with dirt not only in the air we breathe, but in the water we drink. And this water is no worse than the other London waters. Thanks to the kindness of Professor Frankland, I have been furnished with specimens of the water of eight London companies. They are all laden with impurities mechanically suspended. But you will ask whether filtering will not remove the suspended matter? The grosser matter, undoubtedly, but not the more finely divided matter. Here is water which has been passed four times through a filter of bibulous paper, but it is still laden with fine matter. Here also is a bottle kindly sent me by Mr. Lipscomb, and passed once through his charcoal filter. But the track of the beam through it is more luminous than through air, because the quantity of matter suspended in the water is greater than that suspended in air. Here is another specimen courteously sent to me by the Silicated Carbon Company. All the grosser matter has been removed, but it is thick with fine matter. Nine-tenths of the light scattered by these particles is perfectly polarised in a direction at right angles to the beam, and this release of the particles from the ordinary law of polarisation is a demonstration of their smallness. I should say by far the greater number of the particles concerned in this scattering are wholly beyond the range of the microscope, and no ordinary filter can intercept them. There is an æsthetic pleasure in the drinking of a glass of cold sparkling water, and I fear these experiments will destroy this pleasure if you ever enjoyed it. And it is next to impossible by artificial means to produce a pure water. Mr. Hartley, for example, some time ago distilled water while it was surrounded by hydrogen, but the water was not free from floating matter. It is so hard to be clean in the midst of dirt. Here, however, is an approach to pure water. It is from the Lake of Geneva, and the bottle was carefully filled for me by my distinguished friend Soret. The track of the beam through it is of a delicate sky blue ; there is scarcely a trace of grosser matter.

The purest water that I have seen—probably the purest which has been seen hitherto—has been obtained from

z

the fusion of selected specimens of ice. But extra-ordinary precautions are required to obtain this degree of purity. An apparatus was devised and constructed by my assistant for this purpose. Through the plate of an air-pump passes the shank of a large funnel, attached to which below the plate is a glass bulb. In the funnel is placed a block of the most transparent ice, and over the funnel is a glass receiver. This is first exhausted and refilled several times with air, which has been filtered by its passage through cotton wool, the ice being thus sur-rounded by pure moteless air. But the ice has pre-viously been in contact with mote-filled air ; it is therefore necessary to let it wash its own face, and wash the bulb which is to receive the water of liquefaction. The ice is permitted to melt, the bulb is filled and emptied several times, until finally the large block dwindles to a small one. We may be sure that all impurity has been thus removed from the surface of the ice. These two bulbs contain water obtained in this way, the purity of which is the maximum hitherto attained. Still I should hesitate to call the water absolutely pure. When the concentrated beam is sent through it the track of the beam is not invisible, but of the most exquisitely delicate blue. This blue is purer than that of the sky, so that the matter which produces it must be finer than that of the sky. It may be, and indeed has been, contended that this blue is scattered by the very molecules of the water, and not by matter suspended in it. But when we remember that this perfection of blue is approached gradually through stages of less perfect blue ; and when we consider that a blue in all respects similar is demonstrably obtainable from particles mecha-nically suspended, we should hesitate, I think, to conclude that we have arrived here at the last stage of purification. The evidence, I think, points distinctly to the conclusion that could we push the process of purification still further, even this last delicate trace of blue would disappear.

Chalk Water. Clark's Softening Process.

But is it not possible to match the water of the Lake of Geneva here in England? Undoubtedly it is. We have in England a kind of rock which constitutes at once an exceedingly clean recipient and a natural filter, and from which we can obtain water extremely free from mechanical impurities. I refer to the chalk formation, in which large quantities of water are held in store. Our chalk hills are in most cases covered with thin layers of soil, and with very scanty vegetation. Neither opposes much obstacle to the entry of the rain into the chalk, where any organic impurity which the water may carry in is soon oxidised and rendered harmless. Those who have scampered like myself over the downs of Hants and Wilts will remember the scarcity of water in these regions. In fact, the rainfall, instead of washing the surface and collecting in streams, sinks into the fissured chalk and percolates through it, and when this formation is suitably tapped we obtain water of exceeding briskness and purity. Here is a large globe filled with the water of a well near Tring. It is wonderfully free from mechanical impurity; indeed, it stands to reason that water wholly withdrawn from surface contamination and percolating through so clean a substance should be pure. Sending a beam through this glass of water its purity is conspicuous; you see the track of the beam, but it is not the thick and

z 2

muddy track revealed in London waters. It has been a subject much debated whether the supply of excellent water which the chalk holds in store could not be rendered available for London. Many of the most eminent engineers and chemists have ardently recommended this source, and have sought to show that not only is its purity unrivalled but that its quantity is practically inexhaustible. Data sufficient to test this are now, I believe, in existence ; the number of wells sunk in the chalk is so considerable and the quantity of water which they yield is so well known.

But this water, so admirable as regards freedom from mechanical impurity, labours under the disadvantage of being very hard. It is rendered hard by the large quantity of carbonate of lime which it holds in solution. The chalk water in the neighbourhood of Watford holds in solution about seventeen grains of carbonate of lime per gallon. This, in the old terminology, used to be called seventeen degrees of hardness. Now this hard water is bad for tea, bad for washing ; it furs your boilers, because the lime held in solution is precipitated by boiling. If the water be used cold its hardness must be neutralised at the expense of soap before it will give a lather. These are serious objections to the use of chalk water in London. But they are now successfully met by the experimental demonstration that such water can be softened inexpensively, and on a grand scale. I had long known the method of softening water called Clark's process, but not until recently, under the guidance of Mr. Homersham, did I see proof of its larger applications. The chalk water is softened for the supply of the city of Canterbury ; at the Chiltern Hills it is softened for the supply of Tring and Aylesbury. Caterham also enjoys the luxury.

I have visited all these places, and made myself acquainted with the works. At Canterbury there are three reservoirs covered in and protected by a concrete roof and layers of pebbles both from the summer's heat and the winter's cold. Each reservoir contains 120,000 gallons of chalk water. Adjacent to these reservoirs are others containing pure slaked lime—the so-called ' cream of lime.' These are filled with water, the lime and water being thoroughly mixed by air forced in by an engine through apertures in the bottom of the reservoir. The water thus well mixed with the lime soon dissolves all of this substance that it is capable of dissolving. The lime is then allowed to subside to the bottom, leaving a perfectly clear lime water behind.

The object is now to soften the chalk water. Into the empty reservoir is introduced a certain quantity of the clear lime water, and after this about nine times the quantity of the chalk water. The transparency immediately disappears—the mixture of the two clear liquids becomes thickly turbid. The carbonate of lime is precipitated, and the precipitate is permitted to subside; it is crystalline and heavy, and therefore sinks rapidly. In about twelve hours you find a layer of pure white carbonate of lime at the bottom of the reservoir, with a water of extraordinary beauty and purity overhead. A few days ago I pitched some half-pence into a reservoir sixteen feet deep at the Chiltern Hills. The sixteen feet hardly perceptibly dimmed the coin. Had I cast a pin in, it could, I am persuaded, have been seen at the bottom. By this process of softening the water is reduced from about seventeen degrees of hardness to three degrees of hardness. It yields a lather immediately. Its temperature is constant throughout the year. In the hottest summer it is cool, its temperature being twenty degrees above the freezing point; and it does not freeze in winter if conveyed in proper pipes. It is not exposed to the contamination of either earth or air. The reservoirs are covered; a leaf cannot blow into the water, no surface contamination can reach it, it passes direct from the main into the house tap; no cisterns are employed, the supply is always fresh and pure. It is highly charged with air. This is the kind of water which is supplied to the fortunate people of Tring, Caterham, and Canterbury.

Let me in conclusion remind you that I do not consider the floating matter revealed by the electric beam to be all *living* matter. I believe that only in exceptional cases, such as those cited in the excellent reports of Dr. Angus Smith, does the quantity of living matter suspended in the air of our streets and rooms amount to more than a small fraction of the total dust. But I believe it to be perfectly well established that, during epidemics, air and water are charged with the specific 'materies morbi' by which the disease is spread; that these two media are, in fact, the chief vehicles of its dissemination. I believe there are the strongest grounds for holding the contagious matter to be 'particulate,' and further that the particles are to all intents and purposes *germs*; exhibiting as they do the fundamental characteristic of propagating their own kind through countless generations, and over vast geographical areas. Their life and reproduction run parallel to and are an incident of the life of man himself. I do not doubt the ability of these particles to scatter light, nor that the means by which the visible floating dust of our air is arrested, and which demonstrably arrest with it the germs of various forms of fungoid and animalcular life, including those concerned in the phenomena of putrefaction, will also be found effectual in arresting contagium.

The following extract from a private letter written to me by Dr. William Budd, is so important, and its reasoning is so clear, that I asked and obtained the permission of its exceedingly able writer to publish it.

' Another point of great practical importance is, as far as possible, to appraise the respective shares which air and water take in the great act of distribution. That both cholera and typhoid fever are sometimes disseminated by drinking water has been amply proved. I have myself related many instances of the fact, and have in my notes the record of many others still more striking. But that water is the sole or even the chief vehicle of cholera and typhoid is a notion which, if I may trust my own experience, facts do not warrant.

' Limiting myself, for the moment, to the case of typhoid, I am in a position to state that all the worst and most widespread outbreaks of that fever which I have ever witnessed, have occurred among communities supplied by drinking water which was absolutely blameless. Two illustrations will suffice.

' I live in a town in which the divorce between sewage and drinking water has long been consummated. Bristol is supplied with drinking water which from its source in the Mendips to the tap from which it is delivered under high pressure to the consumer, flows through conduits out of all reach of sewage contamination.

' And yet typhoid fever has not only not ceased to exist in Bristol, but about eight or ten years ago (before the appointment of a health officer), there occurred in the Parish of St. James' one of the worst outbreaks of this fever which I have ever seen in the city. In the course of a circuit which I took one morning with the late Dr. Pring (at that time Poor Law Medical Officer) I saw within a comparatively small area more than eighty cases of the disease.

' Now, with the exception of a single household, all the patients were drinking the Mendip water ; the very same water which, as far as fever is concerned, more than 150,000 of their fellow-citizens outside the infected area were drinking with absolute impunity.

' Some four or five years ago I was sent for to advise what measures should be taken to stay an outbreak of typhoid fever which had occurred in a large convent about two miles from Bristol. The inmates were divided into three distinct divisions ; the largest being a reformatory for girls, who occupied the central block of the building. Into this reformatory the fever was brought by a girl already suffering from it, and who had contracted it in a seaside place more than twenty miles off. From this girl the disease spread until, at the date of my visit, more than fifty girls were lying ill of it. From first to last the fever was confined to the reformatory girls, and to persons in immediate attendance upon them.

' Now, the facts as to the drinking water were these.

' 1. The water was proved by examination of the well,

and by chemical analysis, to be entirely free from sewage
contamination.

'2. The inmates of another large division of the convent,
who remained entirely free from fever, drank the same
water as the girls among whom fever was raging like a
plague.

'3. From the very time when disinfection was brought
to bear on the excreta the disease ceased to spread,
although the inmates of the infected division continued to
drink the same water as before.

'Lastly, nothing has since been done to the well—the
water remains what it was—but no fever has occurred in
the convent since.

'The evidence in both these cases is, as you see, of that
crucial, decisive order that admits of no reply.

'They show, at least, that typhoid fever may do its worst
where drinking water takes absolutely no part in the
distribution of the poison.

'But if water be excluded, the air is the only other
possible vehicle by which a poison generated in the living
body can find its way back to other living bodies on a
scale sufficiently large to cause the resulting disease to
assume an epidemic form.

'I may remark further that the infection of the air in
these two cases was obviously not the work of chance,
but only represented the effect of agencies which are
always in operation where this fever prevails.

'The phenomenon is, in fact, merely the expression of a
general law.

'The germs cast off in the liquid excreta of contagious
diseases rise into the air by no power of their own, but
in virtue of the very same physical conditions which cause

the germs of the great tribe of Infusoria, which, as their name bespeaks, breed in liquids, to rise in swarms into the same medium.

' If there were time or need, I could show, by evidence quite as decisive, that all these statements apply equally to cholera also.

' I do not know that there is anything in these data to suggest additional matter for your essay, but I have thought it worth while to bring them under your notice, harmonising as they do with your own investigations which show by such striking phenomena that air and water are equally objects of distrust.

' As to the germ theory itself, that is a matter on which I have long since made up my mind. From the day when I first began to think upon these subjects, I have never had a doubt that the specific cause of contagious fevers must be living organisms.

' It is impossible, in fact, to make any statement bearing upon the essence or distinctive characters of these fevers, without using terms which are of all others *the most distinctive* of life. Take up the writings of the most violent opponent of the germ theory, and, ten to one, you will find them full of such terms as " propagation," " self-propagation," " reproduction," " self-multiplication," and so on. Try as he may—if he has anything to say of these diseases which is characteristic of them—he cannot evade the use of these terms or the exact equivalents of them. While perfectly applicable to life and to living things, these terms express qualities which are not only inapplicable to common chemical agents, but, as far I can see, actually inconceivable of them.'

XII.

LIFE AND LETTERS OF FARADAY.

BY Dr. HENRY BENCE JONES.

A REVIEW.

[*The Academy for May and June*, 1870.]

Fame is the spur that the clear spirit doth raise
(That last infirmity of noble minds)
To scorn delights and live laborious days;
But the fair guerdon when we hope to find,
And think to burst out into sudden blaze,
Comes the blind fury with the abhorred shears,
And slits the thin-spun life.   But not the praise
Phœbus reply'd, and touched my trembling ears;
Fame is no plant that grows on mortal soil,
Nor in the glistering foil
Set off to the world, nor in broad rumour lies;
But lives and spreads aloft by those pure eyes
And perfect witness of all-judging Jove.'

MILTON.

XII.

LIFE AND LETTERS OF FARADAY.

UNDERTAKEN and executed in a reverent and loving spirit, the work of Dr. Bence Jones makes Faraday the virtual writer of his own life. Everybody now knows the story of the philosopher's birth ; that his father was a smith ; that he was born at Newington Butts in 1791 ; that he slid along the London pavements, a bright-eyed errand boy, with a load of brown curls upon his head and a packet of newspapers under his arm ; that the lad's master was a bookseller and bookbinder—a kindly man, who became attached to the little fellow and in due time made him his apprentice without fee ; that during his apprenticeship he found his appetite for knowledge provoked and strengthened by the books he stitched and covered. Thus he grew in wisdom and stature to his year of legal manhood, when he appears in the volumes before us as a writer of letters, which reveal his occupation, acquirements, and tone of mind. His correspondent was Mr. Abbott, a member of the Society of Friends, who, with a forecast of his friend's greatness, preserved his letters and produced them at the proper time.

In later years Faraday always carried in his pocket a blank card on which he jotted down in pencil his thoughts and memoranda. He made his notes in the laboratory, in the theatre, and in the streets. This distrust of his memory reveals itself in his first letter to Abbott. To a proposition that no new enquiry should be started between them before the old one had been exhaustively discussed, Faraday objects. 'Your notion,' he says, 'I can hardly allow, for the following reason : ideas and thoughts spring up in my mind which are irrevocably lost for want of noting at the time.' Gentle as he seemed, he wished to have his own way, and he had it throughout his life. Differences of opinion sometimes arose between the two friends, and then they resolutely faced each other. ' I accept your offer to fight it out with joy, and shall in the battle of experience cause not pain, but, I hope, pleasure.' Faraday notes his own impetuosity, and incessantly checks it. There is at times something mechanical in his self-restraint. In another nature it would have hardened into mere ' correctness ' of conduct ; but his overflowing affections prevented this in his case. The habit became a second nature to him at last, and lent serenity to his later years.

In October, 1812, he was engaged by a Mr. De la Roche as a journeyman bookbinder ; but the situation did not suit him. His master appears to have been an austere and passionate man, and Faraday was to the last degree sensitive. All his life he continued so. He suffered at times from dejection ; and a certain grimness, too, pervaded his moods. ' At present,' he writes to Abbott, ' I am as serious as you can be, and would not scruple to speak a truth to any human being, whatever repugnance it might give rise to. Being in this state of mind, I should have refrained from writing to you, did I not conceive from the general tenor of your letters that your mind is, at proper times, occupied upon serious subjects to the

exclusion of those that are frivolous.' Plainly he had
fallen into that stern Puritan mood which not only crucifies the flesh, affections, and lusts of him who harbours
it, but is often a cause of disturbed digestion to his friends.

About three months after his engagement with De la
Roche, Faraday quitted him and bookbinding together.
He had heard Davy, copied his lectures, and written to
him entreating to be released from Trade which he hated
and enabled to pursue Science. Davy recognised the
merit of his correspondent, kept his eye upon him, and
when occasion offered, drove to his door and sent in a
letter offering him the post of assistant in the laboratory
of the Royal Institution. He was engaged upon March 1,
1812, and on the 8th we find him extracting the sugar
from beet-root. He joined the City Philosophical Society
which had been founded by Mr. Tatum in 1808. ' The
discipline was very sturdy, the remarks very plain, and
the results most valuable.' Faraday derived great profit
from this little association. In the laboratory he had a
discipline sturdier still. Both Davy and himself were at
this time cut and bruised by explosions of chloride of
nitrogen. One explosion was so rapid ' as to blow my
hand open, tear away a part of one nail, and make my
fingers so sore that I cannot use them easily.' In another
experiment ' the tube and receiver were blown to pieces,
I got a cut on the head, and Sir Humphry a bruise on
his hand.' And again speaking of the same substance,
he says, ' when put in the pump and exhausted, it stood
for a moment, and then exploded with a fearful noise.
Both Sir H. and I had masks on, but I escaped this time
the best. Sir H. had his face cut in two places about the
chin, and a violent blow on the forehead struck through

a considerable thickness of silk and leather.' It was this same substance that blew out the eye of Dulong.

Over and over again, even at this early date, we can discern the quality which, compounded with his rare intellectual power, made him a great experimental philosopher. This was his desire to see facts, and not to rest contented with the descriptions of them. He frequently pits the eye against the ear, and affirms the enormous superiority of the organ of vision. Late in life I have heard him say that he could never fully understand an experiment until he had seen it. But he did not confine himself to experiment. He aspired to be a teacher, and reflected and wrote upon the method of scientific exposition. 'A lecturer,' he observes, 'should appear easy and collected, undaunted and unconcerned:' still 'his whole behaviour should evince respect for his audience.' These recommendations were afterwards in great part embodied by himself. I doubt his unconcern, but his fearlessness was often manifested. It used to rise within him as a wave, which carried both him and his audience along with it. On rare occasions also, when he felt himself and his subject hopelessly unintelligible, he suddenly evoked a certain recklessness of thought, and without halting to extricate his bewildered followers, he would dash alone through the jungle into which he had unwittingly led them ; thus saving them from ennui by the exhibition of a vigour which, for the time being, they could neither share nor comprehend.

In October, 1813, he quitted England with Sir Humphry and Lady Davy. During his absence he kept a journal, from which copious and interesting extracts have been made by Dr. Bence Jones. Davy was considerate, preferring at times to be his own servant rather than impose on Faraday duties which he disliked. But Lady Davy was the reverse. She treated him as an underling; he chafed under the treatment, and was often on the point of returning home. They halted at Geneva. De la Rive the elder had known Davy in 1799, and by his writings in the 'Bibliothèque britannique,' had been the first to make the English chemist's labours known abroad. He welcomed Davy to his country residence in 1814. Both were sportsmen, and they often went out shooting together. On these occasions Faraday charged Davy's gun, while De la Rive charged his own. Once the Genevese philosopher found himself by the side of Faraday, and in his frank and genial way entered into conversation with the young man. It was evident that a person possessing such a charm of manner and such high intelligence could be no mere servant. On enquiry De la Rive was somewhat shocked to find that the *soi-disant domestique* was really *préparateur* in the laboratory of the Royal Institution; and he immediately proposed that Faraday thenceforth should join the masters instead of the servants at their meals. To this Davy, probably out of weak deference to his wife, objected; but an arrangement was come to that Faraday thenceforward should have his food in his own room. Rumour states that a dinner in honour of Faraday was given by De la Rive. This is a delusion; there was no such banquet; but Faraday never forgot the kindness of the friend who saw his merit when he was a mere *garçon de laboratoire*.[1]

---

[1] While confined last autumn at Geneva by the effects of a fall in the Alps, my friends, with a kindness I can never forget, did all that friendship could suggest to render my captivity pleasant to me. M. de la Rive then wrote out for me the full account, of which the foregoing is a condensed abstract.

He returned in 1815 to the Royal Institution. Here he helped Davy for years; he worked also for himself, and lectured frequently at the City Philosophical Society. He took lessons in elocution, happily without damage to his natural force, earnestness, and grace of delivery. He was never pledged to theory, and he changed in opinion as knowledge advanced. With him life was growth. In those early lectures we hear him say, ' In knowledge, that man only is to be contemned and despised who is not in a state of transition.' And again—' Nothing is more difficult and requires more caution than philosophical deduction, nor is there anything more adverse to its accuracy than fixity of opinion.' Not that he was wafted about by every wind of doctrine; but that he united flexibility with his strength. In striking contrast with this intellectual expansiveness is his fixity in religion, but this is a subject which cannot be discussed here.

Of all the letters published in these volumes none possess a greater charm than those of Faraday to his wife. Here, as Dr. Bence Jones truly remarks, ' he laid open all his mind and the whole of his character, and what can be made known can scarcely fail to charm every one by its loveliness, its truthfulness, and its earnestness.' Abbott and he sometimes swerved into word-play about love; but up to 1820, or thereabouts, the passion was potential merely. Faraday's journal indeed contains entries which show that he took pleasure in the assertion of his contempt for love; but these very entries became links in his

---

It was at the desire of Dr. Bence Jones that I asked him to do so. The rumour of a banquet at Geneva illustrates the tendency to substitute for the youth of 1814 the Faraday of later years.

destiny. It was through them that he became acquainted with one who inspired him with a feeling which only ended with his life. His biographer has given us the means of tracing the varying moods which preceded his acceptance. They reveal more than the common alternations of light and gloom ; at one moment he wishes that his flesh might melt and he become nothing ; at another he is intoxicated with hope. The impetuosity of his character was then unchastened by the discipline to which it was subjected in after-years. The very strength of his passion proved for a time a bar to its advance, suggesting as it did to the conscientious mind of Miss Barnard doubts of her capability to return it with adequate force. But they met again and again, and at each successive meeting he found his heaven clearer, until at length he was able to say, ' Not a moment's alloy of this evening's happiness occurred. Everything was delightful to the last moment of my stay with my companion, because she was so.' The turbulence of doubt subsided, and a calm and elevating confidence took its place. ' What can I call myself,' he writes to her in a subsequent letter, ' to convey most perfectly my affection and love for you Can I or can truth say more than that for this world I am yours?' Assuredly he made his profession good, and no fairer light falls upon his character than that which reveals his relations to his wife. Never, I believe, existed a manlier, purer, steadier love. Like a burning diamond it continued to shed, for six-and-forty years, its white and smokeless glow.

Faraday was married on June 12, 1821 ; and up to this date Davy appears throughout as his friend. Soon afterwards, however, disunion occurred between them, which, while it lasted, must have given Faraday intense pain. It is impossible to doubt the honesty of conviction with which this subject has been treated by Dr. Bence Jones, and there may be facts known to him, but not appearing in these volumes, which justify his opinion that Davy in those days had become jealous of Faraday. This, which is the prevalent belief, is also reproduced in an excellent article in the March number of 'Fraser's Magazine.' But the best analysis I can make of the data fails to present Davy in this light to me. The facts, as I regard them, are briefly these.

In 1820, Oersted of Copenhagen made the celebrated discovery which connects electricity with magnetism, and immediately afterwards the acute mind of Wollaston perceived that a wire carrying a current ought to rotate round its own axis under the influence of a magnetic pole. In 1821 he tried, but failed, to realise this result in the laboratory of the Royal Institution. Faraday was not present at the moment, but he came in immediately afterwards, and heard the conversation of Wollaston and Davy about the experiment. He had also heard a rumour of a wager that Dr. Wollaston would eventually succeed.

This was in April. In the autumn of the same year Faraday wrote a history of electro-magnetism, and repeated for himself the experiments which he described. It was while thus instructing himself that he succeeded in causing a wire carrying an electric current to rotate round a magnetic pole. This was not the result sought by Wollaston, but it was closely related to it.

The strong tendency of Faraday's mind to look upon the reciprocal actions of natural forces gave birth to his greatest discoveries; and we, who know this, should be justified in concluding that, even had Wollaston not preceded him, the result would have been the same. But in judging Davy we ought to transport ourselves to his time, and carefully exclude from our thoughts and feelings that noble subsequent life which would render simply impossible the ascription to Faraday of anything unfair. It would be unjust to Davy to put our knowledge in the place of his, or to credit him with data which he could not have possessed. Rumour and fact had connected the name of Wollaston with these supposed interactions between magnets and currents. When, therefore, Faraday in October published his successful experiment without any allusion to Wollaston, general, though really ungrounded, criticism followed. I say ungrounded because, firstly, Faraday's experiment was not that of Wollaston, and secondly, Faraday before he published it, had actually called upon Wollaston, and not finding him at home did not feel himself authorised to mention his name.

In December Faraday published a second paper on the same subject, from which, through a misapprehension, the name of Wollaston was also omitted. Warburton and others thereupon affirmed that Wollaston's ideas had been appropriated without acknowledgment, and it is plain that Wollaston himself, though cautious in his utterance, was also hurt. Censure grew till it became intolerable. 'I hear,' writes Faraday to his friend Stodart, 'every day more and more of these sounds, which, though only whispers to me, are, I suspect, spoken aloud among scientific men.' He might have written explanations and defences, but he went straighter to the point. He wished to see the principals face to face—to plead his cause before them personally. There is a certain vehemence in his desire to do this. He saw Wollaston, he saw Davy, he saw Warburton ; and I am inclined to think that it was the irresistible candour and truth of character which these *vivâ voce* defences revealed, as much as the defences themselves, that disarmed resentment at the time.

As regards Davy, another cause of dissension arose in 1823. In the spring of that year Faraday analysed the hydrate of chlorine, a substance once believed to be the element chlorine, but proved by Davy to be a compound of that element and water. The analysis was looked over by Davy, who then and there suggested to Faraday to heat the hydrate in a closed glass tube. This was done, the substance was decomposed, and one of the products of decomposition was proved by Faraday to be chlorine liquefied by its own pressure. On the day of its discovery he communicated this result to Dr. Paris. Davy, on being informed of it, instantly liquefied another gas in the same way. Having struck thus into Faraday's enquiry, ought he not to have left the matter in Faraday's hands? I think he ought. But, considering his relation to both Faraday and the hydrate of chlorine, Davy, I submit, may be excused for thinking differently. A father is not always wise enough to see that his son has ceased to be a boy, and estrangement on this account is not rare ; nor was Davy wise enough to discern that Faraday had passed the mere assistant stage and become a discoverer. It is now hard to avoid magnifying this error. But had Faraday died or ceased to work at this time, or had his subsequent life been devoted to money-getting instead of to research, would anybody now dream of ascribing jealousy to Davy? Assuredly not. Why should

he be jealous? His reputation at this time was almost
without a parallel: his glory was without a cloud. He
had added to his other discoveries that of Faraday, and
after having been his teacher for seven years, his language
to him was this: ' It gives me great pleasure to hear that
you are comfortable at the Royal Institution, and I trust
that you will not only do something good and honourable
for yourself, but also for science.' This is not the language of jealousy, potential or actual. But the chlorine
business introduced irritation and anger, to which, and not
to any ignobler motive, Davy's opposition to the election
of Faraday to the Royal Society is, I am persuaded, to be
ascribed.

These matters are touched upon with perfect candour
and becoming consideration in the volumes of Dr. Bence
Jones; but in ' society ' they are not always so handled.
Here a name of noble intellectual associations is surrounded by injurious rumours which I would willingly
scatter for ever. The pupil's magnitude and the splendour of his position are too great and absolute to need
as a foil the humiliation of his master. Brothers in
intellect, Davy and Faraday, however, could never have
become brothers in feeling; their characters were too
unlike. Davy loved the pomp and circumstance of
fame; Faraday the inner consciousness that he had fairly
won renown. They were both proud men. But with
Davy pride projected itself into the outer world; while
with Faraday it became a steadying and dignifying
inward force. In one great particular they agreed.
Each of them could have turned his science to immense
commercial profit, but neither of them did so. The noble
excitement of research, and the delight of discovery, con-

stituted their reward.   I commend them to the reverence which great gifts greatly exercised ought to inspire. They were both ours ; and through the coming centuries England will be able to point with just pride to the possession of such men.

The first volume of the 'Life and Letters' reveals to us the youth who was to be father to the man. Skilful, aspiring, resolute, he grew steadily in knowledge and in power. Consciously or unconsciously, the relation of Action to Reaction was ever present to Faraday's mind. It had been fostered by his discovery of Magnetic Rotations, and it planted in him more daring ideas of a similar kind. Magnetism he knew could be evoked by electricity, and he thought that electricity, in its turn, ought to be capable of evolution by magnetism. On August 29, 1831, his experiments on this subject began. He had been fortified by previous trials, which, though failures, had begotten instincts directing him towards the truth. He, like every strong worker, might at times miss the outward object, but he always gained the inner light, education and expansion. Of this Faraday's life was a constant illustration. By November he had discovered and colligated a multitude of the most wonderful and unexpected phenomena. He had generated currents by currents ; currents by magnets, permanent and transitory ; and he afterwards generated currents by the earth itself. Arago's 'Magnetism of Rotation,' which had for years offered itself as a challenge to the best scientific intellects of Europe, now fell into his hands. It proved to be a beautiful but still special illustration of the great principle of Magneto electric Induction. Nothing equal to this, in the way of pure experimental enquiry, had previously been achieved.

Electricities from various sources were next examined, and their differences and resemblances revealed. He thus assured himself of their substantial identity. He then took up Conduction, and gave many striking illustrations of the influence of Fusion on Conducting Power. Renouncing professional work, from which at this time he might have derived an income of many thousands a year, he poured his whole momentum into his researches. He was long entangled in Electro-chemistry. The light of law was for a time obscured by the thick umbrage of novel facts; but he finally emerged from his researches with the great principle of Definite Electro-chemical Decomposition in his hands. If his discovery of Magneto-electricity may be ranked with that of the Pile by Volta, this new discovery may almost stand beside that of Definite Combining Proportions in Chemistry. He passed on to Static Electricity—its Conduction, Induction, and Mode of Propagation. He discovered and illustrated the principle of Inductive Capacity; and, turning to theory, he asked himself how electrical attractions and repulsions are transmitted. Are they, like gravity, actions at a distance, or do they require a medium? If the former, then, like gravity, they will act in straight lines; if the latter, then, like sound or light, they may turn a corner. Faraday held, and his views are gaining ground, that his experiments proved the fact of curvilinear propagation, and hence the operation of a medium. Others denied this; but none can deny the profound and philosophic character of his leading thought.[1] The first volume of the researches contains all the papers here referred to.

Faraday had heard it stated that henceforth physical discovery would be made solely by the aid of mathematics ; that we had our data, and needed only to work deductively. Statements of a similar character crop out from time to time in our day. They arise from an imperfect acquaintance with the nature, present condition, and prospective vastness of the field of physical enquiry. The tendency of natural science doubtless is to bring all physical phenomena under the dominion of mechanical laws ; to give them, in other words, mathematical expression. But our approach to this result is asymptotic ; and for ages to come—possibly for all the ages of the human race—nature will find room for both the philosophical experimenter and the mathematician. Faraday entered his protest against the foregoing statement by labelling his investigations 'Experimental Researches in Electricity.' They were completed in 1854, and three volumes of them have been published. For the sake of reference, he numbered every paragraph, the last number being 3362. In 1859 he collected and published a fourth volume of papers under the title, 'Experimental Researches in Chemistry and Physics.' Thus the apostle of experiment magnified his office.

The second volume of the Researches embraces memoirs on the Electricity of the Gymnotus ; on the Source of Power in the Voltaic Pile ; on the Electricity evolved by the Friction of Water and Steam, in which the phenomena

---

[1] In a very remarkable paper published in Poggendorff's *Annalen* for 1857, Werner Siemens developes and accepts Faraday's theory of Molecular Induction.

and principles of Sir William Armstrong's Hydro-electric machine are described and developed; a paper on Magnetic Rotations, and Faraday's letters in relation to the controversy it aroused. The contribution of most permanent value here is that on the Source of Power in the Voltaic Pile. By it the Contact Theory pure and simple was totally overthrown, and the necessity of chemical action to the maintenance of the current demonstrated.

The third volume of the Researches opens with a memoir entitled 'The Magnetisation of Light, and the Illumination of Magnetic Lines of Force.' It is difficult even now to affix a definite meaning to this title; but the discovery of the rotation of the plane of polarisation which it announced seems pregnant with great results. The writings of William Thomson on the theoretic aspects of the discovery; the excellent electro-dynamic measurements of Wilhelm Weber, which are models of experimental completeness and skill; Weber's labours in conjunction with his lamented friend Kohlrausch—above all, the researches of Clerk Maxwell on the Electro-magnetic Theory of Light—point to that wonderful and mysterious medium which is the vehicle of light and radiant heat as the probable basis also of magnetic and electric phenomena. The hope of such a combination was first raised by the discovery here referred to.[1] Faraday himself seemed to

---

[1] A letter addressed to me by Prof. Weber on the 18th of last March contains the following reference to the connection here mentioned: ' Die Hoffnung einer solchen Combination ist durch Faraday's Entdeckung der Drehung der Polarisationsebene durch magnetische Directionskraft zuerst, und sodann durch die Uebereinstimmung derjenigen Geschwindigkeit, welche das Verhältniss der electro-dynamischen Einheit zur electro-statischen ausdrückt, mit der Geschwindigkeit des Lichts angeregt worden; und mir scheint von allen Versuchen, welche zur Verwirklichung dieser Hoffnung gemacht worden sind, das von Herrn Maxwell gemachte am erfolgreichsten.'

cling with particular affection to this discovery. He felt that there was more in it than he was able to unfold. He predicted that it would grow in meaning with the growth of science. This it has done; this it is doing now. Its right interpretation will probably mark an epoch in scientific history.

Rapidly following it is the discovery of Diamagnetism, or the Repulsion of Matter by a magnet. Brugmans had shown that bismuth repelled a magnetic needle. Here he stopped. Le Bailliff proved that antimony did the same. Here he stopped. Seebeck, Becquerel, and others, also touched the discovery. These fragmentary gleams excited a momentary curiosity, and were almost forgotten, when Faraday, independently, alighted on the same facts; and, instead of stopping, made them the inlets to a new and vast region of research. The value of a discovery is to be measured by the intellectual action it calls forth; and it was Faraday's good fortune to strike such lodes of scientific truth as give some of the best intellects of the age occupation.

The salient quality of Faraday's scientific character reveals itself from beginning to end of these volumes: a union of ardour and patience—the one prompting the attack, the other holding him on to it till defeat was final or victory assured. Certainty in one sense or the other was necessary to his peace of mind. The right method of investigation is perhaps incommunicable; it depends on the individual rather than on the system, and the mark is missed when Faraday's researches are pointed to as merely illustrative of the power of the inductive philosophy. The brain may be filled with that philosophy, but without the energy and insight which this man possessed, and which with him were personal and distinctive, we should never rise to the level of his achievements. His power is that of individual genius, rather than of philosophic method; the energy of a strong soul expressing itself after its own fashion, and acknowledging no mediator between it and Nature.

The second volume of the ' Life and Letters,' like the first, is a historic treasury as regards Faraday's work and character, and his scientific and social relations. It contains letters from Humboldt, Herschel, Hachette, De la Rive, Dumas, Liebig, Melloni, Becquerel, Oersted, Plücker, Du Bois-Reymond, Lord Melbourne, Prince Louis Napoleon, and many other distinguished men. I notice with particular pleasure a letter from Sir John Herschel in reply to a sealed packet addressed to him by Faraday, but which he had permission to open if he pleased. The packet referred to one of the many unfulfilled hopes which spring up in the mind of fertile investigators :—

' Go on and prosper, " from strength to strength," like a victor marching with assured step to further conquests; and be certain that no voice will join more heartily in the peans that already begin to rise, and will speedily swell into a shout of triumph, astounding even to yourself, than that of J. F. W. Herschel.'

As an encourager of the scientific worker, this fine spirit is still active.

Faraday's behaviour to Melloni in 1835 merits a word of notice. The young man was a political exile in Paris. He had newly fashioned and applied the thermo-electric pile, and had obtained with it results of the greatest importance. But they were not appreciated. With the sickness of disappointed hope Melloni waited for the report of the Commissioners appointed by the Academy of Sciences to examine his labours. At length he published his researches in the 'Annales de Chimie.' They thus fell into the hands of Faraday, who, discerning at once their extraordinary merit, obtained for their author the Rumford Medal of the Royal Society. A sum of money always accompanies this medal, and the pecuniary help was at this time even more essential than the mark of honour to the young refugee. Melloni's gratitude was boundless :—

'Et vous, monsieur,' he writes to Faraday, 'qui appartenez à une société à laquelle je n'avais rien offert, vous qui me connaissiez à peine le nom ; vous n'avez pas demandé si j'avais des ennemis faibles ou puissants, in calculé quel en était le nombre ; mais vous avez parlé pour l'opprimé étranger, pour celui qui n'avait pas le moindre droit à tant de bienveillance, et vos paroles ont été accueillies favorablement par des collègues consciencieux ! Je reconnais bien là des hommes dignes de leur noble mission, les véritables représentants de la science d'un pays libre et généreux.'

Within the prescribed limits of this article it would be impossible to give even the slenderest summary of Faraday's correspondence, or to carve from it more than the merest fragments of his character. His letters, written to Lord Melbourne and others in 1836, regarding his pension, illustrate his uncompromising independence. The Prime Minister had offended him, but assuredly the apology demanded and given was complete. I think it certain that, notwithstanding the very full account of this transaction given by Dr. Bence Jones, motives and influences were at work which even now are not entirely revealed. The minister was bitterly attacked, but he bore the censure of the press with great dignity. Faraday, while he disavowed having either directly or indirectly furnished the matter of those attacks, did not publicly exonerate his lordship. The Hon. Caroline Fox had proved herself Faraday's ardent friend, and it was she who had healed the breach between the philosopher and the minister. She manifestly thought that Faraday ought to have come forward in Lord Melbourne's defence, and there is a flavour of resentment in one of her letters to him on the subject. No doubt Faraday had good grounds for his reticence, but they are to me unknown.

In 1841 his health broke down utterly, and he went to Switzerland with his wife and brother-in-law. His bodily vigour soon revived, and he accomplished feats of walking respectable even for a trained mountaineer. The published extracts from his Swiss journal contain many beautiful and touching allusions. Amid references to the tints of the Jungfrau, the blue rifts of the glaciers, and the noble Niesen towering over the Lake of Thun, we come upon the charming little scrap which I have elsewhere quoted :—' Clout-nail making goes on here rather considerably, and is a very neat and pretty operation to observe. I love a smith's shop and anything relating to smithery. My father was a smith.' This is from his journal; but he is unconsciously speaking to somebody —perhaps to the world.

His descriptions of the Staub-bach, Giessbach, and of the scenic effects of sky and mountain, are all fine and sympathetic. But amid it all, and in reference to it all, he tells his sister that ' true enjoyment is from within, not from without.'   In those days Agassiz was living under a slab of gneiss on the glacier of the Aar.   Faraday met Forbes at the Grimsel, and arranged with him an excursion to the 'Hôtel des Neuchatelois;' but indisposition put the project out.

From the Fort of Ham, in 1843, Faraday received a letter addressed to him by Prince Louis Napoleon Bonaparte.   He read this letter to me many years ago, and the desire, shown in various ways by the French Emperor, to turn modern science to account have often reminded me of it since.   At the age of thirty-five the prisoner of Ham speaks of 'rendering his captivity less sad by studying the great discoveries' which science owes to Faraday; and he asks a question which reveals his cast of thought at the time: 'What is the most simple combination to give to a voltaic battery, in order to produce a spark capable of setting fire to powder under water or under ground?'   Should the necessity arise, the French Emperor will not lack at the outset the best appliances of modern science; while we, I fear, shall have to learn the magnitude of the resources we are now neglecting amid the pangs of actual war.[1]

One turns with renewed pleasure to Faraday's letters to his wife, published in the second volume.   Here surely the loving essence of the man appears more distinctly than anywhere else.   From the house of Dr. Percy, in Birmingham, he writes thus :—

'Here—even here—the moment I leave the table I wish I were with you IN QUIET.   Oh, what happiness is

---

[1] The 'science' has since been applied with astonishing effect by those who had studied it far more thoroughly than the Emperor of the French.

ours! My runs into the world in this way only serve to make me esteem that happiness the more.'

And again—

'We have been to a grand conversazione in the town-hall, and I have now returned to my room to talk with you, as the pleasantest and happiest thing that I can do. Nothing rests me so much as communion with you. I feel it even now as I·write, and catch myself saying the words aloud as I write them.'

Take this, moreover, as indicative of his love for Nature :—

'After writing, I walk out in the evening hand in hand with my dear wife to enjoy the sunset ; for to me who love scenery, of all that I have seen or can see there is none surpasses that of heaven. A glorious sunset brings with it a thousand thoughts that delight me.'

Of the numberless lights thrown upon him by the 'Life and Letters,' some fall upon his religion. In a letter to a lady, he describes himself as belonging to ' a very small and despised sect of Christians, known, if known at all, as *Sandemanians*, and our hope is founded on the faith that is in Christ.' He adds, ' I do not think it at all necessary to tie the study of the natural sciences and religion together, and in my intercourse with my fellow-creatures, that which is religious, and that which is philosophical, have ever been two distinct things.' He saw clearly the danger of quitting his moorings, and his science became the safeguard of his particular faith. For his investigations so filled his mind as to leave no room for sceptical questionings, thus shielding from the assaults of philosophy the creed of his youth. His religion was constitutional and hereditary. It was implied in the eddies of his blood and in the tremors of his brain ; and however its outward and visible form might have changed, Faraday would still have possessed its elemental constituents—awe, reverence, truth, and love.

It is worth enquiring how so profoundly religious a mind, and so great a teacher, would be likely to regard our present discussions on the subject of education. Faraday would be a ' secularist' were he now alive. He had no sympathy with those who contemn knowledge unless it be accompanied by dogma. A lecture delivered before the City Philosophical Society in 1818, when he was twenty-six years of age, expresses the views regarding education which he entertained to the end of his life. ' First, then,' he says, ' all theological considerations are banished from the society, and of course from my remarks ; and whatever I may say has no reference to a future state, or to the means which are to be adopted in this world in anticipation of it. Next, I have no intention of substituting anything for religion, but I wish to take that part of human nature which is independent of it. Morality, philosophy, commerce, the various institutions and habits of society, are independent of religion, and may exist either with or without it. They are always the same, and can dwell alike in the breasts of those who from opinion are entirely opposed in the set of principles they include in the term religion, or in those who have none.

' To discriminate more closely, if possible, I will observe that we have *no* right to judge religious opinions, but the human nature of this evening is that part of man which we *have* a right to judge; and I think it will be found, on examination, that this humanity—as it may perhaps be called—will accord with what I have before described as being in our own hands so improvable and perfectible.'

Among my old papers I find the following remarks on one of my earliest dinners with Faraday. ' At 2 o'clock he came down for me. He, his niece, and myself, formed the party. " I never give dinners," he said. " I don't know how to give dinners, and I never dine out. But I should not like my friends to attribute this to a wrong cause. I act thus for the sake of securing time for work, and not through religious motives, as some imagine." He said grace. I am almost ashamed to call his prayer a " saying " of grace. In the language of Scripture, it might be described as the petition of a son, into whose heart God had sent the Spirit of His Son, and who with absolute trust asked a blessing from his father. We dined on roast beef, Yorkshire pudding, and potatoes ; drank sherry, talked of research and its requirements, and of his habit of keeping himself free from the distractions of society. He was bright and joyful—boylike, in fact, though he is now sixty-two. His work excites admiration, but contact with him warms and elevates the heart. Here, surely, is a strong man. I love strength, but let me not forget the example of its union with modesty, tenderness, and sweetness in the character of Faraday.'

Faraday's progress in discovery, and the salient points of his character, are well brought out by the wise choice of letters and extracts published in these volumes. I will not call the labours of the biographer final. So great a character will challenge reconstruction. In the coming time some sympathetic spirit, with the requisite strength, knowledge, and solvent power, will, I

doubt not, render these materials plastic, give them more
perfect organic form, and send through them, with less of
interruption, the currents of Faraday's life. 'He was too
good a man,' writes his present biographer, 'for me to
estimate rightly, and too great a philosopher for me to
understand thoroughly.'   That may be, but the reverent
affection to which we owe the discovery, selection, and
arrangement of the materials here placed before us, is
probably a surer guide than mere literary skill.   The
task of the artist who may wish in future times to re-
produce the real though unobtrusive grandeur, the purity,
beauty, and childlike simplicity of him whom we have
lost, will find his chief treasury already provided for him
by Dr. Bence Jones's labour of love.

XIII.

AN
ELEMENTARY LECTURE ON MAGNETISM.

ADDRESSED TO THE TEACHERS OF PRIMARY SCHOOLS AT THE
SOUTH KENSINGTON MUSEUM.

30th April, 1861.

'Next in order I will proceed to discuss by what law of nature it comes to pass that iron can be attracted by that stone which the Greeks call the Magnet from the name of its native place, because it has its origin within the bounds of the country of the Magnesians. This stone is more wondered at because it often produces a chain of [iron] rings hanging down from it. Thus you may see five and more suspended in succession and tossing about in the light airs, one always hanging from the other and attached to its lower side, and each in turn one from the other experiencing the binding power of the stone : with such a continued current its force flies through all.

'In things of this kind many things must be established before you can assign the true law of the thing in question, and it must be approached by a very circuitous road ; wherefore all the more I call for an attentive ear and mind.'—LUCRETIUS, *De Rerum Natura*, Lib. VI., Munro's Translation, p. 317.

This lecture is a plain statement of the elementary facts of magnetism, of one magnetic theory, and of the methods to be pursued in mastering both. It has already circulated among the teachers mentioned on its title-page, and I had some doubts as to the propriety of its insertion here. But on reading it, it seemed so likely to be helpful, that my scruples disappeared.

J. T.

MAGNETIC LINES OF FORCE.

From a Photograph by Professor MAYER, *Lehigh University, United States·*

XIII.

A LECTURE ON MAGNETISM.

WE have no reason to believe that the sheep or the dog, or indeed any of the lower animals, feel an interest in the laws by which natural phenomena are regulated. A herd may be terrified by a thunder-storm; birds may go to roost, and cattle return to their stalls during a solar eclipse; but neither birds nor cattle, as far as we know, ever think of enquiring into the causes of these things. It is otherwise with man. The presence of natural objects, the occurrence of natural events, the varied appearances of the universe in which he dwells, penetrate beyond his organs of sense, and appeal to an inner power of which the senses are the mere instruments and excitants. No fact is to him either final or original. He cannot limit himself to the contemplation of it alone, but endeavours to ascertain its position in a series to which the constitution of his mind assures him it must belong. He regards all that he witnesses in the present as the efflux and sequence of something that has gone before, and as the source of a system of events which is to follow. The notion of spontaneity, by which in his ruder state he accounted for natural events, is abandoned; the idea that nature is an aggregate of independent parts also disappears, as the connection and mutual dependence of physical powers become more and more manifest: until he is finally led, and that chiefly by the science of which

I happen this evening to be the exponent, to regard Nature as an organic whole, as a body each of whose members sympathises with the rest, changing, it is true, from ages to ages, but without one real break of continuity, or a single interruption of the fixed relations of cause and effect,

The system of things which we call Nature is, however, too vast and various to be studied first-hand by any single mind. As knowledge extends there is always a tendency to subdivide the field of investigation, its various parts being taken up by different individuals, and thus receiving a greater amount of attention than could possibly be bestowed on them if each investigator aimed at the mastery of the whole. East, west, north, and south, the human mind pushes its conquests; but the centripetal form in which knowledge, as a whole, advances, spreading ever wider on all sides, is due in reality to the exertions of individuals, each of whom directs his efforts, more or less, along a single line. Accepting, in many respects, his culture from his fellow-men, taking it from spoken words and from written books, in some one direction, the student of nature must actually touch his work. He may otherwise be a distributor of knowledge, but not a creator, and fails to attain that vitality of thought and correctness of judgment which direct and habitual contact with natural truth can alone impart.

One large department of the system of nature which forms the chief subject of my own studies, and to which it is my duty to call your attention this evening, is that of physics, or natural philosophy. This term is large enough to cover the study of nature generally, but it is usually restricted to a department which, perhaps, lies closer to our perceptions than any other. It deals with the phenomena and laws of light and heat—with the phenomena and laws of magnetism and electricity—with those of sound—with the pressures and motions of liquids and gases, whether in a state of translation or of undulation. The science of mechanics is a portion of natural philosophy, though at present so large as to need the exclusive attention of him who would cultivate it profoundly. Astronomy is the application of physics to the motions of the heavenly bodies, the vastness of the field causing it, however, to be regarded as a department in itself. In chemistry physical agents play important parts. By heat and light we cause bodies to combine, and by heat and light we decompose them. Electricity tears asunder the locked atoms of compounds, through their power of separating carbonic acid into its constituents; the solar beams build up the whole vegetable world, and by it the animal, while the touch of the selfsame beams causes hydrogen and chlorine to unite with sudden explosion and form by their combination a powerful acid. Thus physics and chemistry intermingle, physical agents being employed by the chemist as a means to an end; while in physics proper the laws and phenomena of the agents themselves, both qualitative and quantitative, are the primary objects of attention.

My duty here to-night is to spend an hour in telling how this subject is to be studied, and how a knowledge of it is to be imparted to others. When first invited to do this, I hesitated before accepting the responsibility. It would be easy to entertain you with an account of what natural philosophy has accomplished. I might point to those applications of science regarding which from without.' In those days Agassiz was living under a slab of gneiss on the glacier of the Aar. Faraday met Forbes at the Grimsel, and arranged with him an excursion to the 'Hôtel des Neuchatelois;' but indisposition put the project out.

From the Fort of Ham, in 1843, Faraday received a letter addressed to him by Prince Louis Napoleon Bonaparte. He read this letter to me many years ago, and the desire, shown in various ways by the French Emperor, to turn modern science to account have often reminded me of it since. At the age of thirty-five the prisoner of Ham speaks of 'rendering his captivity less sad by studying the great discoveries' which science owes to Faraday; and he asks a question which reveals his cast of thought at the time : 'What is the most simple combination to give to a voltaic battery, in order to produce a spark capable of setting fire to powder under water or under ground?' Should the necessity arise, the French Emperor will not lack at the outset the best appliances of modern science; while we, I fear, shall have to learn the magnitude of the resources we are now neglecting amid the pangs of actual war.[1]

One turns with renewed pleasure to Faraday's letters to his wife, published in the second volume. Here surely the loving essence of the man appears more distinctly than anywhere else. From the house of Dr. Percy, in Birmingham, he writes thus :—

'Here—even here—the moment I leave the table I wish I were with you IN QUIET. Oh, what happiness is

---

[1] The 'science' has since been applied with astonishing effect by those who had studied it far more thoroughly than the Emperor of the French.

ours! My runs into the world in this way only serve to make me esteem that happiness the more.'

And again—

'We have been to a grand conversazione in the town-hall, and I have now returned to my room to talk with you, as the pleasantest and happiest thing that I can do. Nothing rests me so much as communion with you. I feel it even now as I write, and catch myself saying the words aloud as I write them.'

Take this, moreover, as indicative of his love for Nature :—

'After writing, I walk out in the evening hand in hand with my dear wife to enjoy the sunset; for to me who love scenery, of all that I have seen or can see there is none surpasses that of heaven. A glorious sunset brings with it a thousand thoughts that delight me.'

Of the numberless lights thrown upon him by the 'Life and Letters,' some fall upon his religion. In a letter to a lady, he describes himself as belonging to 'a very small and despised sect of Christians, known, if known at all, as *Sandemanians*, and our hope is founded on the faith that is in Christ.' He adds, 'I do not think it at all necessary to tie the study of the natural sciences and religion together, and in my intercourse with my fellow-creatures, that which is religious, and that which is philosophical, have ever been two distinct things.' He saw clearly the danger of quitting his moorings, and his science became the safeguard of his particular faith. For his investigations so filled his mind as to leave no room for sceptical questionings, thus shielding from the assaults of philosophy the creed of his youth. His religion was constitutional and hereditary. It was implied in the

... and in the presence of ...
... and visible from ...
... still hope pursued its ...
... reverence, truth, and love.
... inquiring how so profound ...
... a teacher, would be ...
... on the subject ...
... "socialist" were he ...
... with those who content ...
... by dogma. A lec...
... Philosophical Society in 1...
... of age, express the v...
... be entrusted to the en...
... all theological con...
... the society, and of cou...
... whatever I may say has no r...
... the means which are to b...
... emancipation of it. Next, I h...
... anything for religion, b...
... of human nature which is in...
... philosophy, commerce, the v...
... of society, are independen...
... either with or without i...
... and can dwell alike in t...
... opinion are entirely oppose...
... include in the term religio...
...

... more closely, if possible, ...
... right to judge religious opi...
... of this evening is that part ...
... to judge; and I think it w...
... on, that this humanity—as it ...

be called—will accord with what I have before described
as being in our own hands so improvable and perfectible.'

Among my old papers I find the following remarks on
one of my earliest dinners with Faraday. 'At 2 o'clock
he came down for me. He, his niece, and myself, formed
the party. "I never give dinners," he said. "I don't
know how to give dinners, and I never dine out. But I
should not like my friends to attribute this to a wrong
cause. I act thus for the sake of securing time for work,
and not through religious motives, as some imagine." He
said grace. I am almost ashamed to call his prayer a
" saying " of grace. In the language of Scripture, it
might be described as the petition of a son, into whose
heart God had sent the Spirit of His Son, and who with
absolute trust asked a blessing from his father. We
dined on roast beef, Yorkshire pudding, and potatoes ;
drank sherry, talked of research and its requirements,
and of his habit of keeping himself free from the distractions of society. He was bright and joyful—boylike,
in fact, though he is now sixty-two. His work excites
admiration, but contact with him warms and elevates the
heart. Here, surely, is a strong man. I love strength,
but let me not forget the example of its union with
modesty, tenderness, and sweetness in the character of
Faraday.'

Faraday's progress in discovery, and the salient points
of his character, are well brought out by the wise
choice of letters and extracts published in these volumes.
I will not call the labours of the biographer final.
So great a character will challenge reconstruction. In
the coming time some sympathetic spirit, with the
requisite strength, knowledge, and solvent power, will, I

doubt not, render these materials plastic, give them more perfect organic form, and send through them, with less of interruption, the currents of Faraday's life. 'He was too good a man,' writes his present biographer, 'for me to estimate rightly, and too great a philosopher for me to understand thoroughly.' That may be, but the reverent affection to which we owe the discovery, selection, and arrangement of the materials here placed before us, is probably a surer guide than mere literary skill. The task of the artist who may wish in future times to re-produce the real though unobtrusive grandeur, the purity, beauty, and childlike simplicity of him whom we have lost, will find his chief treasury already provided for him by Dr. Bence Jones's labour of love.

XIII.

AN
ELEMENTARY LECTURE ON MAGNETISM.

ADDRESSED TO THE TEACHERS OF PRIMARY SCHOOLS AT THE
SOUTH KENSINGTON MUSEUM.

30th April, 1861.

...nature it come
...words all the
...origin within
...more wondered
...own from it
...ing about
...accorded to its
...the binding
...this through

...before you can
...reproached by a
...native ear and
...s Translation,

...production of
...serving both
...title-page,
...re. But on
...appeared
J. T.

I happen this evening to be the exponent, to regard Nature as an organic whole, as a body each of whose members sympathises with the rest, changing, it is true, from ages to ages, but without one real break of continuity, or a single interruption of the fixed relations of cause and effect,

The system of things which we call Nature is, however, too vast and various to be studied first-hand by any single mind. As knowledge extends there is always a tendency to subdivide the field of investigation, its various parts being taken up by different individuals, and thus receiving a greater amount of attention than could possibly be bestowed on them if each investigator aimed at the mastery of the whole. East, west, north, and south, the human mind pushes its conquests; but the centripetal form in which knowledge, as a whole, advances, spreading ever wider on all sides, is due in reality to the exertions of individuals, each of whom directs his efforts, more or less, along a single line. Accepting, in many respects, his culture from his fellow-men, taking it from spoken words and from written books, in some one direction, the student of nature must actually touch his work. He may otherwise be a distributor of knowledge, but not a creator, and fails to attain that vitality of thought and correctness of judgment which direct and habitual contact with natural truth can alone impart.

One large department of the system of nature which forms the chief subject of my own studies, and to which it is my duty to call your attention this evening, is that of physics, or natural philosophy. This term is large enough to cover the study of nature generally, but it is usually restricted to a department which, perhaps, lies closer to our perceptions than any other. It deals with the phenomena and laws of light and heat—with the phenomena and laws of magnetism and electricity—with those of sound—with the pressures and motions of liquids and gases, whether in a state of translation or of undulation. The science of mechanics is a portion of natural philosophy, though at present so large as to need the exclusive attention of him who would cultivate it profoundly. Astronomy is the application of physics to the motions of the heavenly bodies, the vastness of the field causing it, however, to be regarded as a department in itself. In chemistry physical agents play important parts. By heat and light we cause bodies to combine, and by heat and light we decompose them. Electricity tears asunder the locked atoms of compounds, through their power of separating carbonic acid into its constituents; the solar beams build up the whole vegetable world, and by it the animal, while the touch of the selfsame beams causes hydrogen and chlorine to unite with sudden explosion and form by their combination a powerful acid. Thus physics and chemistry intermingle, physical agents being employed by the chemist as a means to an end; while in physics proper the laws and phenomena of the agents themselves, both qualitative and quantitative, are the primary objects of attention.

My duty here to-night is to spend an hour in telling how this subject is to be studied, and how a knowledge of it is to be imparted to others. When first invited to do this, I hesitated before accepting the responsibility. It would be easy to entertain you with an account of what natural philosophy has accomplished. I might point to those applications of science regarding which we hear so much in the newspapers, and which we often find mistaken for science itself. I might, of course, ring changes on the steam engine and the telegraph, the electrotype and the photograph, the medical applications of physics, and the million other inlets by which scientific thought filters into practical life. That would be easy compared with the task of informing you how you are to make the study of physics the instrument of your own culture, how you are to possess its facts and make them living seeds which shall take root and grow in the mind, and not lie like dead lumber in the store-house of memory. This is a task much heavier than the mere cataloguing of scientific achievements; and it is one which, feeling my own want of time and power to execute it aright, I might well hesitate to accept.

But let me sink excuses, and attack the work to the best of my ability. First and foremost, then, I would advise you to get a knowledge of facts from actual observation. Facts looked at directly are vital; when they pass into words half the sap is taken out of them. You wish, for example, to get a knowledge of magnetism; well, provide yourself with a good book on the subject, if you can, but do not be content with what the book tells you; do not be satisfied with its descriptive woodcuts; see the actual thing yourself. Half of our book writers describe experiments which they never made, and their descriptions often lack both force and truth; but no matter how clever or conscientious they may be, their written words cannot supply the place of actual observation. Every fact has numerous radiations, which are shorn off by the man who describes it. Go, then, to a philosophical instrument maker, and give, according to your means, for a straight bar-magnet, say, half-a-crown, or, if you can afford it, five shillings for a pair of them; or get a smith to cut a length of ten inches from a bar of steel an inch wide and half an inch thick; file its ends decently, harden it, and get somebody like myself to magnetise it. Two bar-magnets are better than one. Procure some darning needles such as these. Provide yourself also with a little unspun silk; which will give you a suspending fibre void of torsion; make a little loop of paper or of wire, thus, and attach your fibre to it. Do it neatly. In the loop place your darning needle, and bring the two ends or poles, as they are called, of your magnet successively up to either end of the needle. Both the poles, you find, attract both ends of the needle. Replace the needle by a bit of annealed iron wire, the same effects ensue. Suspend successively little rods of lead, copper, silver, or brass, of wood, glass, ivory, or whalebone; the magnet produces no sensible effect upon any of these substances. You thence infer a special property in the case of steel and iron. Multiply your experiments, however, and you will find that some other substances besides iron are acted upon by your magnet. A rod of the metal nickel, or of the metal cobalt, from which the blue colour used by painters is derived, exhibits powers similar to those observed with the iron and steel.

In studying the character of the force you may, however, confine yourself to iron and steel, which are always at hand. Make your experiments with the darning needle over and over again; operate on both ends of the needle; try both ends of the magnet. Do not think the work stupid; you are conversing with Nature, and must acquire a certain grace and mastery over her language;

and these practice can alone impart. Let every movement be made with care, and avoid slovenliness from the onset. In every one of your experiments endeavour to feel the responsibility of a moral agent. Experiment, as I have said, is the language by which we address Nature, and through which she sends her replies; in the use of this language a lack of straightforwardness is as possible and as prejudicial as in the spoken language of the tongue. If you wish to become acquainted with the truth of Nature you must from the first resolve to deal with her sincerely.

Now remove your needle from its loop, and draw it from end to end along one of the ends of the magnet; resuspend it, and repeat your former experiment. You find the result different. You now find that each extremity of the magnet attracts one end of the needle, and repels the other. The simple attraction observed in the first instance is now replaced by a *dual* force. Repeat the experiment till you have thoroughly observed the ends which attract and those which repel each other.

Withdraw the magnet entirely from the vicinity of your needle, and leave the latter freely suspended by its fibre. Shelter it as well as you can from currents of air, and if you have iron buttons on your coat or a steel penknife in your pocket, beware of their action. If you work at night, beware of iron candlesticks, or of brass ones with iron rods inside. Freed from such disturbances, the needle takes up a certain determinate position. It sets its length nearly north and south. Draw it aside from this position and let it go. After several oscillations it will again come to it. If you have obtained your magnet from a philosophical instrument maker, you will see a mark on one of its ends. Supposing, then, that you drew your needle along the end thus marked, and that the eye-end of your needle was the last to quit the magnet, you will find that the eye turns to the south, the point of the needle turning towards the north. Make sure of this, and do not take this statement on my authority.

Now take a second darning needle like the first, and magnetise it in precisely the same manner : freely suspended it also will turn its point to the north and its eye to the south. Your next step is to examine the action of the two needles which you have thus magnetised upon each other.

Take one of them in your hand, and leave the other suspended ; bring the eye-end of the former near the eye-end of the latter ; the suspended needle retreats : it is repelled. Make the same experiment with the two points, you obtain the same result, the suspended needle is repelled. Now cause the dissimilar ends to act on each other—you have attraction—point attracts eye and eye attracts point. Prove the reciprocity of this action by removing the suspended needle, and putting the other in its place. You obtain the same result. The attraction then, is mutual, and the repulsion is mutual, and you have thus demonstrated in the clearest manner the fundamental law of magnetism, that like poles repel, and that unlike poles attract each other. You may say that this is all easily understood without doing ; but *do it*, and your knowledge will not be confined to what I have uttered here.

I have said that one end of your magnet has a mark upon it ; lay several silk fibres together, so as to get sufficient strength, or employ a thin silk ribbon, and form a loop large enough to hold your magnet. Suspend it ;

it turns its marked end towards the north. This marked end is that which in England is called the north pole. If a common smith has made your magnet, it will be convenient to determine its north pole yourself, and to mark it with a file. You vary your experiments by causing your magnetised darning needle to attract and repel your large magnet; it is quite competent to do so. In magnetising the needle, I have supposed the eye-end to be the last to quit the marked end of the magnet; that end of the needle is a south pole. The end which last quits the magnet is always opposed in polarity to the end of the magnet with which it has been in contact. Brought near each other they mutually attract, and thus demonstrate that they are unlike poles.

You may perhaps learn all this in a single hour; but spend several at it, if necessary; and remember understanding it is not sufficient: you must obtain a manual aptitude in addressing Nature. If you speak to your fellow-man you are not entitled to use jargon. Bad experiments are jargon addressed to Nature, and just as much to be deprecated. A manual dexterity in illustrating the interaction of magnetic poles is of the utmost importance at this stage of your progress; and you must not neglect attaining this power over your implements. As you proceed, moreover, you will be tempted to do more than I can possibly suggest. Thoughts will occur to you which you will endeavour to follow out; questions will arise which you will try to answer. The same experiment may be twenty things to twenty people. Having witnessed the action of pole on pole through the air, you will perhaps try whether the magnetic power is not to be screened off. You use plates of glass, wood, slate, pasteboard, or gutta-percha, but find them all pervious to this wondrous force. One magnetic pole acts upon another through these bodies as if they were not present. And should you become a patentee for the regulation of ships' compasses, you will not fall, as some projectors have done, into the error of screening off the magnetism of the ship by the interposition of such substances.

If you wish to teach a class you must contrive that the effects which you have thus far witnessed for yourself shall be witnessed by twenty or thirty pupils. And here your private ingenuity must come into play. You will attach bits of paper to your needles, so as to render their movements visible at a distance, denoting the north and south poles by different colours, say green and red. You may also improve upon your darning needle. Take a strip of sheet steel, the rib of a lady's stays will answer, heat it to vivid redness and plunge it into cold water. It is thereby hardened, rendered, in fact, almost as brittle as glass. Six inches of this, magnetised in the manner of the darning needle, will be better able to carry your paper indexes. Having secured such a strip, you proceed thus :—

Magnetise a small sewing needle and determine its poles ; or, break half an inch or an inch off your magnetised darning needle and suspend it by a fine silk fibre. The sewing needle or the fragment of the darning needle is now to be used as a test needle to examine the distribution of the magnetism in your strip of steel. Hold the strip upright in your left hand, and cause the test needle to approach the lower end of your strip ; one end is attracted, the other is repelled. Raise your needle along the strip; its oscillations, which at first were quick, become slower; opposite the middle of the strip they cease entirely; neither end of the needle is attracted; above the middle the test needle turns suddenly round, its other end being now attracted. Go through the experiment thoroughly; you thus learn that the entire lower half of the strip attracts one end of the needle, while the entire upper half attracts the opposite end. Supposing the north end of your little needle to be that attracted below, you infer that the entire lower half of your magnetised strip exhibits south magnetism, while the entire upper half exhibits north magnetism. So far, then, you have determined the distribution of magnetism in your strip of steel. .

You look at this fact, you think of it; in its suggestiveness the value of the experiment chiefly consists. The thought arises, 'What will occur if I break my strip of steel across in the middle? Shall I obtain two magnets each possessing a single pole?' Try the experiment; break your strip of steel, and test each half as you tested the whole. The mere presentation of its two ends in succession to your test needle suffices to show you that you have *not* a magnet with a single pole, that each half possesses two poles with a neutral point between them. And if you again break the half into two other halves, you will find that each quarter of the original strip exhibits precisely the same magnetic distribution as the strip itself. You may continue the breaking process: no matter how small your fragment may be, it still possesses two opposite poles and a neutral point between them. Well, your hand ceases to break where breaking becomes a mechanical impossibility; but does the mind stop there? No: you follow the breaking process in idea when you can no longer realise it in fact; your thoughts wander amid the very atoms of your steel, and you conclude that each atom is a magnet, and that the force exerted by the strip of steel is the mere summation or resultant of the forces of its ultimate particles.

Here, then, is an exhibition of power which we can call forth or cause to disappear at pleasure. We magnetise our strip of steel by drawing it along the pole of a magnet; we can demagnetise it, or reverse its magnetism, by properly drawing it along the same pole in the opposite direction. What, then, is the real nature of this wondrous change? What is it that takes place among the atoms of the steel when the substance is magnetised? The question leads us beyond the region of sense, and into that of imagination. This faculty, indeed, is the divining rod of the man of science. Not, however, an imagination which catches its creations from the air, but one informed and inspired by facts, capable of seizing firmly on a physical image as a principle, of discerning its consequences, and of devising means whereby these forecasts of thought may be brought to an experimental test. If such a principle be adequate to account for all the phenomena, if from an assumed cause the observed facts necessarily follow, we call the assumption a theory, and once possessing it, we can not only revive at pleasure facts already known, but we can predict others which we have never seen. Thus, then, in the prosecution of physical science, our powers of observation, memory, imagination, and inference, are all drawn upon. We observe facts and store them up; imagination broods upon these memories, and by the aid of reason tries to discern their interdependence. The theoretic principle flashes, or slowly dawns upon the mind, and then the deductive faculty interposes to carry out the principle to its logical consequences. A perfect theory gives dominion over natural facts; and even an assumption which can only partially stand the test of a comparison with facts, may be of eminent use in enabling us to connect and classify groups of phenomena. The theory of magnetic fluids is of this latter character, and with it we must now make ourselves familiar.

With the view of stamping the thing more firmly on your minds, I will make use of a strong and vivid image. In optics, red and green are called complementary colours; their mixture produces *white*. Now I ask you to imagine each of these colours to possess a self-repulsive power; that red repels red, and that green repels green; but that red attracts green and green attracts red, the attraction of the dissimilar colours being equal to the repulsion of the similar ones. Imagine the two colours mixed so as to produce white, and suppose two strips of wood painted with this white; what will be their action upon each other? Suspend one of them freely as we suspended our darning needle, and bring the other near it; what will occur? The red component of the strip you hold in your hand will repel the red component of your suspended strip, but then it will attract the green; and the forces being equal they neutralise each other. In fact, the least reflection shows you that the strips will be as indifferent to each other as two unmagnetised darning needles would be under the same circumstances.

But suppose, instead of mixing the colours, we painted one half of each strip from centre to end red, and the other half green, it is perfectly manifest that the two strips would now behave towards each other exactly as our two magnetised darning needles—the red end would repel the red and attract the green, the green would repel the green and attract the red; so that, assuming two colours thus related to each other, we could by their mixture produce the neutrality of an unmagnetised body, while by their separation we could produce the duality of action of magnetised bodies.

But you have already anticipated a defect in my conception; for if we break one of our strips of wood in the middle we have one half entirely red and the other entirely green, and with these it would be impossible to imitate the action of our broken magnet. How then must we modify our conception? We must evidently suppose *each atom of wood* painted green on one face and red on the opposite one. If this were done the resultant action of all the atoms would exactly resemble the action of a magnet. Here also, if the two opposite colours of each atom could be caused to mix so as to produce white, we should have, as before, perfect neutrality.

Substitute in your minds for these two self-repellant and mutually attractive colours two invisible self-repellant and mutually attractive fluids, which in ordinary steel are mixed to form a neutral compound, but which the act of magnetisation separates from each other, placing the opposite fluids on the opposite faces of each atom, and you have a perfectly distinct conception of the celebrated theory of magnetic fluids. The strength of the magnetism excited is supposed to be proportional to the quantity of neutral fluid decomposed. According to this theory nothing is actually transferred from the exciting magnet to the excited steel. The act of magnetisation consists in the forcible separation of two powers which existed in the steel before it was magnetised, but which then neutralised each other by their coalescence. And if you test your magnet after it has excited a hundred pieces of steel, you will find that it has lost no force—no more, indeed, than I should lose had my words such a magnetic influence on your minds, as to excite in them a strong resolve to study natural philosophy. I should, in fact, be the gainer by my own utterance and by the reaction of your strength; and so also the magnet is the gainer by the reaction of the body which it magnetises.

Look now to your excited piece of steel; figure each atom to your minds with its opposed fluids spread over its opposite faces. How can this state of things be permanent? The fluids, by hypothesis, attract each other; what, then, keeps them apart? Why do they not instantly rush together across the equator of the atom, and thus neutralise each other? To meet this question philosophers have been obliged to infer the existence of a special force which holds the fluids asunder. They call it *coercive force*; and it is found that those kinds of steel which offer most resistance to being magnetised, which require the greatest amount of coercion to tear their fluids asunder, are the very ones which offer the greatest resistance to the re-union of the fluids after they have been once separated. Such kinds of steel are most suited to the formation of *permanent* magnets. It is manifest, indeed, that without coercive force a permanent magnet would not be at all possible.

You have not forgotten that previous to magnetising your darning needle *both* its ends were attracted by your magnet ; and that both ends of your bit of iron wire were acted upon in the same way. Probably also long before this you will have dipped the end of your magnet among iron filings, and observed how they cling to it, or into a nailbox, and found how it drags the nails after it. I know very well that if you are not the slaves of routine, you will have by this time done many things that I have not told you to do, and thus multiplied your experience beyond what I have indicated. You are almost sure to have caused a bit of iron to hang from the end of your magnet, and you have probably succeeded in causing a second piece to attach itself to the first, a third to the second ; until finally the force has become too feeble to bear the weight of more. If you have operated with nails you may have observed that the points and edges hold together with the greatest tenacity ; and that a bit of iron clings more firmly to the corner of your magnet than to one of its flat surfaces. In short, you will in all likelihood have enriched your experience in many ways without any special direction from me.

Well, the magnet attracts the nail, and that nail attracts a second one. This proves that the nail in contact with the magnet has had the magnetic quality developed in it by that contact. If it be withdrawn from the magnet its power to attract its fellow nail ceases. Contact, however, is not necessary. A sheet of glass or paper, or a space of air may exist between the magnet and the nail ; the latter is still magnetised, though not so forcibly as when in actual contact. The nail then presented to the magnet is itself a temporary magnet. That end which is turned towards the magnetic pole has the opposite magnetism of the pole which excites it ; the end most remote from the pole has the same magnetism as the pole itself, and between the two poles the nail, like the magnet, possesses a magnetic equator.

Conversant as you now are with the theory of magnetic fluids, you have already, I doubt not, anticipated me in imagining the exact condition of the iron under the influence of the magnet. You picture the iron as possessing the neutral fluid in abundance, you picture the magnetic pole, when brought near, decomposing the fluid; repelling the fluid of a like kind with itself, and attracting the unlike fluid; thus exciting in the parts of the iron nearest to itself the opposite polarity. But the iron is incapable of becoming a permanent magnet. It only shows its virtue as long as the magnet acts upon it. What then does the iron lack which the steel possesses? It lacks coercive force. Its fluids are separated with ease, but once the separating cause is removed, they flow together again and neutrality is restored. Your imagination must be quite nimble in picturing these changes. You must be able to see the fluids dividing and reuniting according as the magnet is brought near or withdrawn. Fixing a definite pole in your imagination you must picture the precise arrangement of the two fluids with reference to this pole. And you must not only be well drilled in the use of this mental imagery yourself, but you must be able to arouse the same pictures in the minds of your pupils, and satisfy yourself that they possess this power of placing actually before themselves magnets and iron in various positions, and describing the exact magnetic state of the iron in each particular case. The mere facts of magnetism will have their interest immensely augmented by an acquaintance with those hidden principles whereon the facts depend. Still, while you use this theory of magnetic fluids to track out the phenomena and link them together, be sure to tell your pupils that it is to be regarded as a symbol merely,—a symbol, moreover, which is incompetent to cover all the facts,[1] but which does good practical service whilst we are waiting for the actual truth.

This state of excitement into which the soft iron is thrown by the influence of the magnet, is sometimes called 'magnetisation by influence.' More commonly, however, the magnetism is said to be 'induced' in the soft iron, and hence this way of magnetising is called 'magnetic induction.' Now, there is nothing theoretically perfect in nature : there is no iron so soft as not to possess a certain amount of coercive force, and no steel so hard as not to be capable, in some degree, of magnetic induction. The quality of steel is in some measure possessed by iron, and the quality of iron is shared in some degree by steel. It is in virtue of this latter fact that the unmagnetised darning needle was attracted in your first experiment ; and from this you may at once deduce the consequence that after the steel has been magnetised, the repulsive action of a magnet must be always less than its attractive action. For the repulsion is opposed by the inductive action of the magnet on the steel, while the attraction is assisted by the same inductive action. Make this clear to your minds, and verify it by your experiments. In some cases you can actually make the attraction due to the temporary

---

[1] This theory breaks down when applied to diamagnetic bodies, which are repelled by magnets. Like soft iron, such bodies are thrown into a state of temporary excitement in virtue of which they are repelled, but any attempt to explain such a repulsion by the decomposition of a fluid will demonstrate its own futility.

magnetism overbalance the repulsion due to the permanent magnetism, and thus cause two poles of the same
kind apparently to attract each other. When, however,
good hard magnets act on each other from a sufficient
distance, the inductive action practically vanishes, and the
repulsion of like poles is sensibly equal to the attraction
of unlike ones.

I dwell thus long on elementary principles, because
they are of the first importance, and it is the temptation
of this age of unhealthy cramming to neglect them. Now
follow me a little further. In examining the distribution
of magnetism in your strip of steel you raised the needle
slowly from bottom to top, and found what we called a
neutral point at the centre. Now does the magnet really
exert no influence on the pole presented to its centre?
Let us see.

FIG. 1.

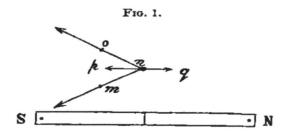

Let s n, fig. 1, be your magnet and let *n* represent a
particle of north magnetism placed exactly opposite the
middle of the magnet. Of course this is an imaginary
case, as you can never in reality thus detach your north
magnetism from its neighbour. What is the action of the
two poles of the magnet on *n*? Your reply will of course
be that the pole s attracts *n* while the pole N repels it. Let
the magnitude and direction of the attraction be expressed

by the line *n m*, and the magnitude and direction of the repulsion by the line *n o*. Now the particle *n* being equally distant from s and N, the line *n o*, expressing the repulsion, will be equal to *m n*, which expresses the attraction, and the particle *n*, acted upon by two such forces, must evidently move in the direction *p n*, exactly midway between *m n* and *n o*. Hence you see that although there is no tendency of the particle *n* to move towards the magnetic equator, there is a tendency on its part to move parallel to the magnet. If instead of a particle of north magnetism we placed a particle of south magnetism opposite to the magnetic equator, it would evidently be urged along the line *n q*; and if instead of two separate particles of magnetism we place a little magnetic needle, containing both north and south magnetism, opposite the magnetic equator, its south pole being urged along *n q*, and its north along *n p*, the little needle will be compelled to set itself parallel to the magnet s N. Make the experiment and satisfy yourselves that this is the case.

Substitute for your magnetic needle a bit of iron wire, devoid of permanent magnetism, and it will set itself exactly as the needle does. Acted upon by the magnet, the wire, as you know, becomes a magnet and behaves as such; it will, of course, turn its north pole towards *p*, and south pole towards *q*, just like the needle.

But supposing you shift the position of your particle of north magnetism, and bring it nearer to one end of your magnet than to the other, the forces acting on the particle are no longer equal; the nearest pole of the magnet will act more powerfully on the particle than the more distant one. Let s N, fig. 2, be the magnet and *n* the particle of north magnetism in its new position. Well, it is repelled by N, and attracted by s. Let the repulsion be represented in magnitude and direction by the line *n o*, and the attraction by the shorter line *n m*. The resultant of these two forces will be found by completing the parallelogram *m n o p*, and drawing its diagonal *n p*. Along *n p*, then, a particle of north magnetism would be urged by the simultaneous action of s and N. Substituting a

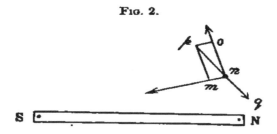

Fig. 2.

particle of south magnetism for *n*, the same reasoning would lead to the conclusion that the particle would be urged along *n q*, and if we place at *n* a short magnetic needle, its north pole will be urged along *n p*, its south pole along *n q*, and the only position possible to the needle, thus acted on, is along the line *p q*, which, as you see, is no longer parallel to the magnet. Verify this by actual experiment.

In this way we might go round the entire magnet, and considering its two poles as two centres from which the force emanates, we could, in accordance with ordinary mechanical principles, assign a definite direction to the magnetic needle at every particular place. And substituting, as before, a bit of iron wire for the magnetic needle, the positions of both will be the same.

Now, I think, without further preface, you will be able to comprehend for yourselves, and explain to others, one of the most interesting effects in the whole domain of magnetism. Iron filings you know are particles of iron, irregular in shape, being longer in some directions than in others. For the present experiment, moreover, instead of the iron filings, very small scraps of thin iron wire might be employed. I place a sheet of paper over the magnet; it is all the better if the paper be stretched on a wooden frame, as this enables us to keep it quite level. I scatter the filings, or the scraps of wire from a sieve upon the paper, and tap the latter gently, so as to liberate the particles for a moment from its friction. The magnet acts on the filings through the paper, and see how it arranges them! They embrace the magnet in a series of beautiful curves, which are technically called magnetic curves, or lines of magnetic force. Does the meaning of these lines yet flash upon you? Set your magnetic needle or your suspended bit of wire at any point of one of the curves and you will find the direction of the needle or of the wire to be exactly that of the particle of iron, or of the magnetic curve at the point. Go round and round the magnet; the direction of your needle always coincides with the direction of the curve on which it is placed. These, then, are the lines along which a particle of south magnetism, if you could detach it, would move to the north pole, and a bit of north magnetism to the south pole; they are the lines along which the decomposition of the neutral fluid takes place, and in the case of the magnetic needle, one of its poles being urged in one direction, and the other pole in the opposite direction, the needle must necessarily set itself as a *tangent* to the curve. I will not seek to simplify this subject further. If there be anything obscure or confused

or incomplete in my statement, you ought now, by patient
thought, to be able to clear away the obscurity, to reduce
the confusion to order, and to supply what is needed to
render the explanation complete. Do not quit the subject
until you thoroughly understand it ; and if you are able
to look with your mind's eye at the play of forces around
a magnet, and see distinctly the operation of those forces
in the production of the magnetic curves, the time which
we have spent together has not been spent in vain.

In this thorough manner we must master our materials,
reason upon them, and by determined study, attain to
clearness of conception. Facts thus dealt with exercise
an expansive force upon the boundaries of thought ;—
they widen the mind to generalisation. We soon re-
cognise a brotherhood between the larger phenomena of
nature and the minute effects which we have observed
in our private chambers. Why, we enquire, does the
magnetic needle set north and south? Evidently it is
compelled to do so by the earth ; the great globe which
we inherit is itself a magnet. Let us learn a little more
about it. By means of a bit of wax or otherwise attach
your silk fibre to your magnetic needle by a single point
at its middle, the needle will thus be uninterfered with by
the paper loop, and will enjoy to some extent a power of
dipping its point or its eye below the horizon. Lay your
magnet on a table, and hold the needle over the equator
of the magnet. The needle sets horizontal. Move it
towards the north end of the magnet ; the south end of
the needle dips, the dip augmenting as you approach
the north pole, over which the needle, if free to move,
will set itself exactly vertical. Move it back to the
centre, it resumes its horizontality ; pass it on towards

the south pole, its north end now dips, and directly over the south pole the needle becomes vertical, its north end being now turned downwards. Thus we learn that on the one side of the magnetic equator the north end of the needle dips; on the other side the south end dips, the dip varying from nothing to ninety degrees. If we go to the equatorial regions of the earth with a suitably suspended needle we shall find there the position of the needle horizontal. If we sail north one end of the needle dips; if we sail south the opposite end dips; and over the north or south terrestrial magnetic pole the needle sets vertical. The south magnetic pole has not yet been found, but Sir James Ross discovered the north magnetic pole on the 1st of June, 1831. In this manner we establish a complete parallelism between the action of the earth and that of an ordinary magnet.

The terrestrial magnetic poles do not coincide with the geographical ones; nor does the earth's magnetic equator quite coincide with the geographical equator. The direction of the magnetic needle in London, which is called the magnetic meridian, incloses an angle of 24 degrees with the true astronomical meridian, this angle being called the Declination of the needle for London. The north pole of the needle now lies to the west of the true meridian; the declination is westerly. In the year 1660, however, the declination was nothing, while before that time it was easterly. All this proves that the earth's magnetic constituents are gradually changing their distribution. This change is very slow; it is technically called the *secular change*, and the observation of it has not yet extended over a sufficient period of time to enable us to guess, even approximately, at its laws.

Having thus discovered, to some extent, the secret of the earth's power, we can turn it to account. I hold in my hand a poker formed of good soft iron ; it is now in the line of dip, a tangent, in fact, to the earth's line of magnetic force. The earth, acting as a magnet, is at this moment constraining the two fluids of the poker to separate, making the lower end of the poker a north pole, and the upper end a south pole. Mark the experiment : —I hold the knob uppermost, and it attracts the north end of a magnetic needle. I now reverse the poker, bringing its knob undermost ; the knob is now a north pole and attracts the south end of a magnetic needle. Get such a poker and carefully repeat this experiment ; satisfy yourselves that the fluids shift their position according to the manner in which the poker is presented to the earth. It has already been stated that the softest iron possesses a certain amount of coercive force. The earth, at this moment, finds in this force an antagonist which opposes the full decomposition of the neutral fluid. The component fluids may be figured as meeting an amount of friction, or possessing an amount of adhesion, which prevents them from gliding over the atoms of the poker. Can we assist the earth in this case ? If we wish to remove the residue of a powder from the interior surface of a glass to which the powder clings, we invert the glass, tap it, loosen the hold of the powder, and thus enable the force of gravity to pull it down. So also by tapping the end of the poker we loosen the adhesion of the fluids to the atoms and enable the earth to pull them apart. But, what is the consequence ? The portion of fluid which has been thus forcibly dragged over the atoms refuses to return when the poker has been removed from the line of dip ; the iron, as you see, has become a permanent magnet. By reversing its position and tapping it again we reverse its magnetism. A thoughtful and competent teacher will well know how to place these remarkable facts before his pupils in a manner which will excite their interest ; he will know, and if not, will try to learn, how by the use of sensible images, more or less gross, to give those he teaches definite conceptions, purifying these conceptions more and more as the minds of his pupils become more capable of abstraction. He will cause his logic to run like a line of light through these images, and by thus acting he will cause his boys to march at his side with a profit and a joy, which the mere exhibition of facts without principles, or the appeal to the bodily senses and the power of memory alone, could never inspire.

As an expansion of the note at p. 393, the following extract may find a place here :—

'It is well known that a voltaic current exerts an attractive force upon a second current, flowing in the same direction ; and that when the directions are opposed to each other the force exerted is a repulsive one. By coiling wires into spirals, Ampère was enabled to make them produce all the phenomena of attraction and repulsion exhibited by magnets, and from this it was but a step to his celebrated theory of molecular currents. He supposed the molecules of a magnetic body to be surrounded by such currents, which, however, in the natural state of the body mutually neutralised each other, on account of their confused grouping. The act of magnetisation he supposed to consist in setting these molecular currents parallel to each other ; and starting from this principle, he reduced all the phenomena of magnetism to the mutual action of electric currents.

'If we reflect upon the experiments recorded in the foregoing pages from first to last, we can hardly fail to be convinced that diamagnetic bodies operated on by magnetic forces possess a polarity "the same in kind as, but the reverse in direction of that acquired by magnetic bodies." But if this be the case, how are we to conceive the *physical mechanism* of this polarity ? According to Coulomb's and Poisson's theory, the act of magnetisation consists in the decomposition of a neutral magnetic fluid ; the north pole of a magnet, for example, possesses an attraction for the south fluid of a piece of soft iron submitted to its influence, draws the said fluid towards it, and with it the material particles with which the fluid is associated. To account for diamagnetic phenomena this theory seems to fail altogether ; according to it, indeed, the oft-used phrase, "a north pole exciting a north pole, and a south pole a south pole," involves a contradiction. For if the north fluid be supposed to be *attracted* towards the influencing north pole, it is absurd to suppose that its presence there could produce *repulsion*. The theory of Ampère is equally at a loss to explain diamagnetic action ; for if we suppose the particles of bismuth surrounded by molecular currents, then, according to all that is known of electro–dynamic laws, these currents would set themselves parallel to, and in the same direction as those of the magnet, and hence attraction, and not repulsion, would be the result. The fact, however, of this not being the case proves that these molecular currents are not the mechanism by which diamagnetic induction is effected. The consciousness of this, I doubt not, drove M. Weber to the assumption that the phenomena of diamagnetism are produced by molecular currents, not *directed*, but actually *excited* in the bismuth by the magnet. Such induced currents would, according to known laws, have a direction *opposed* to those of the inducing magnet, and hence would produce the phenomena of repulsion. To carry out the assumption here made, M. Weber is obliged to suppose that the molecules of diamagnetic bodies are surrounded by channels, in which the induced molecular currents, once excited, continue to flow without resistance.'—*Diamagnetism and Magne-crystallic Action*, p. 136–7.

XIV.

SHORTER ARTICLES.

SLATES. .

PART OF A LECTURE

DELIVERED IN THE ROYAL INSTITUTION OF GREAT BRITAIN,

6th June, 1856.

SLATES.

WHEN the student of physical science has to investigate the character of any natural force, his first care must be to purify it from the mixture of other forces, and thus study its simple action. If, for example, he wishes to know how a mass of liquid would shape itself, if at liberty to follow the bent of its own molecular forces, he must see that these forces have free and undisturbed exercise. We might perhaps refer him to the dew-drop for a solution of the question; but here we have to do, not only with the action of the molecules of the liquid upon each other, but also with the action of gravity upon the mass, which pulls the drop downwards and elongates it. If he would examine the problem in its purity, he must do as Plateau has done, detach the liquid mass from the action of gravity; he would then find the shape to be a perfect sphere. Natural processes come to us in a mixed manner, and to the uninstructed mind are a mass of unintelligible confusion. Suppose half-a-dozen of the best musical performers to be placed in the same room, each playing his own instrument to perfection, but no two playing the same tune; though each individual instrument might be a source of perfect music, still the mixture of all would produce mere noise. Thus it is with the processes of nature. Here mechanical and molecular laws intermingle and create apparent confusion. Their mixture constitutes what may be called the *noise* of natural laws, and it is the vocation of the man of science to resolve this noise into its components, and thus to detect the 'music' in which the foundations of nature are laid.

The necessity of this detachment of one force from all other forces is nowhere more strikingly exhibited than in the phenomena of crystallisation. Here, for example, is a solution of common sulphate of soda or Glauber salt. Looking into it mentally, we see the molecules of that liquid, like disciplined squadrons under a governing eye, arranging themselves into battalions, gathering round distinct centres, and forming themselves into solid masses, which after a time assume the visible shape of the crystal now held in my hand. I may, like an ignorant meddler wishing to hasten matters, introduce confusion into this order. This may be done by plunging a glass rod into the vessel ; the consequent action is not the pure expression of the crystalline forces ; the molecules rush together with the confusion of an unorganised mob, and not with the steady accuracy of a disciplined host. In this mass of bismuth also we have an example of confused crystallisation ; but in the crucible behind me a slower process is going on : here there is an architect at work 'who makes no chips, no din,' and who is now building the particles into crystals, similar in shape and structure to those beautiful masses which we see upon the table. By permitting alum to crystallise in this slow way, we obtain these perfect octahedrons; by allowing carbonate of lime to crystallise, nature produces these beautiful rhomboids ; when silica crystallises, we have formed these hexagonal prisms capped at the ends by pyramids ; by allowing saltpetre to crystallise we have these prismatic masses, and when carbon crystallises, we have the diamond. If we wish to obtain a perfect crystal we must allow the molecular forces free play : if the crystallising mass be permitted to rest upon a surface it will be flattened, and to prevent this a small crystal must be so suspended as to be surrounded on all sides by the liquid, or, if it rest upon the surface, it must be turned daily so as to present all its faces in succession to the working builder.

In building up crystals these little atomic bricks often arrange themselves into layers which are perfectly parallel to each other, and which can be separated by mechanical means ; this is called the cleavage of the crystal. The crystal of sugar I hold in my hand thus far escaped the solvent and abrading forces which sooner or later determine the fate of sugar-candy. I readily discover that it cleaves with peculiar facility in one direction. Again I lay my knife upon this piece of rocksalt, and with a blow cleave it in one direction. Laying the knife at right angles to its former position, the crystal cleaves again ; and finally placing the knife at right angles to the two former positions, we find a third cleavage. Rocksalt cleaves in three directions, and the resulting solid is this perfect cube, which may be broken up into any number of smaller cubes. Iceland spar also cleaves in three directions, not at right angles, but oblique to each other, the resulting solid being a rhomboid. In each of these cases the mass cleaves with equal facility in all three directions. For the sake of completeness I may say that many crystals cleave with unequal facility in different directions : heavy spar presents an example of this kind of cleavage.

Turn we now to the consideration of some other phenomena to which the term cleavage may be applied. Beech, deal, and other woods cleave with facility along the fibre, and this cleavage is most perfect when the edge of the axe is laid across the rings which mark the growth of the tree. If you look at this bundle of hay severed from a rick, you will see a sort of cleavage in it also; the stalks lie in parallel planes, and only a small force is required to separate them laterally. But we cannot regard the cleavage of the tree as the. same in character as that of the hayrick. In the one case it is the molecules arranging themselves according to organic laws which produce a cleavable structure, in the other case the easy separation in one direction is due to the mechanical arrangement of the coarse sensible masses of the stalks of hay.

This sandstone rock was once a powder, more or less coarse, held in mechanical suspension by water. The powder was composed of two distinct parts, fine grains of sand and small plates of mica. Imagine a wide strand covered by a tide, or an estuary with water which holds such powder in suspension : how will it sink? The rounded grains of sand will reach the bottom first, because they encounter least resistance, the mica afterwards, and when the tide recedes we have the little plates shining like spangles upon the surface of the sand. Each successive tide brings its charge of mixed powder, deposits its duplex layer day after day, and finally masses of immense thickness are piled up, which by preserving the alternations of sand and mica tell the tale of their formation. Take the sand and mica, mix them together in water, and allow them to subside ; they will arrange themselves in the manner indicated, and by repeating the process you can actually build up a mass which shall be the exact counterpart of that presented by nature. Now this structure cleaves with readiness along the planes in which the particles of mica are strewn. Specimens of such a rock sent to me from Halifax, and other masses from the quarries of Over Darwen in Lancashire, are here before you. With a hammer and chisel I can cleave them into flags ; indeed these flags are employed for roofing purposes in the districts from which the specimens have come, and receive the name of ' slatestone.' But you will discern without a word from me, that this cleavage is not a crystalline cleavage any more than that of a hayrick is. It is molar, not molecular.

This, so far as I am aware of, has never been imagined, and it has been agreed among geologists not to call such splitting as this cleavage at all, but to restrict the term to a phenomenon of a totally different character.

Those who have visited the slate quarries of Cumberland and North Wales will have witnessed the phenomenon to which I refer. We have long drawn our supply of roofing-slates from such quarries; school-boys ciphered on these slates, they were used for tombstones in church-yards, and for billiard-tables in the metropolis ; but not until a comparatively late period did men begin to enquire how their wonderful structure was produced. What is the agency which enables us to split Honister Crag, or the cliffs of Snowdon, into laminæ from crown to base ? This question is at the present moment one of the great diffi-culties of geologists, and occupies their attention perhaps more than any other. You may wonder at this. Looking into the quarry of Penrhyn, you may be disposed to offer the explanation I heard given two years ago. 'These planes of cleavage,' said a friend who stood beside me on the quarry's edge, 'are the planes of stratification which have been lifted by some convulsion into an almost vertical position.' But this was a mistake, and indeed here lies the grand difficulty of the problem. The planes of cleavage stand in most cases at a high angle to the bedding. Thanks to Sir Roderick Murchison, I am able to place the proof of this before you. Here is a specimen of slate in which both the planes of cleavage and of bedding are distinctly marked, one of them making a large angle with the other. This is common. The cleavage of slates then is not a question of stratification; what then is its cause?

In an able and elaborate essay published in 1835, Prof. Sedgwick proposed the theory that cleavage is due to the action of crystalline or polar forces subsequent to the consolidation of the rock. 'We may affirm,' he says, 'that no retreat of the parts, no contraction of dimensions in passing to a solid state, can explain such phenomena. They appear to me only resolvable on the supposition that crystalline or polar forces acted upon the whole mass simultaneously in one direction and with adequate force.' And again, in another place: 'Crystalline forces have re-arranged whole mountain masses, producing a beautiful crystalline cleavage, passing alike through all the strata.'[1] The utterance of such a man struck deep, as it ought to do, into the minds of geologists, and at the present day there are few who do not entertain this view either in whole or in part.[2] The boldness of the theory,

---

[1] *Transactions of the Geological Society*, ser. ii. vol. iii. p. 477.

[2] In a letter to Sir Charles Lyell, dated from the Cape of Good Hope February 20, 1836, Sir John Herschel writes as follows:—'If rocks have

indeed, has, in some cases, caused speculation to run riot, and we have books published on the action of polar forces and geologic magnetism, which rather astonish those who know something about the subject. According to this theory whole districts of North Wales and Cumberland, mountains included, are neither more nor less than the parts of a gigantic crystal. These masses of slate were originally fine mud, composed of the broken and abraded particles of older rocks. They contain silica, alumina, potash, soda, and mica mixed mechanically together. In the course of ages the mixture became consolidated, and the theory before us assumes that a process of crystallisation afterwards re-arranged the particles and developed in it a single plane of cleavage. Though a bold, and I think inadmissible, stretch of analogies, this hypothesis has done good service. Right or wrong, a thoughtfully uttered theory has a dynamic power which operates against intellectual stagnation; and even by provoking opposition is eventually of service to the cause of truth. It would, however, have been remarkable if, among the ranks of geologists themselves, men were not found to seek an explanation of slate-cleavage involving a less hardy assumption.

The first step in an enquiry of this kind is to seek facts. This has been done, and the labours of Daniel Sharpe (the late President of the Geological Society, who, to the

been so heated as to allow of a commencement of crystallisation, that is to say, if they have been heated to a point at which the particles can begin to move amongst themselves, or at least on their own axes, some general law must then determine the position in which these particles will rest on cooling. Probably that position will have some relation to the direction in which the heat escapes. Now when all or a majority of particles of the same nature have a general tendency to one position, that must of course determine a cleavage plane.'

loss of science and the sorrow of all who knew him, has so suddenly been taken away from us), Mr. Henry Clifton Sorby, and others have furnished us with a body of facts associated with slaty cleavage, and having a most important bearing upon the question.

Fossil shells are found in these slate-rocks. I have here several specimens of such shells in the actual rock, and occupying various positions in regard to the cleavage planes. They are squeezed, distorted, and crushed; in all cases the distortion leads to the inference that the rock which contains these shells has been subjected to enormous pressure in a direction at right angles to the planes of cleavage. The shells are all flattened and spread out in these planes. Compare this fossil trilobite of normal proportions with these others which have suffered distortion. Some have lain across, some along, and some oblique to the cleavage of the slate in which they are found; but in all cases the distortion is such as required for its production a compressing force acting at right angles to the planes of cleavage. As the trilobites lay in the mud, the jaws of a gigantic vice appear to have closed upon them and squeezed them into the shapes you see.

We sometimes find a thin layer of coarse gritty material, between two layers of finer rock, through which and across the gritty layer pass the planes of lamination. The coarse layer is found bent by the pressure into sinuosities like a contorted ribbon. Mr. Sorby has described a striking case of this kind. This crumpling can be experimentally imitated; the amount of compression might, moreover, be roughly estimated by supposing the contorted bed to be stretched out, its length measured and compared with the shorter distance into which it has been squeezed. We find in this way that the yielding of the mass has been considerable.

Let me now direct your attention to another proof of pressure; you see the varying colours which indicate the bedding on this mass of slate. The dark portion is gritty, being composed of comparatively coarse particles, which, owing to their size, shape, and gravity, sink first and constitute the bottom of each layer. Gradually, from bottom to top the coarseness diminishes, and near the upper surface we have a layer of exeedingly fine mud. It is the mud thus consolidated from which are derived the German razor-stones, so much prized for the sharpening of surgical instruments. When a bed is thin, the fine white mud is permitted to rest upon a slab of the coarser slate in contact with it: when the bed is thick, it is cut into slices which are cemented to pieces of ordinary slate, and thus rendered stronger. The mud thus deposited is, as might be expected, often rolled up into nodular masses, carried forward, and deposited among coarser material by the rivers from which the slate-mud has subsided. Here are such nodules enclosed in sandstone. Everybody, moreover, who has ciphered upon a school-slate must remember the whitish-green spots which sometimes dotted the surface of the slate, and over which the pencil usually slid as if the spots were greasy. Now these spots are composed of the finer mud, and they could not, on account of their fineness, *bite* the pencil like the surrounding gritty portions of the slate. Here is a beautiful example of these spots: you observe them on the cleavage surface in broad round patches. But turn the slate edgeways and the section of each nodule is seen to be a sharp oval with its longer axis parallel to the cleavage.

... has been ... by Mr. Sorby. I ... ... the quarries of Wales and Cumberland ... ... the same ... of London, and ... ... This we deduce a common ... ... evidence of the highest ... ... important geological pro... ... the magnetic development of these slates, I ... ... contain a less amount of ... ... slate. An analysis was ... ... the laboratory of Dr. ... ... These were the following result:—

... of Slate

Dark slate, no ...

...	.	5·65
. .	.	. 6·13
	Mean	. 5·89

... Green ...

...	.	3·24
. .	.	. 3·12
	Mean	. 3·18

... ... the quantity of iron in the ... ... in the greenish spot is ... ... contained in the spot itself. ... ... which the magnetic experiments ...

... ... the facts brought before ... ... is the representative of a class. ... ... contains the unhappy trilobites ... ... nodules of greenish marl ... ... these sources of independent testimony ... ... same conclusion, namely, that slate-rocks have been subjected to enormous pressure in a direction at right angles to the planes of cleavage.

In reference to Mr. Sorby's contorted bed, I have said that by supposing it to be stretched out and its length measured, it would give us an idea of the amount of yielding of the mass above and below the bed. Such a measurement, however, would not give the exact amount of yielding. I hold in my hand a specimen of slate with its bedding marked upon it; the lower portions of each layer being composed of a comparatively coarse gritty material something like what you may suppose the contorted bed to be composed of. Now in crossing these gritty portions, the cleavage turns, as if tending to cross the bedding at another angle. When the pressure began to act, the intermediate bed, which is not entirely unyielding, suffered longitudinal pressure; as it bent, the pressure became gradually more lateral, and the direction of its cleavage is exactly such as you would infer from an action of this kind—it is neither quite across the bed, nor yet in the same direction as the cleavage of the slate above and below it, but intermediate between both. Supposing the cleavage to be at right angles to the pressure, this is the direction which it ought to take across these more unyielding strata.

Thus we have established the concurrence of the phenomena of cleavage and pressure—that they accompany each other; but the question still remains, Is the pressure sufficient to account for the cleavage? A single geologist, as far as I am aware, answers boldly in the affirmative. This geologist is Sorby, who has attacked the question in the true spirit of a physical investigator. Call to mind the cleavage of the flags of Halifax and

Over Darwen, which is caused by the interposition of layers of mica between the gritty strata. Mr. Sorby finds plates of mica to be also a constituent of slate-rock. He asks himself, what will be the effect of pressure upon a mass containing such plates confusedly mixed up in it? It will be, he argues, and he argues rightly, to place the plates with their flat surfaces more or less perpendicular to the direction in which the pressure is exerted. He takes scales of the oxide of iron, mixes them with a fine powder, and on squeezing the mass finds that the tendency of the scales is to set themselves at right angles to the line of pressure. Along the planes of weakness produced by the scales the mass cleaves.

By tests of a different character from those applied by Mr. Sorby, it might be shown how true his conclusion is, that the effect of pressure on elongated particles, or plates, will be such as he describes it. But while the scales must be regarded as a true cause, I should not ascribe to them a large share in the production of the cleavage. I believe that even if the plates of mica were wholly absent the cleavage of slate-rocks would be much the same as it is at present.

Here is a mass of pure white wax: it contains no mica particles, no scales of iron, or anything analogous to them. Here is the selfsame substance submitted to pressure. I would invite the attention of the eminent geologists now before me to the structure of this wax. No slate ever exhibited so clean a cleavage; it splits into laminæ of surpassing tenuity, and proves at a single stroke that pressure is sufficient to produce cleavage, and that this cleavage is independent of intermixed plates or scales. I have purposely mixed this wax with elongated particles, and am unable to say at the present moment that the cleavage is sensibly affected by their presence—if anything, I should say they rather impair its fineness and clearness than promote it.

The finer the slate is the more perfect will be the resemblance of its cleavage to that of the wax. Compare the surface of the wax with the surface of this slate from Borrodale in Cumberland. You have precisely the same features in both : you see flakes clinging to the surfaces of each, which have been partially torn away in cleaving. Let any close observer compare these two effects, he will, I am persuaded, be led to the conclusion that they are the product of a common cause.[1]

But you will ask me how, according to my view, does pressure produce this remarkable result. This may be stated in a very few words.

There is no such thing in nature as a body of perfectly homogeneous structure. I break this clay which seems so uniform, and find that the fracture presents to my eyes innumerable surfaces along which it has given way, and it has yielded along those surfaces because in them the cohesion of the mass is less than elsewhere. I break this marble, and even this wax, and observe the same result ; look at the mud at the bottom of a dried pond ; look to some of the ungravelled walks in Kensington Gardens on drying after rain,—they are cracked and split, and other circumstances being equal, they crack and split where the cohesion is least. Take then a mass of partially con-

---

[1] I have usually softened the wax by warming it, kneaded it with the fingers, and pressed it between thick plates of glass previously wetted. At the ordinary summer temperature the pressed wax is soft, and tears rather than cleaves ; on this account I cool my compressed specimens in a mixture of pounded ice and salt, and when thus cooled they split beautifully.

solidated mud. Such a mass is divided and subdivided by interior surfaces along which the cohesion is comparatively small. Penetrate the mass in idea, and you will see it composed of numberless irregular polyhedra bounded by surfaces of weak cohesion. Imagine such a mass subjected to pressure,—it yields and spreads out in the direction of least resistance;[1] the little polyhedra become converted into laminæ, separated from each other by surfaces of weak cohesion, and the infallible result will be a tendency to cleave at right angles to the line of pressure.

Further, a mass of dried mud is full of cavities and fissures. If you break dried pipe-clay you see them in great numbers, and there are multitudes of them so small that you cannot see them. A flattening of these cavities must take place in squeezed mud, and this must to some extent facilitate the cleavage of the mass in the direction indicated.

Although the time at my disposal has not permitted me duly to develope these thoughts, yet for the last twelve months the subject has presented itself to me almost daily under one aspect or another. I have never eaten a biscuit during this period without remarking the cleavage developed by the rollingpin. You have only to break a biscuit across, and to look at the fracture, to see the laminated structure. We have here the means of

---

[1] It is scarcely necessary to say that if the mass were squeezed equally in all directions no laminated structure could be produced; it must have room to yield in a lateral direction. Mr. Warren De la Rue informs me that he once wished to obtain white-lead in a fine granular state, and to accomplish this he first compressed it. The mould was conical, and permitted the lead to spread out a little laterally. The lamination was as perfect as that of slate, and it quite defeated him in his effort to obtain a granular powder.

pushing the analogy further. I invite you to compare the structure of this slate, which was subjected to a high temperature during the conflagration of Mr. Scott Russell's premises, with that of a biscuit. Air or vapour within the slate has caused it to swell, and the mechanical structure it reveals is precisely that of a biscuit. During these enquiries I have received much instruction in the manufacture of puff-paste. Here is some such paste baked under my own superintendence. The cleavage of our hills is accidental cleavage, but this is cleavage with intention. The volition of the pastrycook has entered into its formation. It has been his aim to preserve a series of surfaces of structural weakness, along which the dough divides into layers. Puff-paste in preparation must not be handled too much; it ought, moreover, to be rolled on a cold slab, to prevent the butter from melting, and diffusing itself, thus rendering the paste more homogeneous and less liable to split. Puff-paste is, then, simply an exaggerated case of slaty cleavage.

The principle which I have enunciated is so simple as to be almost trivial; nevertheless, it embraces not only the cases mentioned, but, if time permitted, it might be shown you that the principle has a much wider range of application. When iron is taken from the puddling furnace it is more or less spongy, an aggregate in fact of small nodules : it is at a welding heat, and at this temperature is submitted to the process of rolling. Bright smooth bars are the result. But notwithstanding the high heat the nodules do not perfectly blend together. The process of rolling draws them into fibres. Here is a mass acted upon by dilute sulphuric acid, which exhibits in a striking manner this fibrous structure. The experiment was made by my friend Dr. Percy, without any reference to the question of cleavage.

Break a piece of ordinary iron and you have a granular fracture ; beat the iron, you elongate these granules, and finally render the mass fibrous.   Here are pieces of rails along which the wheels of locomotives have slidden ; the granules have yielded and become plates.   They exfoliate or come off in leaves ; all these effects belong, I believe, to the great class of phenomena of which slaty cleavage forms the most prominent example.[1]

[I would now lay more stress on the lateral yielding, referred to in the note at the bottom of page 418, accompanied as it is by tangential sliding, than I was prepared to do when this lecture was given.   This sliding is, I think, the principal cause of the planes of weakness, both in pressed wax and slate rock.   J. T.   1871.]

[1] For some further observations on this subject by Mr. Sorby and myself, see *Philosophical Magazine* for August 1856.

DEATH BY LIGHTNING.

PEOPLE in general imagine, when they think at all about the matter, that an impression upon the nerves—a blow, for example, or the prick of a pin—is felt at the moment it is inflicted. But this is not the case. The seat of sensation is the brain, and to it the intelligence of any impression made upon the nerves has to be transmitted before this impression can become manifest in consciousness. The transmission, moreover, requires *time,* and the consequence is, that a wound inflicted on a portion of the body distant from the brain is more tardily appreciated than one inflicted adjacent to the brain. By an extremely ingenious experimental arrangement, Helmholtz has determined the velocity of this nervous transmission, and finds it to be about one hundred feet a second, or less than one-tenth of the velocity of sound in air. If, therefore, a whale fifty feet long were wounded in the tail, it would not be conscious of the injury till half a second after the wound had been inflicted.[1] But this is not the only ingredient in the delay. There can scarcely be a doubt that to every act of consciousness belongs a determinate molecular arrangement of the brain—that every

---

[1] A most admirable lecture on the velocity of nervous transmission has been published by Dr. Du Bois-Raymond in the *Proceedings of the Royal Institution* for 1866, vol. iv. p. 575.

thought or feeling has its physical correlative in that organ; and nothing can be more certain than that every physical change, whether molecular or mechanical, requires time for its accomplishment. So that, besides the interval of transmission, a still further time is necessary for the brain to put itself in order—for its molecules to take up the motions or positions necessary to the completion of consciousness. Helmholtz considers that one-tenth of a second is demanded for this purpose. Thus, in the case of the whale above supposed, we have first half a second consumed in the transmission of the intelligence through the sensor nerves to the head, one-tenth of a second consumed by the brain in completing the arrangements necessary to consciousness, and, if the velocity of transmission through the motor be the same as that through the sensor nerves, half a second in sending a command to the tail to defend itself. Thus one second and a tenth would elapse before an impression made upon its caudal nerves could be responded to by a whale fifty feet long.

Now, it is quite conceivable that an injury might be inflicted which would render the nerves unfit to be the conductors of the motion which results in sensation; and if such a thing occurred, no matter how severe the injury might be, we should not be conscious of it. Or it may be, that long before the time required by the brain to complete the arrangement necessary to consciousness, its power of arrangement might be wholly suspended. In such a case also, though the injury might be of a nature to cause death, this would occur without feeling of any kind. Death in this case would be simply the sudden negation of life, without any intervention of consciousness whatever.

Doubtless there are many kinds of death of this character. The passage of a musket-bullet through the brain is a case in point ; and the placid aspect of a man thus killed is in perfect accordance with the conclusion which might be drawn *à priori* from the experiments of Helmholtz. Cases of insensibility, moreover, are not uncommon which do not result in death, and after which the persons affected have been able to testify that no pain was felt prior to the loss of consciousness.

The time required for a rifle-bullet to pass clean through a man's head may be roughly estimated at a thousandth of a second. Here, therefore, we should have no room for sensation, and death would be painless. But there are other actions which far transcend in rapidity that of the rifle-bullet. A flash of lightning cleaves a cloud, appearing and disappearing in less than a hundred-thousandth of a second, and the velocity of electricity is such as would carry it in a single second over a distance almost equal to that which separates the earth and moon. It is well known that a luminous impression once made upon the retina endures for about one-sixth of a second, and that this is the reason why we see a ribbon of light when a glowing coal is caused to pass rapidly through the air. A body illuminated by an instantaneous flash continues to be seen for the sixth of a second after the flash has become extinct ; and if the body thus illuminated be in motion, it appears at rest at the place where the flash falls upon it. The colour-top is familiar to most of us. By this instrument a disk with differently-coloured sectors is caused to rotate rapidly ; the colours blend together, and, if they are chosen in the proper proportions, when the motion is sufficiently rapid the disk appears white. Such a top, rotating in a dark room and illuminated by an electric spark, appears motionless, each distinct colour being clearly seen. Professor Dove has found that a flash of lightning produces the same effect. During a thunder-storm he put a colour-top in exceedingly rapid motion, and found that every flash revealed the top as a motionless object with its colours distinct. If illuminated solely by a flash of lightning, the motion of all bodies on the earth's surface would, as Dove has remarked, appear suspended. A cannon-ball, for example, would have its flight apparently arrested, and would seem to hang motionless in space as long as the luminous impression which revealed the ball remained upon the eye.

If, then, a rifle-bullet move with sufficient rapidity to destroy life without the interposition of sensation, much more is a flash of lightning competent to produce this effect. Accordingly, we have well-authenticated cases of people being struck senseless by lightning who, on recovery, had no memory of pain. The following circumstantial case is described by Hemmer :—

On June 30, 1788, a soldier in the neighbourhood of Mannheim, being overtaken by rain, placed himself under a tree, beneath which a woman had previously taken shelter. He looked upwards to see whether the branches were thick enough to afford the required protection, and, in doing so, was struck by lightning, and fell senseless to the earth. The woman at his side experienced the shock in her foot, but was not struck down. Some hours afterwards the man revived, but remembered nothing about what had occurred, save the fact of his looking up at the branches. This was his last act of consciousness, and he passed from the conscious to the unconscious condition without pain. The visible marks of a lightning stroke are usually insignificant : the hair is sometimes burnt ; slight wounds are observed ; while, in some instances, a red streak marks the track of the discharge over the skin.

Under ordinary circumstances, the discharge from a small Leyden jar is exceedingly unpleasant to myself. Some time ago I happened to stand in the presence of a numerous audience, with a battery of fifteen large Leyden jars charged beside me. Through some awkwardness on my part, I touched a wire leading from the battery, and the discharge went through my body. Life was absolutely blotted out for a very sensible interval, without a trace of pain. In a second or so consciousness returned ; I saw myself in the presence of the audience and apparatus, and, by the help of these external appearances, immediately concluded that I had received the battery discharge. The *intellectual* consciousness of my position was restored with exceeding rapidity, but not so the *optical* consciousness. To prevent the audience from being alarmed, I observed that it had often been my desire to receive accidentally such a shock, and that my wish had at length been fulfilled. But while making this remark, the appearance which my body presented to myself was that of a number of separate pieces. The arms, for example, were detached from the trunk, and seemed suspended in the air. In fact, memory and the power of reasoning appeared to be complete long before the optic nerve was restored to healthy action. But what I wish chiefly to dwell upon here is, the absolute painlessness of the shock ; and there cannot be a doubt, that to a person struck dead by lightning, the passage from life to death occurs without consciousness being in the least degree implicated. It is an abrupt stoppage of sensation, unaccompanied by a pang.

July 8, 1865.

SCIENCE AND SPIRITS.

THEIR refusal to investigate ' spiritual phenomena ' is often urged as a reproach to scientific men. I here propose to give a sketch of an attempt to apply to the ' phenomena ' those methods of enquiry which are found available in dealing with natural truth.

Some time ago, when the spirits were particularly active in this country, a celebrated philosopher was invited, or rather entreated, by one of his friends to meet and question them. He had, however, already made their acquaintance, and did not wish to renew it. I had not been so privileged, and he therefore kindly arranged a transfer of the invitation to me. The spirits themselves named the time of meeting, and I was conducted to the place at the day and hour appointed.

Absolute unbelief in the facts was by no means my condition of mind. On the contrary, I thought it probable that some physical principle, not evident to the spiritualists themselves, might underlie their manifestations. Extraordinary effects are produced by the accumulation of small impulses. Galileo set a heavy pendulum in motion by the well-timed puffs of his breath. Ellicot set one clock going by the ticks of another, even when the two clocks were separated by a wall. Preconceived notions can, moreover, vitiate, to an extraordinary degree, the testimony of even veracious persons. Hence my desire to witness those extraordinary phenomena, the existence of which seemed placed beyond a doubt by the known veracity of those who had witnessed and described them. The meeting took place at a private residence in the neighbourhood of London. My host, his intelligent wife, and a gentleman who may be called X., were in the house when I arrived. I was informed that the 'medium' had not yet made her appearance; that she was sensitive, and might resent suspicion. It was therefore requested that the tables and chairs should be examined before her arrival, in order to be assured that there was no trickery in the furniture. This was done; and I then first learned that my hospitable host had arranged that the *séance* should be a dinner-party. This was to me an unusual form of investigation; but I accepted it, as one of the accidents of the occasion.

The 'medium' arrived—a delicate-looking young lady, who appeared to have suffered much from ill health. I took her to dinner and sat close beside her. Facts were absent for a considerable time, a series of very wonderful narratives supplying their place. The duty of belief on testimony was frequently insisted on. X. appeared to be a chosen spiritual agent, and told us many surprising things. He affirmed that, when he took a pen in his hand, an influence ran from his shoulder downwards, and impelled him to write oracular sentences. I listened for a time, offering no observation. 'And now,' continued X., 'this power has so risen as to reveal to me the thoughts of others. Only this morning I told a friend what he was thinking of, and what he intended to do during the day.' Here, I thought, is something that can be at once tested. I said immediately to X.:—' If

you wish to win to your cause an apostle, who will proclaim your principles to the world without fear, tell me
what I am now thinking of.' X. reddened, and did *not* tell
me my thought.

Some time previously I had visited Baron Reichenbach, in Vienna, and I now asked the young lady who
sat beside me, whether she could see any of the curious
things which he describes—the light emitted by crystals,
for example? Here is the conversation which followed,
as extracted from my notes, written on the day following
the *séance.*

Medium.—'Oh, yes; but I see light around all
bodies.'

I.—'Even in perfect darkness?'

Medium.—'Yes; I see luminous atmospheres round
all people. The atmosphere which surrounds Mr. R. C.
would fill this room with light.'

I.—'You are aware of the effects ascribed by Baron
Reichenbach to magnets?'

Medium.—'Yes; but a magnet makes me terribly ill.'

I.—'Am I to understand that, if this room were
perfectly dark, you could tell whether it contained a
magnet, without being informed of the fact?'

Medium.—'I should know of its presence on entering
the room.'

I.—'How?'

Medium.—'I should be rendered instantly ill.'

I.—'How do you feel to-day?'

Medium.—'Particularly well; I have not been so well
for months.'

I.—'Then, may I ask you whether there is, at the
present moment, a magnet in my possession?'

The young lady looked at me, blushed, and stammered, 'No; I am not *en rapport* with you.'

I sat at her right hand, and a left-hand pocket, within six inches of her person, contained a magnet.

Our host here deprecated discussion, as it 'exhausted the medium.' The wonderful narratives were resumed; but I had narratives of my own quite as wonderful. These spirits, indeed, seemed clumsy creations, compared with those with which my own researches had made me familiar. I therefore began to match the wonders related to me by other wonders. A lady present discoursed on spiritual atmospheres, which she could see as beautiful colours when she closed her eyes. I professed myself able to see similar colours, and more than that, to be able to see the interior of my own eyes. The medium affirmed that she could see actual waves of light coming from the sun. I retorted that men of science could tell the exact number of waves emitted in a second, and also their exact length. The medium spoke of the performances of the spirits on musical instruments. I said that such performance was gross, in comparison with a kind of music which had been discovered some time previously by a scientific man. Standing at a distance of twenty feet from a jet of gas, he could command the flame to emit a melodious note; it would obey, and continue its song for hours. So loud was the music emitted by the gas-flame, that it might be heard by an assembly of a thousand people. These were acknowledged to be as great marvels as any of those of spiritdom. The spirits were then consulted, and I was pronounced to be a first-class medium.

During this conversation a low knocking was heard

from time to time under the table. These were the spirits' knocks. I was informed that one knock, in answer to a question, meant ' No ; ' that two knocks meant 'Not yet ; ' and that three knocks meant ' Yes.' In answer to the question whether I was a medium, the response was three brisk and vigorous knocks. I noticed that the knocks issued from a particular locality, and therefore requested the spirits to be good enough to answer from another corner of the table. They did not comply ; but I was assured that they would do it, and much more, by-and-bye. The knocks continuing, I turned a wine-glass upside down, and placed my ear upon it, as upon a stethescope. The spirits seemed disconcerted by the act ; they lost their playfulness, and did not quite recover it for a considerable time.

Somewhat weary of the proceedings, I once threw myself back against my chair and gazed listlessly out of the window. While thus engaged, the table was rudely pushed. Attention was drawn to the wine, still oscillating in the glasses, and I was asked whether that was not convincing. I readily granted the fact of motion, and began to feel the delicacy of my position. There were several pairs of arms upon the table, and several pairs of legs under it ; but how was I, without offence, to express the conviction which I really entertained ? To ward off the difficulty, I again turned a wine-glass upside down and rested my ear upon it. The rim of the glass was not level, and the hair on touching it, caused it to vibrate, and produce a peculiar buzzing sound. A perfectly candid and warm-hearted old gentleman at the opposite side of the table, whom I may call A., drew attention to the sound, and expressed his entire belief that it was spiritual. I, however, informed him that it was the moving hair acting on the glass. The explanation was not well received; and X., in a tone of severe pleasantry, demanded whether it was the hair that had moved the table. The promptness of my negative probably satisfied him that my notion was a very different one.

The superhuman power of the spirits was next dwelt upon. The strength of man, it was stated, was unavailing in opposition to theirs. No human power could prevent the table from moving when they pulled it. During the evening this pulling of the table occurred, or rather was attempted, three times. Twice the table moved when my attention was withdrawn from it; on a third occasion, I tried whether the act could be provoked by an assumed air of inattention. Grasping the table firmly between my knees, I threw myself back in the chair, and waited, with eyes fixed on vacancy, for the pull. It came. For some seconds it was pull spirit, hold muscle; the muscle, however, prevailed, and the table remained at rest. Up to the present moment, this interesting fact is known only to the particular spirit in question and myself.

A species of mental scene-painting, with which my own pursuits had long rendered me familiar, was employed to figure the changes and distribution of spiritual power. The spirits were provided with atmospheres, which combined with and interpenetrated each other, considerable ingenuity being shown in demonstrating the necessity of *time* in effecting the adjustment of the atmospheres. In fact, just as in science, the senses, time, and space constituted the conditions of the phenomena. A re-arrangement of our positions was proposed and carried out; and soon afterwards my attention was drawn to a scarcely sensible vibration on the part of the table. Several persons were leaning on the table at the time, and I asked permission to touch the medium's hand. 'Oh, I know I tremble,' was her reply. Throwing one leg across the other, I accidentally nipped a muscle, and produced thereby an involuntary vibration of the free leg. This vibration, I knew, must be communicated to the floor, and thence to the chairs of all present. I therefore intentionally promoted it. My attention was promptly drawn to the motion; and a gentleman beside me, whose value as a witness I was particularly desirous to test, expressed his belief, that it was out of the compass of human power to produce so strange a tremor. 'I believe,' he added earnestly, 'that it is entirely the spirits' work.' 'So do I,' added, with heat, the candid and warmhearted old gentleman A. 'Why, sir,' he continued, 'I feel them at this moment shaking my chair.' I stopped the motion of the leg. 'Now, sir,' A. exclaimed, 'they are gone.' I began again, and A. once more ejaculated. I could, however, notice that there were doubters present, who did not quite know what to think of the manifestations. I saw their perplexity; and, as there was sufficient reason to believe that the disclosure of the secret would simply provoke anger, I kept it to myself.

Again a period of conversation intervened, during which the spirits became animated. The evening was confessedly a dull one, but matters appeared to brighten towards its close. The spirits were requested to spell the name by which I am known in the heavenly world. Our

F F

host commenced repeating the alphabet, and when he reached the letter 'P' a knock was heard. He began again, and the spirits knocked at the letter 'O.' I was puzzled, but waited for the end. The next letter knocked down was 'E.' I laughed, and remarked that the spirits were going to make a poet of me. Admonished for my levity, I was informed that the frame of mind proper for the occasion ought to have been superinduced by a perusal of the Bible immediately before the *séance*. The spelling, however, went on, and sure enough I came out a poet. But matters did not end here. Our host continued his repetition of the alphabet, and the next letter of the name proved to be 'O.' Here was manifestly an unfinished word ; and the spirits were apparently in their most communicative mood. The knocks came from under the table, but no person present evinced the slightest desire to look under it. I asked whether I might go underneath ; the permission was granted ; so I crept under the table. Some tittered ; but the candid old A. exclaimed, ' He has a right to look into the very dregs of it, to convince himself.' Having pretty well assured myself that no sound could be produced under the table without its origin being revealed, I requested our host to continue his questions. He did so, but in vain. He adopted a tone of tender entreaty ; but the ' dear spirits ' had become dumb dogs, and refused to be entreated. I continued under that table for at least a quarter of an hour, after which, with a feeling of despair as regards the prospects of humanity never before experienced, I regained my chair. Once there, the spirits resumed their loquacity, and dubbed me ' Poet of Science.'

This, then, is the result of an attempt made by a

scientific man to look into these spiritual phenomena. It is not encouraging ; and for this reason. The present promoters of spiritual phenomena divide themselves into two classes, one of which needs no demonstration, while the other is beyond the reach of proof. The victims like to believe, and they do not like to be undeceived. Science is perfectly powerless in the presence of this frame of mind. It is, moreover, a state perfectly compatible with extreme intellectual subtlety and a capacity for devising hypotheses which only require the hardihood engendered by strong conviction, or by callous mendacity, to render them impregnable. The logical feebleness of science is not sufficiently borne in mind. It keeps down the weed of superstition, not by logic, but by slowly rendering the mental soil unfit for its cultivation. When science appeals to uniform experience, the spiritualist will retort, ' How do you know that a uniform experience will continue uniform ? You tell me that the sun has risen for 6,000 years : that is no proof that it will rise to-morrow ; within the next twelve hours it may be puffed out by the Almighty.' Taking this ground, a man may maintain the story of ' Jack and the Bean-stalk ' in the face of all the science in the world. You urge, in vain, that science has given us all the knowledge of the universe which we now possess, while spiritualism has added nothing to that knowledge. The drugged soul is beyond the reach of reason. It is in vain that impostors are exposed, and the special demon cast out. He has but slightly to change his shape, return to his house, and find it ' empty, swept, and garnished.'

December 10, 1864.

VITALITY.

THE origin, growth, and energies of living things are subjects which have always engaged the attention of thinking men. To account for them it was usual to assume a special agent, to a great extent free from the limitations observed among the powers of inorganic nature. This agent was called the *vital force*; and, under its influence, plants and animals were supposed to collect their materials and to assume determinate forms. Within the last twenty years, however, our ideas of vital processes have undergone profound modifications; and the interest, and even disquietude, which the change has excited in some minds are amply evidenced by the discussions and protests which are now common regarding the phenomena of vitality. In tracing out these phenomena through all their modifications the most advanced philosophers of the present day declare that they ultimately arrive at a single source of power, from which all vital energy is derived; and the disquieting circumstance is that this source is not the direct fiat of a supernatural agent, but a reservoir of what, if we do not accept the creed of Zoroaster, must be regarded as *inorganic* force. In short, it is considered as proved that all the energy which we derive from plants and animals is drawn from the sun.

A few years ago, when the sun was affirmed to be the source of life, nine out of ten of those who are alarmed by the form which this assertion has latterly assumed would have assented, in a general way, to its correctness. Their assent, however, was more poetical than scientific, and they were by no means prepared to see a rigid mechanical signification attached to their words. This, however, is the peculiarity of modern conclusions :—that there is no *creative* energy whatever in the vegetable or animal organism, but that all the power which we obtain from the muscles of men and animals, as much as that which we develope by the combustion of wood or coal, has been produced at the sun's expense. The sun is so much colder that we may have our fires ; he is also so much colder that we may have our horse-racing and Alpine climbing. It is, for example, certain that the sun has beeen chilled to an extent capable of being accurately expressed in numbers, in order to furnish the power which lifted this year a certain number of tourists from the vale of Chamouni to the summit of Mont Blanc.

To most minds, however, the energy of light and heat presents itself as a thing totally distinct from ordinary mechanical energy. But either of them can be derived from the other. By the friction of wood a savage can raise it to the temperature of ignition ; by properly striking a piece of iron a skilful blacksmith can cause it to glow, and thus, by the rude agency of his hammer, he generates light and heat. This action, if carried far enough, would produce the light and heat of the sun. In fact the sun's light and heat have actually been referred to the fall of meteoric matter upon his surface ; and whether the sun is thus supported or not, it is perfectly certain that he *might be* thus supported. Whether, moreover, the whilom molten condition of our planet was, as supposed by eminent men, due to the collision of cosmic masses or not, it is perfectly certain that the molten condition *might be* thus brought about. If, then, solar light and heat can be produced by the impact of dead matter, and if from the light and heat thus produced we can derive the energies which we have been accustomed to call *vital*, it indubitably follows that vital energy may have a proximately mechanical origin.

In what sense, then, is the sun to be regarded as the origin of the energy derivable from plants and animals? Let us try to give an intelligible answer to this question. Water may be raised from the sea-level to a high elevation, and then permitted to descend. In descending it may be made to assume various forms—to fall in cascades, to spurt in fountains, to boil in eddies, or to flow tranquilly along a uniform bed. It may, moreover, be caused to set complex machinery in motion, to turn millstones, throw shuttles, work saws and hammers, and drive piles. But every form of power here indicated would be derived from the original power expended in raising the water to the height from which it fell. There is no energy *generated* by the machinery; the work performed by the water in descending is merely the parcelling out and distribution of the work expended in raising it. In precisely this sense is all the energy of plants and animals the parcelling out and distribution of a power originally exerted by the sun. In the case of the water, the source of the power consists in the forcible separation of a quantity of the liquid from a low level of the earth's surface and its elevation to a higher position, the power thus expended being returned by the water in its descent. In the case of vital phenomena, the source of power consists in the forcible separation of the atoms of compound substances by the sun. We name the force which draws the water earthward 'gravity,' and that which draws atoms together 'chemical affinity;' but these different names must not mislead us regarding the qualitative identity of the two forces. They are both *attractions*, and, to the intellect, the falling of carbon atoms against oxygen atoms is not more difficult of conception than the falling of water to the earth.

The building up of the vegetable, then, is effected by the sun through the reduction of chemical compounds. *The phenomena of animal life are more or less complicated reversals of these processes of reduction.* We eat the vegetable, and we breathe the oxygen of the air, and in our bodies the oxygen which had been *lifted* from the carbon and hydrogen by the action of the sun again falls towards them, producing animal heat and developing animal forms. Through the most complicated phenomena of vitality this law runs :—the vegetable is produced while a weight rises, the animal is produced while a weight falls. But the question is not exhausted here. The water employed in our first illustration generates all the motion displayed in its descent, but the *form* of the motion depends on the character of the machinery interposed in the path of the water. In a similar way the primary action of the sun's rays is qualified by the atoms and molecules among which their energy is distributed. Molecular forces determine the form which the solar energy will assume. In the separation of the carbon and oxygen this energy may be so conditioned as to result in one case in the formation of a cabbage, and in another case in the formation of an oak. So also as regards the reunion of the carbon and the oxygen, the molecular machinery through which the combining energy acts may, in one case, weave the texture of a frog, while in another it may weave the texture of a man.

The matter of the animal body is that of inorganic nature. There is no substance in the animal tissues which is not primarily derived from the rocks, the water, and the air. Are the forces of organic matter, then, different in kind from those of inorganic matter? The philosophy of the present day negatives the question. It is the compounding in the organic world of forces belonging equally to the inorganic that constitutes the mystery and the miracle of vitality. Every portion of every animal body may be reduced to purely inorganic matter. A perfect reversal of this process of reduction would carry us from the inorganic to the organic; and such a reversal is at least conceivable. The tendency, indeed, of modern science is to break down the wall of partition between organic and inorganic, and to reduce both to the operation of forces which are the same in kind, but whose combinations differ in complexity.

Consider now the question of personal identity, in relation to this of molecular form. Twenty-six years ago Mayer, of Heilbronn, with that power of genius which breathes large meanings into scanty facts, pointed out that the blood was 'the oil of life,' the combustion of which, like that of coal in grosser cases, sustained muscular action. The muscles are the machinery by which the dynamic power of the blood is brought into play. Thus the blood is consumed. But the whole body, though more slowly than the blood, wastes also, so that after a certain number of years it is entirely renewed. How is the sense of personal identity maintained across this flight of molecules? To man, as we know him, *matter* is necessary to consciousness, but the matter of any period may be all changed, while consciousness exhibits no solution of continuity. Like changing sentinels, the oxygen, hydrogen, and carbon that depart seem to whisper their secret to their comrades that arrive, and thus, while the Non-ego shifts the Ego remains intact. Constancy of *form* in the grouping of the molecules, and not constancy of the molecules themselves, is the correlative of this constancy of perception. Life is a *wave* which in no two consecutive moments of its existence is composed of the same particles.

Supposing, then, the molecules of the human body instead of replacing others, and thus renewing a pre-existing form, to be gathered first hand from nature and put together in the same relative positions as those which they occupy in the body; that they have the selfsame forces and distribution of forces, the selfsame motions and distribution of motions—would this organised concourse of molecules stand before us as a sentient thinking being? There seems no valid reason to believe that it would not. Or, supposing a planet carved from the sun, and set spinning round an axis, and revolving round the sun at a distance from him equal to that of our earth, would one of the consequences of its refrigeration be the development of organic forms? I lean to the affirmative. *Structural* forces are certainly in the mass, whether or not those forces reach to the extent of forming a plant or an animal. In an amorphous drop of water lie latent all the marvels of crystalline force; and who will set limits to the possible play of molecules in a cooling planet? If these statements startle, it is because matter has been defined and maligned by philosophers and theologians who were equally unaware that it is, at bottom, essentially mystical and transcendental.

Questions such as these derive their present interest in great part from their audacity, which is sure, in due time, to disappear. And the sooner the public dread is abolished with reference to such questions the better for the cause of truth. As regards knowledge, physical science is polar. In one sense it knows, or is destined to know, everything. In another sense it knows nothing. Science knows much of this intermediate phase of things that we call nature, of which it is the product; but science knows nothing of the origin or destiny of nature. Who or what made the sun, and gave his rays their alleged power? Who or what made and bestowed upon the ultimate particles of matter their wondrous power of varied interaction? Science does not know : the mystery, though pushed back, remains unaltered. To many of us who feel that there are more things in heaven and earth than are dreamt of in the present philosophy of science, but who have been also taught, by baffled efforts, how vain is the attempt to grapple with the Inscrutable, the ultimate frame of mind is that of Goethe :

> Who dares to name His name,
> Or belief in Him proclaim,
> Veiled in mystery as He is, the All-enfolder?
> Gleams across the mind His light,
> Feels the lifted soul His might,
> Dare it then deny His reign, the All-upholder?

One or two interpolations excepted, the foregoing brief article was written on an Alpine slope in the summer of 1863. Seven years afterwards I was singularly interested to learn, that nearly 300 years ago, in explaining the actions and energies of the human body, Descartes employed similar imagery and expressed similar views as far as the knowledge of his time allowed. Professor Huxley, who possesses a reading faculty which I can but envy, has published in his ' Lay Sermons ' the following remarkable extracts from the ' Traité de l'Homme :'—

' In proportion as these spirits (the animal spirits) enter the cavities of the brain, they pass thence into the pores of its substance, and from these pores into the nerves ; where, according as they enter, or even only tend to enter, more or less, into one than into another, they have the power of altering the figure of the muscles into which the nerves are inserted, and by this means of causing all the limbs to move. Thus, as you may have seen in the grottoes and the fountains in royal gardens, the force with which the water issues from its reservoir is sufficient to move various machines, and even to make them play instruments, or pronounce words, according to the different disposition of the pipes which lead the water.

' And, in truth, the nerves of the machine which I am describing may very well be compared to the pipes of these waterworks ; its muscles and its tendons to the other various engines and springs which seem to move them ; its animal spirits to the water which impels them, of which the heart is the fountain ; while the cavities of the brain are the central office. Moreover, respiration and other such actions as are natural and usual in the body, and which depend on the course of the spirits, are like the movements of a clock, or of a mill, which may be kept up by the ordinary flow of water.

' The external objects which, by their mere presence, act upon the organs of the senses ; and which, by this means, determine the corporal machine to move in many different ways, according as the parts of the brain are arranged, are like the strangers who, entering into some of the grottoes of those waterworks, unconsciously cause the movements which take place in their presence. For they cannot enter without treading upon certain planks so arranged that, for example, if they approach a bathing Diana, they cause her to hide among the reeds; and if they attempt to follow her, they see approaching a Neptune, who threatens them with his trident; or if they try some other way, they cause some monster who vomits water into their faces, to dart out; or like contrivances, according to the fancy of the engineers who have made them. And lastly, when the *rational soul* is lodged in this machine, it will have its principal seat in the brain, and will take the place of the engineer, who ought to be in that part of the works with which all the pipes are connected, when he wishes to increase, or to slacken, or in some way to alter, their movements.'

' All the functions which I have attributed to this machine (the body), as the digestion of food, the pulsation of the heart and of the arteries; the nutrition and the growth of the limbs; respiration, wakefulness, and sleep; the reception of light, sounds, odours, flavours, heat, and such like qualities, in the organs of the external senses; the impression of the ideas of these in the organ of common sense and in the imagination; the retention, or the impression, of these ideas on the memory; the internal movements of the appetites and the passions; and lastly, the external movements of all the limbs, which follow so aptly, as well the action of the objects which are presented to the senses, as the impressions which meet in the memory, that they imitate as nearly as possible those of a real man: I desire, I say, that you should consider that these functions in the machine naturally proceed from the mere arrangement of its organs, neither more nor less than do the movements of a clock, or other automaton, from that of its weights and its wheels; so that, so far as these are concerned, it is not necessary to conceive any other vegetative or sensitive soul, nor any other principle of motion, or of life, than the blood and the spirits agitated by the fire which burns continually in the heart, and which is no wise essentially different from all the fires which exist in inanimate bodies.'

ADDITIONAL REMARKS ON MIRACLES.

AMONG the scraps of manuscript written at the time when Mr. Mozley's work occupied my attention I find the following reflections :—

With regard to the influence of modern science which Mr. Mozley rates so low, one effect of it is certainly to enhance the magnitude of many of the recorded miracles, and to increase proportionably the difficulties of belief. The ancients knew but little of the vastness of the universe. The Rev. Mr. Kirkman, for example, has shown what inadequate notions the Jews entertained regarding the ' firmament of heaven ; ' and Professor Airy refers to the case of a Greek philosopher who was persecuted for hazarding the assertion, then deemed monstrous, that the sun might be as large as the whole country of Greece. The concerns of a universe, regarded from this point of view, were vastly more commensurate with man and his concerns than those of the universe which science now reveals to us ; and hence that to suit man's purposes, or in compliance with his prayers, changes should occur in the order of the universe, was more easy of belief in the ancient world than it can be now. In the very magnitude which it assigns to natural phenomena, science has augmented the distance between them and man, and increased the popular belief in their orderly progression.

As a natural consequence the demand for evidence is more exacting than it used to be, whenever it is affirmed that such order has been disturbed.

Let us take as an illustration the miracle by which the victory of Joshua over the Amorites was rendered complete, where the sun is reported to have stood still for ' a whole day ' upon Gibeon, and the moon in the valley of Ajalon. An Englishman of average education at the present day would naturally demand a greater amount of evidence to prove that this occurrence took place than would have satisfied an Israelite in the age succeeding that of Joshua. For to the one the miracle probably consisted of the stoppage of a ball of fire less than a yard in diameter, while to the other it would be the stoppage of an orb fourteen hundred thousand times the earth in size. And even accepting the interpretation which instructed divines now put upon this text, that Joshua dealt with what was apparent merely, but that what really occurred was the suspension of the earth's rotation, I think a greater reserve in accepting the miracle, and a right to demand stronger evidence in support of it, will be conceded to a modern man of science than would have sufficed for an ancient Jew.

There is a scientific imagination as well as a historic imagination, and when by the exercise of the former the stoppage of the earth's rotation is clearly realised, the event assumes proportions so vast in comparison with the result to be obtained by it that belief reels under the reflection. The energy here involved is equal to that of six trillions of horses working for the whole of the time employed by Joshua in the destruction of his foes. The amount of power thus expended would be sufficient to supply every individual of an army a thousand times the strength of that of Joshua, with a thousand times the fighting power of each of Joshua's soldiers, not for the few hours necessary to the extinction of a handful of Amorites, but for millions of years. All this wonder is silently passed over by the sacred historian, confessedly because he knew nothing about it. Whether therefore we consider the miracle as purely evidential, or as a practical means of vengeance, the same lavish squandering of energy stares the scientific man in the face. If evidential, the energy was wasted, because the Israelites knew nothing of its amount; if simply destructive, then the ratio of the quantity lost to that employed may be inferred from the foregoing figures.

To other miracles similar remarks apply. Transferring thought from our little sand-grain of an earth to the immeasurable heavens, where countless worlds, with their freights of life, probably revolve unseen, the very suns which warm them being barely seen by us across abysmal space ; reflecting that beyond these sparks of solar fire suns innumerable may lie, whose light can never stir the optic nerve at all ; and bringing this conception face to face with the idea that the Builder and Sustainer of it all should contract himself to a burning bush, or behave in other familiar ways ascribed to him—it is easy to understand how astounding the incongruity must appear to the scientific man. Did this credulous prattle of the ancients about miracles stand alone ; were it not locally associated with words of imperishable wisdom, and with examples of moral grandeur unmatched elsewhere in the history of the human race, both the miracles and their ' evidences ' would have long since ceased to be the trans-

mitted inheritance of intelligent men. Under the pressure
of the awe which this universe inspires, well may we
exclaim in David's spirit, if not in David's words :—
' When I consider the heavens the work of thy fingers,
the moon, and the stars, which thou hast ordained ;
what is man that thou shouldst be mindful of him, or
the son of man that thou shouldst so regard him.'

If you ask me who is to limit the outgoings of Almighty
power, my answer is, not I. If you should urge that if
the Builder and the Maker of this universe chose to stop
the rotation of the earth, or to take the form of a burning bush, there is nothing to prevent him from doing so,
I am not prepared to contradict you. I neither agree
with you nor differ from you, for it is a subject of which
I know nothing. But I observe that in such questions
regarding Almighty power, your enquiries relate, not to
that power as it is actually displayed in the universe, but
to the power of your own imagination. Your question
is, not has the Omnipotent done so and so ? or is it in the
least likely that the Omnipotent should do so and so ?
but, is my imagination competent to picture a being able
and willing to do so and so ? I am not prepared to
deny your competence. To the human mind belongs the
faculty of enlarging and diminishing, of distorting and
combining indefinitely the objects revealed by the senses,
or by its own consciousness. It can imagine a mouse as
large as an elephant, an elephant as large as a mountain,
and a mountain as high as the stars. It can separate
congruities and unite incongruities. We see a fish and
we see a woman ; we can drop one half of each, and
unite in idea the other two halves to a mermaid. We
see a horse and we see a man ; we are able to drop one

half of each, and unite the other two halves to a centaur. Thus also the pictorial representations of the Deity, the bodies and wings of cherubs and seraphs, the hoofs, horns, and tail of the Evil One, the joys of the blessed, and the torments of the damned, have been elaborated from materials furnished to the imagination by the senses. And it behoves you and me to take care that our notions of the Power which rules the universe are not mere fanciful or ignorant enlargements of human power. The capabilities of what you call your reason are not denied. By the exercise of the power here adverted to, and which may be called the *mythologic imagination*, you can picture to yourself a being able and willing to do any and every conceivable thing. You are right in saying that in opposition to this power science is of no avail. Mr. Mozley would call it 'a weapon of air.' The man of science, however, while accepting the figure, would probably reverse its parts, thinking that it is not science which is here the thing of air, but the unsubstantial figment of the imagination to which its solidity is opposed.

THE END.

LONDON: PRINTED BY
SPOTTISWOODE AND CO., NEW-STREET SQUARE
AND PARLIAMENT STREET

HEAT A MODE OF MOTION.

Fourth Edition, with Alterations and Additions.   Plate and 108 Woodcuts. Crown 8vo. price 10s. 6d.

'We see, with not a little satisfaction, that this work has actually gone through four separate editions since the year 1863, and so far we have every reason to be satisfied. It must be confessed that it is the only work which the English student finds convenient. But this is not all; it is carefully brought up in each edition to the state of actual knowledge of the time, and even the present edition contains matters and facts not to be found in any of the earlier issues. . . . We take the foregoing as one

of the most striking of the novelties in this admirable work, and also one of the most interesting of the recent facts recorded by the author. But the whole work is full of such, and we cannot do better than recommend those of our readers who are already unfamiliar with it to procure it for themselves at once. It is not only the best work on the subject in the English language, but it is in itself especially valuable as an eloquent and comprehensive treatise.'
POPULAR SCIENCE REVIEW.

SOUND;

A Course of Eight Lectures delivered at the Royal Institution of Great Britain. Second Edition, revised; with a Portrait of M. Chladni and 169 Woodcut Illustrations.   Crown 8vo. price 9s.

'The contents of Professor TYNDALL'S book are of so attractive a nature, and recommend themselves so strongly, not only to the dilettante lover of knowledge, but to those who are earnestly engaged in the cultivation of science, that we are not surprised a second edition has been speedily called for. Having already noticed at length, in our review of the first edition, the characteristic features of the work, the number and ingenuity of the experiments (in which Professor TYNDALL stands without a rival), the felicitous explanations and varied illustrations, we need here make no further remark than to say that the present is a reprint of the former

edition with the exception of a chapter containing a summary of the recent researches of M. REGNAULT, written by himself. This contains some interesting observations on the propagation of sound in closed tubes, in which it is shewn that the diameter of the tubes makes a considerable difference in the intensity with which the wave is propagated through it, diminishing rapidly the smaller the section of the tube. . . . There are other interesting facts described in reference to the velocity of the propagation of waves, which we have not space to give, but which will well repay perusal.'   The LANCET.

FARADAY AS A DISCOVERER.

New and Cheaper Edition, with Two Portraits.   Fcp. 8vo. price 3s. 6d.

RESEARCHES ON DIAMAGNETISM AND MAGNE-

CRYSTALLIC ACTION; including the question of Diamagnetic Polarity. With Six Copper Plates and numerous Woodcut Illustrations.   8vo. price 14s.

'Dr. FARADAY, by a series of inductive researches of the most perfect character, established DIAMAGNETISM as a force of almost universal influence upon matter, but possessing principles which broadly distinguished it from that magnetism which peculiarly belongs to iron, but which is manifested in a less degree by some three or four other metals. M. PLUCKER discovered the action of a magnet upon crystallised bodies, and gave the name of Magne-crystallic force to it, finding it to be distinct from either magnetism or diamagnetism by its giving a determined position to the mass under its influence. M. PLUCKER'S investigations led him to believe that the direction assumed by a crystal under magnetic influence was determined by the optic axis of the crystal; and Dr. FARADAY, concurring in this view, called it the optic axis force. Dr. TYNDALL took up the inquiry at this point, and was led to a somewhat different conclusion. He appeared to prove that the position of the optic axis is not necessarily the line of magne-crystallic force, and that the force which determined the position of the optic axis in the

magnetic field was not independent of the magnetism or diamagnetism of the mass of the crystal. Beyond this Dr. TYNDALL has shewn that the lines of cleavage seem to influence the position of the crystal in the magnetic field, as they will be axial in a magnetic and equatorial in a diamagnetic crystal; and everything that tends to destroy the cleavages tends also to destroy the directive power. This volume is devoted to a republication of the papers in which these important researches were recorded. . . . There are many advantages in thus collecting together a set of researches of this nature. We have now in a collected form, for the use of the scientific student, a record of the experimental evidence upon which the conclusions above referred to are based. The book is plentifully illustrated with carefully-drawn woodcuts, which will be found of great use to those who may desire to investigate further those magnetic phenomena which appear to lead to an elucidation of the mysteries involved in the atomic constitution of matter.'   ATHENÆUM.

London : LONGMANS and CO. Paternoster Row.

39 Paternoster Row, E.C.

London: *January* 1871.

GENERAL LIST OF WORKS

PUBLISHED BY

Messrs. LONGMANS, GREEN, READER, and DYER.

History, Politics, Historical Memoirs, &c.

The History of England from the fall of Wolsey to the Defeat of the Spanish Armada. By James Anthony Froude, M.A.

> Cabinet Edition, 12 vols. cr. 8vo. £3 12s.
> Library Edition, 12 vols. 8vo. £8 18s.

The History of England from the Accession of James II. By Lord Macaulay.

> Library Edition, 5 vols. 8vo. £4.
> Cabinet Edition, 8 vols. post 8vo. 48s.
> People's Edition, 4 vols. crown 8vo. 16s.

Lord Macaulay's Works. Complete and uniform Library Edition. Edited by his Sister, Lady Trevelyan. 8 vols. 8vo. with Portrait, price £5 5s. cloth, or £8 8s. bound in tree-calf by Rivière.

An Essay on the History of the English Government and Constitution, from the Reign of Henry VII. to the Present Time. By John Earl Russell. Fourth Edition, revised. Crown 8vo. 6s.

Selections from Speeches of Earl Russell, 1817 to 1841, and from Despatches, 1859 to 1865; with Introductions. 2 vols. 8vo. 28s.

Varieties of Vice-Regal Life. By Major-General Sir William Denison, K.C.B. late Governor-General of the Australian Colonies, and Governor of Madras. With Two Maps. 2 vols. 8vo. 28s.

On Parliamentary Government in England : its Origin, Development, and Practical Operation. By Alpheus Todd, Librarian of the Legislative Assembly of Canada. 2 vols. 8vo. price £1 17s.

The Constitutional History of England since the Accession of George III. 1760—1860. By Sir Thomas Erskine May, K.C.B. Second Edit. 2 vols. 8vo. 33s.

A Historical Account of the Neu- trality of Great Britain during the American Civil War. By Montague Bernard, M.A. Royal 8vo. price 16s.

The History of England, from the Earliest Times to the Year 1866. By C. D. Yonge, Regius Professor of Modern History in the Queen's University, Belfast. New Edition. Crown 8vo. 7s. 6d.

A History of Wales, derived from Authentic Sources. By Jane Williams, Ysgafell, Author of a Memoir of the Rev. Thomas Price, and Editor of his Literary Remains. 8vo. 14s.

A

Lectures on the History of England, from the Earliest Times to the Death of King Edward II. By WILLIAM LONGMAN. With Maps and Illustrations. 8vo. 15s.

The History of the Life and Times of Edward the Third. By WILLIAM LONGMAN. With 9 Maps, 8 Plates, and 16 Woodcuts. 2 vols. 8vo. 28s.

History of Civilization in England and France, Spain and Scotland. By HENRY THOMAS BUCKLE. New Edition of the entire work, with a complete INDEX. 3 vols. crown 8vo. 24s.

Realities of Irish Life. By W. STEUART TRENCH, Land Agent in Ireland to the Marquess of Lansdowne, the Marquess of Bath, and Lord Digby. Fifth Edition. Crown 8vo. 6s.

The Student's Manual of the History of Ireland. By M. F. CUSACK, Authoress of the 'Illustrated History of Ireland, from the Earliest Period to the Year of Catholic Emancipation.' Crown 8vo. price 6s.

A Student's Manual of the History of India, from the Earliest Period to the Present. By Colonel MEADOWS TAYLOR, M.R.A.S. M.R.I.A. Crown 8vo. with Maps, 7s. 6d.

The History of India, from the Earliest Period to the close of Lord Dalhousie's Administration. By JOHN CLARK MARSHMAN. 3 vols. crown 8vo. 22s. 6d.

Indian Polity: a View of the System of Administration in India. By Lieut.-Col. GEORGE CHESNEY. Second Edition, revised, with Map. 8vo. 21s.

Home Politics: being a Consideration of the Causes of the Growth of Trade in relation to Labour, Pauperism, and Emigration. By DANIEL GRANT. 8vo. 7s.

Democracy in America. By ALEXIS DE TOCQUEVILLE. Translated by HENRY REEVE. 2 vols. 8vo. 21s.

Waterloo Lectures: a Study of the Campaign of 1815. By Colonel CHARLES C. CHESNEY, R.E. late Professor of Military Art and History in the Staff College. Second Edition. 8vo. with Map, 10s. 6d.

The Military Resources of Prussia and France, and Recent Changes in the Art of War. By Lieut.-Col. CHESNEY, R.E. and HENRY REEVE, D.C.L. Crown 8vo. 7s. 6d.

The Overthrow of the Germanic Confederation by Prussia in 1866. By Sir A. MALET, Bart. K.B.C. late H.B.M. Envoy and Minister at Frankfort. With 5 Maps. 8vo. 18s.

The Oxford Reformers—John Colet, Erasmus, and Thomas More; being a History of their Fellow-Work. By FREDERIC SEEBOHM. Second Edition. 8vo. 14s.

History of the Reformation in Europe in the Time of Calvin. By J. H. MERLE D'AUBIGNÉ, D.D. VOLS. I. and II. 8vo. 28s. VOL. III. 12s. VOL. IV. price 16s. and VOL. V. price 16s.

Chapters from French History; St. Louis, Joan of Arc, Henri IV. with Sketches of the Intermediate Periods. By J. H. GURNEY, M.A. New Edition. Fcp. 8vo. 6s. 6d.

The History of Greece. By C. THIRLWALL, D.D. Lord Bishop of St. David's. 8 vols. fcp. 28s.

The Tale of the Great Persian War, from the Histories of Herodotus. By GEORGE W. COX, M.A. late Scholar of Trin. Coll. Oxon. Fcp. 3s. 6d.

Greek History from Themistocles to Alexander, in a Series of Lives from Plutarch. Revised and arranged by A. H. CLOUGH. Fcp. with 44 Woodcuts, 6s.

Critical History of the Language and Literature of Ancient Greece. By WILLIAM MURE, of Caldwell. 5 vols. 8vo. £3 9s.

History of the Literature of Ancient Greece. By Professor K. O. MÜLLER. Translated by LEWIS and DONALDSON. 3 vols. 8vo. 21s.

The History of Rome. By WILHELM IHNE. Translated and revised by the Author. VOLS. I. and II. 8vo. [Just ready.

History of the City of Rome from its Foundation to the Sixteenth Century of the Christian Era. By THOMAS H. DYER, LL.D. 8vo. with 2 Maps, 15s.

History of the Romans under the Empire. By Very Rev. CHARLES MERIVALE, D.C.L. Dean of Ely. 8 vols. post 8vo. price 48s.

The Fall of the Roman Republic; a Short History of the Last Century of the Commonwealth. By the same Author. 12mo. 7s. 6d.

Historical and Chronological Encyclopædia; comprising Chronological Notices of all the Great Events of Universal History, including Treaties, Alliances, Wars, Battles, &c.; Incidents in the Lives of Eminent Men, Scientific and Geographical Discoveries, Mechanical Inventions, and Social, Domestic, and Economical Improvements. By the late B. B. Woodward, B.A. and W. L. R. Cates. 1 vol. 8vo. [In the press.

History of European Morals from Augustus to Charlemagne. By W. E. H. Lecky, M.A. 2 vols. 8vo. price 28s.

History of the Rise and Influence of the Spirit of Rationalism in Europe. By the same Author. Cabinet Edition (the Fourth). 2 vols. crown 8vo. price 16s.

God in History; or, the Progress of Man's Faith in the Moral Order of the World. By the late Baron Bunsen. Translated from the German by Susanna Winkworth; with a Preface by Dean Stanley 3 vols. 8vo. 42s.

Socrates and the Socratic Schools. Translated from the German of Dr. E. Zeller, with the Author's approval, by the Rev. Oswald J. Reichel, B.C.L. and M.A. Crown 8vo. 8s. 6d.

The Stoics, Epicureans, and Sceptics. Translated from the German of Dr. E. Zeller, with the Author's approval, by Oswald J. Reichel, B.C.L. and M.A. Crown 8vo. 14s.

The History of Philosophy, from Thales to Comte. By George Henry Lewes. Third Edition, rewritten and enlarged. 2 vols. 8vo. 30s.

The Mythology of the Aryan Nations. By George W. Cox, M.A. late Scholar of Trinity College, Oxford. 2 vols. 8vo. price 28s.

The English Reformation. By F. C. Massingberd, M.A. Chancellor of Lincoln. 4th Edition, revised. Fcp. 7s. 6d.

Maunder's Historical Treasury; comprising a General Introductory Outline of Universal History, and a Series of Separate Histories. Fcp. 6s.

Critical and Historical Essays contributed to the *Edinburgh Review* by the Right Hon. Lord Macaulay:—

Cabinet Edition, 4 vols. 24s.
Library Edition, 3 vols. 8vo. 36s.
People's Edition, 2 vols. crown 8vo. 8s.
Student's Edition, crown 8vo. 6s.

History of the Early Church, from the First Preaching of the Gospel to the Council of Nicæa, A.D. 325. By the Author of 'Amy Herbert.' New Edition. Fcp. 4s. 6d.

Sketch of the History of the Church of England to the Revolution of 1688. By the Right Rev. T. V. Short, D.D. Lord Bishop of St. Asaph. Eighth Edition. Crown 8vo. 7s. 6d.

History of the Christian Church, from the Ascension of Christ to the Conversion of Constantine. By E. Burton, D.D late Regius Prof. of Divinity in the University of Oxford. Fcp. 8s. 6d.

Biographical Works.

The Life of Isambard Kingdom Brunel, Civil Engineer. By Isambard Brunel, B.C.L. of Lincoln's Inn, Chancellor of the Diocese of Ely. With Portrait, Plates, and Woodcuts. 8vo. 21s.

The Life and Letters of the Rev. Sydney Smith. Edited by his Daughter, Lady Holland, and Mrs. Austin. New Edition, complete in One Volume. Crown 8vo. price 6s.

A Memoir of G. E. L. Cotton, D.D. late Lord Bishop of Calcutta; with Selections from his Journals and Letters. Edited by Mrs. Cotton. With Portrait. 8vo. [Just ready.

Some Memorials of R. D. Hampden, Bishop of Hereford. Edited by his Daughter, Henrietta Hampden. With Portrait. 8vo. [Just ready.

The Life and Travels of George Whitefield, M.A. of Pembroke College, Oxford, Chaplain to the Countess of Huntingdon. By J. P. Gledstone. Post 8vo. [Just ready.

Memoir of Pope Sixtus the Fifth. By Baron Hübner. Translated from the Original in French, with the Author's sanction, by Hubert E. H. Jerningham. 2 vols. 8vo. [In the press.

The Life and Letters of Faraday. By Dr. BENCE JONES, Secretary of the Royal Institution. Second Edition, with Portrait and Woodcuts. 2 vols. 8vo. 28s.

Faraday as a Discoverer. By JOHN TYNDALL, LL.D. F.R.S. Professor of Natural Philosophy in the Royal Institution. New and Cheaper Edition, with Two Portraits. Fcp. 8vo. 3s. 6d.

Lives of the Lord Chancellors and Keepers of the Great Seal of Ireland, from the Earliest Times to the Reign of Queen Victoria. By J. R. O'FLANAGAN, M.R.I.A. Barrister. 2 vols. 8vo. 36s.

Dictionary of General Biography; containing Concise Memoirs and Notices of the most Eminent Persons of all Countries, from the Earliest Ages to the Present Time. Edited by WILLIAM L. R. CATES. 8vo. price 21s.

Memoirs of Baron Bunsen, drawn chiefly from Family Papers by his Widow, FRANCES Baroness BUNSEN. Second Edition, abridged; with 2 Portraits and 4 Woodcuts. 2 vols. post 8vo. 21s.

The Letters of the Right Hon. Sir George Cornewall Lewis to various Friends. Edited by his Brother, the Rev. Canon Sir G. F. LEWIS, Bart. 8vo. with Portrait, 14s.

Life of the Duke of Wellington. By the Rev. G. R. GLEIG, M.A. Popular Edition, carefully revised; with copious Additions. Crown 8vo. with Portrait, 5s.

Father Mathew: a Biography. By JOHN FRANCIS MAGUIRE, M.P. Popular Edition, with Portrait. Crown 8vo. 8s. 6d.

History of my Religious Opinions. By J. H. NEWMAN, D.D. Being the Substance of Apologia pro Vitâ Suâ. Post 8vo. price 6s.

Letters and Life of Francis Bacon, including all his Occasional Works. Collected and edited, with a Commentary, by J. SPEDDING. VOLS. I. & II. 8vo. 24s. VOLS. III. & IV. 24s. VOL. V. 12s.

Felix Mendelssohn's Letters from *Italy and Switzerland,* and *Letters* from 1833 to 1847, translated by Lady WALLACE. With Portrait. 2 vols. crown 8vo. 5s. each.

Memoirs of Sir Henry Havelock, K.C.B. By JOHN CLARK MARSHMAN. People's Edition, with Portrait. Crown 8vo. price 3s. 6d.

Essays in Ecclesiastical Biography. By the Right Hon. Sir J. STEPHEN, LL.D. Cabinet Edition. Crown 8vo. 7s. 6d.

The Earls of Granard: a Memoir of the Noble Family of Forbes. Written by Admiral the Hon. JOHN FORBES, and Edited by GEORGE ARTHUR HASTINGS, present Earl of Granard, K.P. 8vo. 10s.

Vicissitudes of Families. By Sir J. BERNARD BURKE, C.B. Ulster King of Arms. New Edition, remodelled and enlarged. 2 vols. crown 8vo. 21s.

Lives of the Tudor Princesses, including Lady Jane Grey and her Sisters. By AGNES STRICKLAND. Post 8vo. with Portrait, &c. 12s. 6d.

Lives of the Queens of England. By AGNES STRICKLAND. Library Edition, newly revised; with Portraits of every Queen, Autographs, and Vignettes. 8 vols. post 8vo. 7s. 6d. each.

Maunder's Biographical Treasury. Thirteenth Edition, reconstructed and partly re-written, with above 1,000 additional Memoirs, by W. L. R. CATES. Fcp. 6s.

Criticism, Philosophy, Polity, &c.

The Subjection of Women. By JOHN STUART MILL. New Edition. Post 8vo. 5s.

On Representative Government. By JOHN STUART MILL. Third Edition. 8vo. 9s. crown 8vo. 2s.

On Liberty. By the same Author. Fourth Edition. Post 8vo. 7s. 6d. Crown 8vo. 1s. 4d.

Principles of Political Economy. By the same. Sixth Edition. 2 vols. 8vo. 30s. or in 1 vol. crown 8vo. 5s.

Utilitarianism. By the same. 3d Edit. 8vo. 5s.

Dissertations and Discussions. By the same Author. Second Edition. 3 vols. 8vo. 36s.

Examination of Sir W. Hamilton's Philosophy, and of the principal Philosophical Questions discussed in his Writings. By the same. Third Edition. 8vo. 16s.

Inaugural Address delivered to the University of St. Andrews. By JOHN STUART MILL. 8vo. 5s. Crown 8vo. 1s.

Analysis of the Phenomena of the Human Mind. By JAMES MILL. A New Edition, with Notes, Illustrative and Critical, by ALEXANDER BAIN, ANDREW FINDLATER, and GEORGE GROTE. Edited, with additional Notes, by JOHN STUART MILL. 2 vols. 8vo. price 28s.

The Elements of Political Eco- nomy. By HENRY DUNNING MACLEOD, M.A. Barrister-at-Law. 8vo. 16s.

A Dictionary of Political Economy; Biographical, Bibliographical, Historical, and Practical. By the same Author. VOL. I. royal 8vo. 30s.

Lord Bacon's Works, collected and edited by R. L. ELLIS, M.A. J. SPEDDING, M.A. and D. D. HEATH. New and Cheaper Edition. 7 vols. 8vo. price £3 13s. 6d.

A System of Logic, Ratiocinative and Inductive. By JOHN STUART MILL. Seventh Edition. 2 vols. 8vo. 25s.

Analysis of Mr. Mill's System of Logic. By W. STEBBING, M.A. New Edition. 12mo. 3s. 6d.

The Institutes of Justinian; with English Introduction, Translation, and Notes. By T. C. SANDARS, M.A. Barrister-at-Law. New Edition. 8vo. 15s.

The Ethics of Aristotle; with Essays and Notes. By Sir A. GRANT, Bart. M.A. LL.D. Second Edition, revised and completed. 2 vols. 8vo. price 28s.

The Nicomachean Ethics of Aris- totle. Newly translated into English. By R. WILLIAMS, B.A. Fellow and late Lecturer Merton College, Oxford. 8vo. 12s.

Bacon's Essays, with Annotations. By R. WHATELY, D.D. late Archbishop of Dublin. Sixth Edition. 8vo. 10s. 6d.

Elements of Logic. By R. WHATELY, D.D. late Archbishop of Dublin. New Edition. 8vo. 10s. 6d. crown 8vo. 4s. 6d.

Elements of Rhetoric. By the same Author. New Edition. 8vo. 10s. 6d. Crown 8vo. 4s. 6d.

English Synonymes. By E. JANE WHATELY. Edited by Archbishop WHATELY. 5th Edition. Fcp. 3s.

An Outline of the Necessary Laws of Thought: a Treatise on Pure and Applied Logic. By the Most Rev. W. THOMSON, D.D. Archbishop of York. Ninth Thousand. Crown 8vo. 5s. 6d.

The Election of Representatives, Parliamentary and Municipal; a Treatise. By THOMAS HARE, Barrister-at-Law. Third Edition, with Additions. Crown 8vo. 6s.

Speeches of the Right Hon. Lord MACAULAY, corrected by Himself. People's Edition, crown 8vo. 3s. 6d.

Lord Macaulay's Speeches on Parliamentary Reform in 1831 and 1832. 16mo. price ONE SHILLING.

Walker's Pronouncing Diction- ary of the English Language. Thoroughly revised Editions, by B. H. SMART. 8vo. 12s. 16mo. 6s.

A Dictionary of the English Language. By R. G. LATHAM, M.A. M.D. F.R.S. Founded on the Dictionary of Dr. S. JOHNSON, as edited by the Rev. H. J. TODD, with numerous Emendations and Additions. 4 vols. 4to. price £7.

Thesaurus of English Words and Phrases, classified and arranged so as to facilitate the expression of Ideas, and assist in Literary Composition. By P. M. ROGET, M.D. New Edition. Crown 8vo. 10s. 6d.

The Debater; a Series of Complete Debates, Outlines of Debates, and Questions for Discussion. By F. ROWTON. Fcp. 6s.

Lectures on the Science of Lan- guage, delivered at the Royal Institution. By MAX MÜLLER, M.A. &c. Foreign Member of the French Institute. 2 vols. 8vo. price 30s.

Chapters on Language. By F. W. FARRAR, M.A. F.R.S. late Fellow of Trin. Coll. Cambridge. Crown 8vo. 8s. 6d.

A Book about Words. By G. F. GRAHAM. Fcp. 8vo. 3s. 6d.

Southey's Doctor, complete in One Volume, edited by the Rev. J. W. WARTER, B.D. Square crown 8vo. 12s. 6d.

Historical and Critical Commen- tary on the Old Testament; with a New Translation. By M. M. KALISCH, Ph.D. Vol. I. *Genesis*, 8vo. 18s. or adapted for the General Reader, 12s. Vol. II. *Exodus*, 15s. or adapted for the General Reader, 12s. Vol III. *Leviticus*, Part I. 15s. or adapted for the General Reader, 8s.

A Hebrew Grammar, with Exercises. By the same. Part I. *Outlines with Exercises*, 8vo. 12s. 6d. KEY, 5s. Part II. *Exceptional Forms and Constructions*, 12s. 6d.

Manual of English Literature, Historical and Critical : with a Chapter on English Metres. By THOMAS ARNOLD, M.A. Second Edition. Crown 8vo. 7s. 6d.

A Latin-English Dictionary. By J. T. WHITE, D.D. of Corpus Christi College, and J. E. RIDDLE, M.A. of St. Edmund Hall, Oxford. Third Edition, revised. 2 vols. 4to. pp. 2,128, price 42s.

White's College Latin-English Dictionary (Intermediate Size), abridged from the Parent Work for the use of University Students. Medium 8vo. pp. 1,048, price 18s.

White's Junior Student's Complete Latin-English and English-Latin Dictionary. Revised Edition. Square 12mo. pp. 1,058, price 12s.

Separately {
ENGLISH-LATIN, 5s. 6d.
LATIN-ENGLISH, 7s. 6d.
}

An English-Greek Lexicon, containing all the Greek Words used by Writers of good authority. By C. D. YONGE, B.A. New Edition. 4to. 21s.

Mr. Yonge's New Lexicon, English and Greek, abridged from his larger work (as above). Square 12mo. 8s. 6d.

The Mastery of Languages; or, the Art of Speaking Foreign Tongues Idiomatically. By THOMAS PRENDERGAST, late of the Civil Service at Madras. Second Edition. 8vo. 6s.

A Greek-English Lexicon. Compiled by H. G. LIDDELL, D.D. Dean of Christ Church, and R. SCOTT, D.D. Dean of Rochester. Sixth Edition. Crown 4to. price 36s.

A Lexicon, Greek and English, abridged for Schools from LIDDELL and SCOTT'S *Greek-English Lexicon.* Twelfth Edition. Square 12mo. 7s. 6d.

A Practical Dictionary of the French and English Languages. By Professor LÉON CONTANSEAU, many years French Examiner for Military and Civil Appointments, &c. New Edition, carefully revised. Post 8vo. 10s. 6d.

Contanseau's Pocket Dictionary, French and English, abridged from the Practical Dictionary, by the Author. New Edition. 18mo. price 3s. 6d.

A Sanskrit-English Dictionary. The Sanskrit words printed both in the original Devanagari and in Roman letters ; with References to the Best Editions of Sanskrit Authors, and with Etymologies and comparisons of Cognate Words chiefly in Greek, Latin, Gothic, and Anglo-Saxon. Compiled by T. BENFEY. 8vo. 52s. 6d.

New Practical Dictionary of the German Language; German-English, and English-German. By the Rev. W. L. BLACKLEY, M.A. and Dr. CARL MARTIN FRIEDLÄNDER. Post 8vo. 7s. 6d.

Staff College Essays. By Lieutenant EVELYN BARING, Royal Artillery. 8vo. with Two Maps, 8s. 6d.

Miscellaneous Works and *Popular Metaphysics.*

The Essays and Contributions of A. K. H. B. Author of 'The Recreations of a Country Parson.' Uniform Editions :—

Recreations of a Country Parson. FIRST and SECOND SERIES, 3s. 6d. each.

The Commonplace Philosopher in Town and Country. Crown 8vo. 3s. 6d.

Leisure Hours in Town; Essays Consolatory, Æsthetical, Moral, Social, and Domestic. Crown 8vo. 3s. 6d.

The Autumn Holidays of a Country Parson. Crown 8vo. 3s. 6d.

The Graver Thoughts of a Country Parson. FIRST and SECOND SERIES, crown 8vo. 3s. 6d. each.

Critical Essays of a Country Parson, selected from Essays contributed to *Fraser's Magazine.* Crown 8vo. 3s. 6d.

Sunday Afternoons at the Parish Church of a Scottish University City. Crown 8vo. 3s. 6d.

Lessons of Middle Age, with some Account of various Cities and Men. Crown 8vo. 3s. 6d.

Counsel and Comfort Spoken from a City Pulpit. Crown 8vo. 3s. 6d.

Changed Aspects of Unchanged Truths; Memorials of St. Andrews Sundays. Crown 8vo. 3s. 6d.

Present-Day Thoughts; Memorials of St. Andrews Sundays. Crown 8vo. 3s. 6d.

Short Studies on Great Subjects. By JAMES ANTHONY FROUDE, M.A. late Fellow of Exeter College, Oxford. Third Edition. 8vo. 12s.

Lord Macaulay's Miscellaneous Writings:—

LIBRARY EDITION, 2 vols. 8vo. Portrait, 21s.
PEOPLE'S EDITION, 1 vol. crown 8vo. 4s. 6d.

The Rev. Sydney Smith's Miscellaneous Works; including his Contributions to the *Edinburgh Review*. 1 vol. crown 8vo. 6s.

The Wit and Wisdom of the Rev. SYDNEY SMITH: a Selection of the most memorable Passages in his Writings and Conversation. Crown 8vo. 3s. 6d.

The Silver Store. Collected from Mediæval Christian and Jewish Mines. By the Rev. S. BARING-GOULD, M.A. Crown 8vo. 3s. 6d.

Traces of History in the Names of Places; with a Vocabulary of the Roots out of which Names of Places in England and Wales are formed. By FLAVELL EDMUNDS. Crown 8vo. 7s. 6d.

The Eclipse of Faith; or, a Visit to a Religious Sceptic. By HENRY ROGERS. Twelfth Edition. Fcp. 5s.

Defence of the Eclipse of Faith, by its Author. Third Edition. Fcp. 3s. 6d.

Selections from the Correspondence of R. E. H. Greyson. By the same Author. Third Edition. Crown 8vo. 7s. 6d.

Families of Speech, Four Lectures delivered at the Royal Institution of Great Britain. By the Rev. F. W. FARRAR, M.A. F.R.S. Post 8vo. with 2 Maps, 5s. 6d.

Chips from a German Workshop; being Essays on the Science of Religion, and on Mythology, Traditions, and Customs. By MAX MÜLLER, M.A. &c. Foreign Member of the French Institute. 3 vols. 8vo. £2.

Word Gossip; a Series of Familiar Essays on Words and their Peculiarities. By the Rev. W. L. BLACKLEY, M.A. Fcp. 8vo. 5s.

An Introduction to Mental Philosophy, on the Inductive Method. By J. D. MORELL, M.A. LL.D. 8vo. 12s.

Elements of Psychology, containing the Analysis of the Intellectual Powers. By the same Author. Post 8vo. 7s. 6d.

The Secret of Hegel: being the Hegelian System in Origin, Principle, Form, and Matter. By JAMES HUTCHISON STIRLING. 2 vols. 8vo. 28s.

Sir William Hamilton; being the Philosophy of Perception: an Analysis. By the same Author. 8vo. 5s.

The Senses and the Intellect. By ALEXANDER BAIN, LL.D. Prof. of Logic in the Univ. of Aberdeen. Third Edition. 8vo. 15s.

The Emotions and the Will, by the same Author. Second Edition. 8vo. 15s.

On the Study of Character, including an Estimate of Phrenology. By the same Author. 8vo. 9s.

Mental and Moral Science: a Compendium of Psychology and Ethics. By the same Author. Second Edition. Crown 8vo. 10s. 6d.

Strong and Free; or, First Steps towards Social Science. By the Author of ' My Life and What shall I do with it ? ' 8vo. 10s. 6d.

The Philosophy of Necessity; or, Natural Law as applicable to Mental, Moral, and Social Science. By CHARLES BRAY. Second Edition. 8vo. 9s.

The Education of the Feelings and Affections. By the same Author. Third Edition. 8vo. 3s. 6d.

On Force, its Mental and Moral Correlates. By the same Author. 8vo. 5s.

Time and Space; a Metaphysical Essay. By SHADWORTH H. HODGSON. (This work covers the whole ground of Speculative Philosophy.) 8vo. price 16s.

The Theory of Practice; an Ethical Inquiry. By the same Author. (This work, in conjunction with the foregoing, completes a system of Philosophy.) 2 vols. 8vo. price 24s.

A Treatise on Human Nature; being an Attempt to Introduce the Experimental Method of Reasoning into Moral Subjects. By DAVID HUME. Edited, with Notes, &c. by T. H. GREEN, Fellow, and T. H. GROSE, late Scholar, of Balliol College, Oxford. [*In the press.*

Essays Moral, Political, and Literary. By DAVID HUME. By the same Editors. [*In the press.*

. The above will form a new edition of DAVID HUME'S *Philosophical Works*, complete in Four Volumes, but to be had in Two separate Sections as announced.

Astronomy, Meteorology, Popular Geography, &c.

Outlines of Astronomy. By Sir J. F. W. HERSCHEL, Bart. M.A. New Edition, revised; with Plates and Woodcuts. 8vo. 18s.

Other Worlds than Ours; the Plurality of Worlds Studied under the Light of Recent Scientific Researches. By R. A. PROCTOR, B.A. F.R.A.S. Second Edition, revised and enlarged; with 14 Illustrations. Crown 8vo. 10s. 6d.

The Sun; Ruler, Light, Fire, and Life of the Planetary System. By the same Author. With 10 Plates (7 coloured) and 107 Woodcuts. Crown 8vo. price 14s.

Saturn and its System. By the same Author. 8vo. with 14 Plates, 14s.

The Handbook of the Stars. By the same Author. Square fcp. 8vo. with 3 Maps, price 5s.

Celestial Objects for Common Telescopes. By T. W. WEBB, M.A. F.R.A.S. Second Edition, revised and enlarged, with Map of the Moon and Woodcuts. 16mo. price 7s. 6d.

Navigation and Nautical As- tronomy (Practical, Theoretical, Scientific) for the use of Students and Practical Men. By J. MERRIFIELD, F.R.A.S. and H. EVERS. 8vo. 14s.

A General Dictionary of Geo- graphy, Descriptive, Physical, Statistical, and Historical; forming a complete Gazetteer of the World. By A. KEITH JOHNSTON, F.R.S.E. New Edition. 8vo. price 31s. 6d.

M'Culloch's Dictionary, Geogra- phical, Statistical, and Historical, of the various Countries, Places, and principal Natural Objects in the World. Revised Edition, with the Statistical Information throughout brought up to the latest returns By FREDERICK MARTIN. 4 vols. 8vo. with coloured Maps, £4 4s.

A Manual of Geography, Physical, Industrial, and Political. By W. HUGHES, F.R.G.S. Prof. of Geog. in King's Coll. and in Queen's Coll. Lond. With 6 Maps. Fcp. 7s. 6d.

The States of the River Plate: their Industries and Commerce, Sheep Farming, Sheep Breeding, Cattle Feeding, and Meat Preserving; the Employment of Capital, Land and Stock and their Values, Labour and its Remuneration. By WILFRID LATHAM, Buenos Ayres. Second Edition. 8vo. 12s.

Maunder's Treasury of Geogra- phy, Physical, Historical, Descriptive, and Political. Edited by W. HUGHES, F.R.G.S. With 7 Maps and 16 Plates. Fcp. 6s.

Natural History and Popular Science.

Ganot's Elementary Treatise on Physics, Experimental and Applied, for the use of Colleges and Schools. Translated and Edited with the Author's sanction by E. ATKINSON, Ph.D. F.C.S. New Edition, revised and enlarged; with a Coloured Plate and 620 Woodcuts. Post 8vo. 15s.

The Elements of Physics or Natural Philosophy. By NEIL ARNOTT, M.D. F.R.S. Physician-Extraordinary to the Queen. Sixth Edition, re-written and completed. 2 Parts, 8vo. 21s.

The Forces of the Universe. By GEORGE BERWICK, M.D. Post 8vo. 5s.

Dove's Law of Storms, considered in connexion with the ordinary Movements of the Atmosphere. Translated by R. H. SCOTT, M.A. T.C.D. 8vo. 10s. 6d.

Sound: a Course of Eight Lectures delivered at the Royal Institution of Great Britain. By Professor JOHN TYNDALL, LL.D. F.R.S. New Edition, with Portrait and Woodcuts. Crown 8vo. 9s.

Heat a Mode of Motion. By Professor JOHN TYNDALL, LL.D. F.R.S. Fourth Edition. Crown 8vo. with Woodcuts, price 10s. 6d.

Researches on Diamagnetism and Magne-Crystallic Action; including the Question of Diamagnetic Polarity. By Professor TYNDALL. With 6 Plates and many Woodcuts. 8vo. 14s.

Notes of a Course of Nine Lec- tures on Light, delivered at the Royal Institution, A.D. 1869. By Professor TYNDALL. Crown 8vo. 1s. sewed, or 1s. 6d. cloth.

Notes of a Course of Seven Lectures on Electrical Phenomena and Theories, delivered at the Royal Institution, A.D. 1870. By Professor TYNDALL. Crown 8vo. 1s. sewed, or 1s. 6d. cloth.

Professor Tyndall's Essays on the Use and Limit of the Imagination in Science. Being the Second Edition, with Additions, of a Discourse on the Scientific Use of the Imagination. 8vo. 3s.

Light: its Influence on Life and Health. By FORBES WINSLOW, M.D. D.C.L. Oxon. (Hon.) Fcp. 8vo. 6s.

A Treatise on Electricity, in Theory and Practice. By A. DE LA RIVE, Prof. in the Academy of Geneva. Translated by C. V. WALKER, F.R.S. 3 vols. 8vo. with Woodcuts, £3 13s.

The Correlation of Physical Forces. By W. R. GROVE, Q.C. V.P.R.S. Fifth Edition, revised, and Augmented by a Discourse on Continuity. 8vo. 10s. 6d. The *Discourse*, separately, price 2s. 6d.

The Beginning: its When and its How. By MUNGO PONTON, F.R.S.E. Post 8vo. with very numerous Illustrations.

Manual of Geology. By S. HAUGHTON, M.D. F.R.S. Fellow of Trin. Coll. and Prof. of Geol. in the Univ. of Dublin. Second Edition, with 66 Woodcuts. Fcp. 7s. 6d.

Van Der Hoeven's Handbook of ZOOLOGY. Translated from the Second Dutch Edition by the Rev. W. CLARK, M.D. F.R.S. 2 vols. 8vo. with 24 Plates of Figures, 60s.

Professor Owen's Lectures on the Comparative Anatomy and Physiology of the Invertebrate Animals. Second Edition, with 235 Woodcuts. 8vo. 21s.

The Comparative Anatomy and Physiology of the Vertebrate Animals. By RICHARD OWEN, F.R.S. D.C.L. With 1,472 Woodcuts. 3 vols. 8vo. £3 13s. 6d.

The Origin of Civilisation and the Primitive Condition of Man; Mental and Social Condition of Savages. By Sir JOHN LUBBOCK, Bart. M.P. F.R.S. Second Edition, revised, with 25 Woodcuts. 8vo. price 16s.

The Primitive Inhabitants of Scandinavia. Containing a Description of the Implements, Dwellings, Tombs, and Mode of Living of the Savages in the North of Europe during the Stone Age. By SVEN NILSSON. 8vo. Plates and Woodcuts, 18s.

Homes without Hands: a Description of the Habitations of Animals, classed according to their Principle of Construction. By Rev. J. G. WOOD, M.A. F.L.S. With about 140 Vignettes on Wood. 8vo. 21s.

Bible Animals; being a Description of Every Living Creature mentioned in the Scriptures, from the Ape to the Coral. By the Rev. J. G. WOOD, M.A. F.L.S. With about 100 Vignettes on Wood. 8vo. 21s.

The Harmonies of Nature and Unity of Creation. By Dr. G. HARTWIG. 8vo. with numerous Illustrations, 18s.

The Sea and its Living Wonders. By the same Author. Third Edition, enlarged. 8vo. with many Illustrations, 21s.

The Tropical World. By the same Author. With 8 Chromoxylographs and 172 Woodcuts. 8vo. 21s.

The Polar World: a Popular Description of Man and Nature in the Arctic and Antarctic Regions of the Globe. By the same Author. With 8 Chromoxylographs, 3 Maps, and 85 Woodcuts. 8vo. 21s.

A Familiar History of Birds. By E. STANLEY, D.D. late Lord Bishop of Norwich. Fcp. with Woodcuts, 3s. 6d.

Kirby and Spence's Introduction to Entomology, or Elements of the Natural History of Insects. Crown 8vo. 5s.

Maunder's Treasury of Natural History, or Popular Dictionary of Zoology. Revised and corrected by T. S. CONBOLD, M.D. Fcp. with 900 Woodcuts, 6s.

The Elements of Botany for Families and Schools. Tenth Edition, revised by THOMAS MOORE, F.L.S. Fcp. with 154 Woodcuts, 2s. 6d.

The Treasury of Botany, or Popular Dictionary of the Vegetable Kingdom; with which is incorporated a Glossary of Botanical Terms. Edited by J. LINDLEY, F.R.S. and T. MOORE, F.L.S. assisted by eminent Contributors. Pp. 1,274, with 274 Woodcuts and 20 Steel Plates. Two PARTS, fcp. 8vo. 12s.

The British Flora; comprising the Phænogamous or Flowering Plants and the Ferns. By Sir W. J. HOOKER, K.H. and G. A. WALKER-ARNOTT, LL.D. 12mo. with 12 Plates, 14s.

The Rose Amateur's Guide. By THOMAS RIVERS. New Edition. Fcp. 4s.

Loudon's Encyclopædia of Plants; comprising the Specific Character, Description, Culture, History, &c. of all the Plants found in Great Britain. With upwards of 12,000 Woodcuts. 8vo. 42s.

B

Maunder's Scientific and Literary Treasury; a Popular Encyclopædia of Science, Literature, and Art. New Edition, thoroughly revised and in great part re-written, with above 1,000 new articles, by J. Y. Johnson, Corr. M.Z.S. Fcp. 6s.

A Dictionary of Science, Literature, and Art. Fourth Edition, re-edited by the late W. T. Brande (the Author) and George W. Cox, M.A. 3 vols. medium 8vo. price 63s. cloth.

Chemistry, Medicine, Surgery, and the Allied Sciences.

A Dictionary of Chemistry and the Allied Branches of other Sciences. By Henry Watts, F.C.S. assisted by eminent Scientific and Practical Chemists. 5 vols. medium 8vo. price £7 3s.

Elements of Chemistry, Theoretical and Practical. By William A. Miller, M.D. LL.D. Professor of Chemistry, King's College, London. Fourth Edition. 3 vols. 8vo. £3.
> Part I. Chemical Physics, 15s.
> Part II. Inorganic Chemistry, 21s.
> Part III. Organic Chemistry, 24s.

A Manual of Chemistry, De-scriptive and Theoretical. By William Odling, M.B. F.R.S. Part I. 8vo. 9s. Part II. nearly ready.

A Course of Practical Chemistry, for the use of Medical Students. By W. Odling, M.B. F.R.S. New Edition, with 70 new Woodcuts. Crown 8vo. 7s. 6d.

Outlines of Chemistry; or, Brief Notes of Chemical Facts. By the same Author. Crown 8vo. 7s. 6d.

Lectures on Animal Chemistry Delivered at the Royal College of Physicians in 1865. By the same Author. Crown 8vo. 4s. 6d.

Lectures on the Chemical Changes of Carbon, delivered at the Royal Institution of Great Britain. By the same Author. Crown 8vo. 4s. 6d.

Chemical Notes for the Lecture Room. By Thomas Wood, F.C.S. 2 vols. crown 8vo. I. on Heat, &c. price 3s. 6d. II. on the Metals, price 5s.

A Treatise on Medical Elec-tricity, Theoretical and Practical; and its Use in the Treatment of Paralysis, Neuralgia, and other Diseases. By Julius Althaus, M.D. &c. Second Edition, revised and partly re-written; with Plate and 62 Woodcuts. Post 8vo. price 15s.

The Diagnosis, Pathology, and Treatment of Diseases of Women; including the Diagnosis of Pregnancy. By Graily Hewitt, M.D. &c. President of the Obstetrical Society of London. Second Edition, enlarged; with 116 Woodcuts. 8vo. 24s.

Lectures on the Diseases of In-fancy and Childhood. By Charles West, M.D. &c. Fifth Edition. 8vo. 16s.

On the Surgical Treatment of Children's Diseases. By T. Holmes, M.A. &c. late Surgeon to the Hospital for Sick Children. Second Edition, with 9 Plates and 112 Woodcuts. 8vo. 21s.

A System of Surgery, Theoretical and Practical, in Treatises by Various Authors. Edited by T. Holmes, M.A. &c. Surgeon and Lecturer on Surgery at St. George's Hospital, and Surgeon-in-Chief to the Metropolitan Police. Second Edition, thoroughly revised, with numerous Illustrations. 5 vols. 8vo. £5 5s.

Lectures on the Principles and Practice of Physic. By Sir Thomas Watson, Bart. M.D. Physician-in-Ordinary to the Queen. New Edition in the press.

Lectures on Surgical Pathology. By James Paget, F.R.S. Third Edition, revised and re-edited by the Author and Professor W. Turner, M.B. 8vo. with 131 Woodcuts, 21s.

Cooper's Dictionary of Practical Surgery and Encyclopædia of Surgical Science. New Edition, brought down to the present time. By S. A. Lane, Surgeon to St. Mary's Hospital, &c. assisted by various Eminent Surgeons. Vol. II. 8vo. completing the work. [Early in 1871.

On Chronic Bronchitis, especially as connected with Gout, Emphysema, and Diseases of the Heart. By E. Headlam Greenhow, M.D. F.R.C.P. &c. 8vo. 7s. 6d.

The Climate of the South of France as Suited to Invalids; with Notices of Mediterranean and other Winter Stations. By C. T. Williams, M.A. M.D. Oxon. Assistant-Physician to the Hospital for Consumption at Brompton. Second Edition. Crown 8vo. 6s.

Pulmonary Consumption; its Nature, Treatment, and Duration exemplified by an Analysis of One Thousand Cases selected from upwards of Twenty Thousand. By C. J. B. WILLIAMS, M.D. F.R.S. Consulting Physician to the Hospital for Consumption at Brompton; and C. T. WILLIAMS, M.A. M.D. Oxon.
[*Nearly ready.*

Clinical Lectures on Diseases of the Liver, Jaundice, and Abdominal Dropsy. By C. MURCHISON, M.D. Physician and Lecturer on the Practice of Medicine, Middlesex Hospital. Post 8vo. with 25 Woodcuts, 10s. 6d.

Anatomy, Descriptive and Surgical. By HENRY GRAY, F.R.S. With about 410 Woodcuts from Dissections. Fifth Edition, by T. HOLMES, M.A. Cantab. With a New Introduction by the Editor. Royal 8vo. 28s.

Clinical Notes on Diseases of the Larynx, investigated and treated with the assistance of the Laryngoscope. By W. MARCET, M.D. F.R.S. Crown 8vo. with 5 Lithographs, 6s.

The House I Live in; or, Popular Illustrations of the Structure and Functions of the Human Body. Edited by T. G. GIRTIN. New Edition, with 25 Woodcuts. 16mo. price 2s. 6d.

Outlines of Physiology, Human and Comparative. By JOHN MARSHALL, F.R.C.S. Professor of Surgery in University College, London, and Surgeon to the University College Hospital. 2 vols. crown 8vo. with 122 Woodcuts, 32s.

Physiological Anatomy and Physiology of Man. By the late R. B. TODD, M.D. F.R.S. and W. BOWMAN, F.R.S. of King's College. With numerous Illustrations. VOL. II. 8vo. 25s.

VOL. I. New Edition by Dr. LIONEL S. BEALE, F.R.S. in course of publication; PART I. with 8 Plates, 7s. 6d.

Copland's Dictionary of Practical Medicine, abridged from the larger work, and throughout brought down to the present state of Medical Science. 8vo. 36s.

A Manual of Materia Medica and Therapeutics, abridged from Dr. PEREIRA'S *Elements* by F. J. FARRE, M.D. assisted by R. BENTLEY, M.R.C.S. and by R. WARINGTON, F.R.S. 1 vol. 8vo. with 90 Woodcuts, 21s.

Thomson's Conspectus of the British Pharmacopœia. Twenty-fifth Edition, corrected by E. LLOYD BIRKETT, M.D. 18mo. 6s.

Essays on Physiological Subjects. By GILBERT W. CHILD, M.A. F.L.S. F.C.S. Second Edition. Crown 8vo. with Woodcuts, 7s. 6d.

The Fine Arts, and Illustrated Editions.

In Fairyland; Pictures from the Elf-World. By RICHARD DOYLE. With a Poem by W. ALLINGHAM. With Sixteen Plates, containing Thirty-six Designs printed in Colours. Folio, 31s. 6d.

Life of John Gibson, R.A. Sculptor. Edited by Lady EASTLAKE. 8vo. 10s. 6d.

Materials for a History of Oil Painting. By Sir CHARLES LOCKE EASTLAKE, sometime President of the Royal Academy. 2 vols. 8vo. 30s.

Albert Durer, his Life and Works; including Autobiographical Papers and Complete Catalogues. By WILLIAM B. SCOTT. With Six Etchings by the Author and other Illustrations. 8vo. 16s.

Half-Hour Lectures on the History and Practice of the Fine and Ornamental Arts. By. W. B. SCOTT. Second Edition. Crown 8vo. with 50 Woodcut Illustrations, 8s. 6d.

The Lord's Prayer Illustrated by F. R. PICKERSGILL, R.A. and HENRY ALFORD, D.D. Dean of Canterbury. Imp. 4to. 21s.

The Chorale Book for England: the Hymns Translated by Miss C. WINKWORTH; the Tunes arranged by Prof. W. S. BENNETT and OTTO GOLDSCHMIDT. Fcp. 4to. 12s. 6d.

Six Lectures on Harmony. Delivered at the Royal Institution of Great Britain. By G. A. MACFARREN. 8vo. 10s.6d.

Lyra Germanica, the Christian Year. Translated by CATHERINE WINKWORTH; with 125 Illustrations on Wood drawn by J. LEIGHTON, F.S.A. Quarto, 21s.

Lyra Germanica, the Christian Life. Translated by CATHERINE WINKWORTH; with about 200 Woodcut Illustrations by J. LEIGHTON, F.S.A. and other Artists. Quarto, 21s.

The New Testament, illustrated with Wood Engravings after the Early Masters, chiefly of the Italian School. Crown 4to. 63s. cloth, gilt top ; or £5 5s. morocco.

The Life of Man Symbolised by the Months of the Year in their Seasons and Phases. Text selected by RICHARD PIGOT. 25 Illustrations on Wood from Original Designs by JOHN LEIGHTON, F.S.A. Quarto, 42s.

Cats' and Farlie's Moral Emblems ; with Aphorisms, Adages, and Proverbs of all Nations : comprising 121 Illustrations on Wood by J. LEIGHTON, F.S.A. with an appropriate Text by R. PIGOT. Imperial 8vo. 31s. 6d.

Shakspeare's Midsummer Night's Dream, illustrated with 24 Silhouettes or Shadow Pictures by P. KONEWKA, engraved on Wood by A. VOGEL. Folio, 31s. 6d.

Sacred and Legendary Art. By Mrs. JAMESON. 6 vols. square crown 8vo. price £5 15s. 6d.

Legends of the Saints and Martyrs. Fifth Edition, with 19 Etchings and 187 Woodcuts.   2 vols. price 31s. 6d.

Legends of the Monastic Orders. Third Edition, with 11 Etchings and 88 Woodcuts. 1 vol. price 21s.

Legends of the Madonna. Third Edition, with 27 Etchings and 165 Woodcuts. 1 vol. price 21s.

The History of Our Lord, with that of His Types and Precursors. Completed by Lady EASTLAKE.   Revised Edition, with 13 Etchings and 281 Woodcuts.   2 vols. price 42s.

The Useful Arts, Manufactures, &c.

Gwilt's Encyclopædia of Architecture, with above 1,600 Woodcuts. Fifth Edition, with Alterations and considerable Additions, by WYATT PAPWORTH. 8vo. 52s. 6d.

A Manual of Architecture : being a Concise History and Explanation of the principal Styles of European Architecture, Ancient, Mediæval, and Renaissance ; with their Chief Variations and a Glossary of Technical Terms. By THOMAS MITCHELL. With 150 Woodcuts. Crown 8vo. 10s. 6d.

Italian Sculptors : being a History of Sculpture in Northern, Southern, and Eastern Italy. By C. C. PERKINS. With 30 Etchings and 13 Wood Engravings. Imperial 8vo. 42s.

Tuscan Sculptors, their Lives, Works, and Times. By the same Author. With 45 Etchings and 28 Woodcuts from Original Drawings and Photographs. 2 vols. imperial 8vo. 63s.

Hints on Household Taste in Furniture, Upholstery, and other Details. By CHARLES L. EASTLAKE, Architect. Second Edition, with about 90 Illustrations. Square crown 8vo. 18s.

The Engineer's Handbook ; explaining the principles which should guide the young Engineer in the Construction of Machinery. By C. S. LOWNDES. Post 8vo. 5s.

Lathes and Turning, Simple, Mechanical, and Ornamental. By W. HENRY NORTHCOTT. With about 240 Illustrations on Steel and Wood. 8vo. 18s.

Principles of Mechanism, designed for the use of Students in the Universities, and for Engineering Students generally. By R. WILLIS, M.A. F.R.S. &c. Jacksonian Professor in the Univ. of Cambridge. Second Edition, enlarged ; with 374 Woodcuts. 8vo. 18s.

Handbook of Practical Telegraphy, published with the sanction of the Chairman and Directors of the Electric and International Telegraph Company, and adopted by the Department of Telegraphs for India. By R. S. CULLEY. Third Edition. 8vo. 12s. 6d.

Ure's Dictionary of Arts, Manufactures, and Mines. Sixth Edition, rewritten and greatly enlarged by ROBERT HUNT, F.R.S. assisted by numerous Contributors. With 2,000 Woodcuts. 3 vols. medium 8vo. £4 14s. 6d.

Treatise on Mills and Millwork. By Sir W. FAIRBAIRN, Bart. With 18 Plates and 322 Woodcuts. 2 vols. 8vo. 32s.

Useful Information for Engineers. By the same Author. FIRST, SECOND, and THIRD SERIES, with many Plates and Woodcuts. 3 vols. crown 8vo. 10s. 6d. each.

The Application of Cast and Wrought Iron to Building Purposes. By the same Author. Fourth Edition, with 6 Plates and 118 Woodcuts. 8vo. 16s.

Iron Ship Building, its History and Progress, as comprised in a Series of Experimental Researches. By W. FAIRBAIRN, Bart. F.R.S. With 4 Plates and 130 Woodcuts, 8vo. 18s.

Encyclopædia of Civil Engineering, Historical, Theoretical, and Practical. By E. CRESY, C.E. With above 3,000 Woodcuts. 8vo. 42s.

A Treatise on the Steam Engine, in its various Applications to Mines, Mills, Steam Navigation, Railways, and Agriculture. By J. BOURNE, C.E. New Edition; with Portrait, 37 Plates, and 546 Woodcuts. 4to. 42s.

Catechism of the Steam Engine, in its various Applications to Mines, Mills, Steam Navigation, Railways, and Agriculture. By JOHN BOURNE, C.E. New Edition, with 89 Woodcuts. Fcp. 6s.

Recent Improvements in the Steam-Engine. By JOHN BOURNE, C.E. being a SUPPLEMENT to his 'Catechism of the Steam-Engine.' New Edition, including many New Examples, with 124 Woodcuts. Fcp. 8vo. 6s.

Bourne's Examples of Modern Steam, Air, and Gas Engines of the most Approved Types, as employed for Pumping, for Driving Machinery, for Locomotion, and for Agriculture, minutely and practically described. In course of publication, to be completed in Twenty-four Parts, price 2s. 6d. each, forming One Volume, with about 50 Plates and 400 Woodcuts.

A Treatise on the Screw Propeller, Screw Vessels, and Screw Engines, as adapted for purposes of Peace and War. By JOHN BOURNE, C.E. Third Edition, with 54 Plates and 287 Woodcuts. Quarto, 63s.

Handbook of the Steam Engine. By JOHN BOURNE, C.E. forming a KEY to the Author's Catechism of the Steam Engine. With 67 Woodcuts. Fcp. 9s.

A History of the Machine-Wrought Hosiery and Lace Manufactures. By WILLIAM FELKIN, F.L.S. F.S.S. With several Illustrations. Royal 8vo. 21s.

Mitchell's Manual of Practical Assaying. Third Edition for the most part re-written, with all the recent Discoveries incorporated. By W. CROOKES, F.R.S. With 188 Woodcuts. 8vo. 28s.

Reimann's Handbook of Aniline and its Derivatives; a Treatise on the Manufacture of Aniline and Aniline Colours. Revised and edited by WILLIAM CROOKES, F.R.S. 8vo. with 5 Woodcuts, 10s. 6d.

On the Manufacture of Beet-Root Sugar in England and Ireland. By WILLIAM CROOKES, F.R.S. With 11 Woodcuts. 8vo. 8s. 6d.

Practical Treatise on Metallurgy, adapted from the last German Edition of Professor KERL's *Metallurgy* by W. CROOKES, F.R.S. &c. and E. RÖHRIG, Ph.D. M.E. 3 vols. 8vo. with 625 Woodcuts, price £4 19s.

The Art of Perfumery; the History and Theory of Odours, and the Methods of Extracting the Aromas of Plants. By Dr. PIESSE, F.C.S. Third Edition, with 53 Woodcuts. Crown 8vo. 10s. 6d.

Chemical, Natural, and Physical Magic, for Juveniles during the Holidays. By the same Author. With 38 Woodcuts. Fcp. 6s.

Loudon's Encyclopædia of Agriculture: comprising the Laying-out, Improvement, and Management of Landed Property, and the Cultivation and Economy of the Productions of Agriculture. With 1,100 Woodcuts. 8vo. 21s.

Loudon's Encyclopædia of Gardening: comprising the Theory and Practice of Horticulture, Floriculture, Arboriculture, and Landscape Gardening. With 1,000 Woodcuts. 8vo. 21s.

Bayldon's Art of Valuing Rents and Tillages, and Claims of Tenants upon Quitting Farms, both at Michaelmas and Lady-Day. Eighth Edition, revised by J. C. MORTON. 8vo. 10s. 6d.

Religious and *Moral Works.*

An Exposition of the 39 Articles, Historical and Doctrinal. By E. HAROLD BROWNE, D.D. Lord Bishop of Ely. Eighth Edition. 8vo. 16s.

Examination-Questions on Bishop Browne's Exposition of the Articles. By the Rev. J. GORLE, M.A. Fcp. 3s. 6d.

The Life and Epistles of St. Paul. By the Rev. W. J. CONYBEARE, M.A. and the Very Rev. J. S. HOWSON, D.D. Dean of Chester.

LIBRARY EDITION, with all the Original Illustrations, Maps, Landscapes on Steel, Woodcuts, &c. 2 vols. 4to. 48s.

INTERMEDIATE EDITION, with a Selection of Maps, Plates, and Woodcuts. 2 vols. square crown 8vo. 31s. 6d.

STUDENT'S EDITION, revised and condensed, with 46 Illustrations and Maps. 1 vol. crown 8vo. 9s.

The Voyage and Shipwreck of St. Paul ; with Dissertations on the Ships and Navigation of the Ancients. By JAMES SMITH, F.R.S. Crown 8vo. Charts, 10s. 6d.

Evidence of the Truth of the Christian Religion derived from the Literal Fulfilment of Prophecy. By ALEXANDER KEITH, D.D. 37th Edition, with numerous Plates, in square 8vo. 12s. 6d.; also the 39th Edition, in post 8vo. with 5 Plates, 6s.

The History and Destiny of the World and of the Church, according to Scripture. By the same Author. Square 8vo. with 40 Illustrations, 10s.

The History and Literature of the Israelites, according to the Old Testament and the Apocrypha. By C. DE ROTHSCHILD and A. DE ROTHSCHILD. With 2 Maps. 2 vols. post 8vo. price 12s. 6d.

VOL. I. *The Historical Books,* 7s. 6d.
VOL. II. *The Prophetic and Poetical Writings,* price 5s.

Ewald's History of Israel to the Death of Moses. Translated from the German. Edited, with a Preface and an Appendix, by RUSSELL MARTINEAU, M.A. Second Edition. 2 vols. 8vo. 24s.

History of the Karaite Jews. By WILLIAM HARRIS RULE, D.D. Post 8vo. price 7s. 6d.

The Life of Margaret Mary Hallahan, better known in the religious world by the name of Mother Margaret. By her RELIGIOUS CHILDREN. Second Edition. 8vo. with Portrait, 10s.

The See of Rome in the Middle Ages. By the Rev. OSWALD J. REICHEL, B.C.L. and M.A. 8vo. 18s.

The Evidence for the Papacy as derived from the Holy Scriptures and from Primitive Antiquity. By the Hon. COLIN LINDSAY. 8vo. 12s. 6d.

The Pontificate of Pius the Ninth; being the Third Edition, enlarged and continued, of 'Rome and its Ruler.' By J. F. MAGUIRE, M.P. Post 8vo. Portrait, price 12s. 6d.

Ignatius Loyola and the Early Jesuits. By STEWART ROSE. New Edition, in the press.

An Introduction to the Study of the New Testament, Critical, Exegetical, and Theological. By the Rev. S. DAVIDSON, D.D. LL.D. 2 vols. 8vo. 30s.

A Critical and Grammatical Com-mentary on St. Paul's Epistles. By C. J. ELLICOTT, D.D. Lord Bishop of Gloucester and Bristol. 8vo.

Galatians, Fourth Edition, 8s. 6d.
Ephesians, Fourth Edition, 8s. 6d.
Pastoral Epistles, Fourth Edition, 10s. 6d.
Philippians, Colossians, and Philemon, Third Edition, 10s. 6d.
Thessalonians, Third Edition, 7s. 6d.

Historical Lectures on the Life of Our Lord Jesus Christ : being the Hulsean Lectures for 1859. By C. J. ELLICOTT, D.D. Lord Bishop of Gloucester and Bristol. Fifth Edition. 8vo. 12s.

The Greek Testament; with Notes, Grammatical and Exegetical. By the Rev. W. WEBSTER, M.A. and the Rev. W. F. WILKINSON, M.A. 2 vols. 8vo. £2 4s.

Horne's Introduction to the Cri-tical Study and Knowledge of the Holy Scriptures. Twelfth Edition ; with 4 Maps and 22 Woodcuts and Facsimiles. 4 vols. 8vo. 42s.

Compendious Introduction to the Study of the Bible. Edited by the Rev. JOHN AYRE, M.A. With Maps, &c. Post 8vo. 6s.

The Treasury of Bible Know-ledge ; being a Dictionary of the Books, Persons, Places, Events, and other Matters of which mention is made in Holy Scripture. By Rev. J. AYRE, M.A. With Maps, 15 Plates, and numerous Woodcuts. Fcp. 6s.

Every-day Scripture Difficulties explained and illustrated. By J. E. PRESCOTT, M.A. VOL. I. *Matthew* and *Mark*; VOL. II. *Luke* and *John.* 2 vols. 8vo. price 9s. each.

The Pentateuch and Book of Joshua Critically Examined. By the Right Rev. J. W. COLENSO, D.D. Lord Bishop of Natal. Crown 8vo. price 6s.

The Four Cardinal Virtues (Fortitude, Justice, Prudence, Temperance) in relation to the Public and Private Life of Catholics : Six Sermons for the Day. With Preface, Appendices, &c. By the Rev. ORBY SHIPLEY, M.A. Crown 8vo. with Frontispiece, 7s. 6d.

The Formation of Christendom. By T. W. ALLIES. PARTS I. and II. 8vo. price 12s. each.

Four Discourses of Chrysostom, chiefly on the parable of the Rich Man and Lazarus. Translated by F. ALLEN, B.A. Crown 8vo. 3s. 6d.

Christendom's Divisions; a Philosophical Sketch of the Divisions of the Christian Family in East and West. By EDMUND S. FFOULKES. Post 8vo. 7s. 6d.

Christendom's Divisions, PART II. Greeks and Latins. By the same Author. Post 8vo. 15s.

The Hidden Wisdom of Christ and the Key of Knowledge; or, History of the Apocrypha. By ERNEST DE BUNSEN. 2 vols. 8vo. 28s.

The Keys of St. Peter; or, the House of Rechab, connected with the History of Symbolism and Idolatry. By the same Author. 8vo. 14s.

The Power of the Soul over the Body. By GEO. MOORE, M.D. M.R.C.P.L. &c. Sixth Edition. Crown 8vo. 8s. 6d.

The Types of Genesis briefly considered as Revealing the Development of Human Nature. By ANDREW JUKES. Second Edition. Crown 8vo. 7s. 6d.

The Second Death and the Restitution of All Things, with some Preliminary Remarks on the Nature and Inspiration of Holy Scripture. By the same Author. Second Edition. Crown 8vo. 3s. 6d.

Thoughts for the Age. BY ELIZABETH M. SEWELL, Author of 'Amy Herbert.' New Edition. Fcp. 8vo. price 5s.

Passing Thoughts on Religion. By the same Author. Fcp. 5s.

Self-examination before Confirmation. By the same Author. 32mo. 1s. 6d.

Thoughts for the Holy Week, for Young Persons. By the same Author. New Edition. Fcp. 8vo. 2s.

Readings for a Month Preparatory to Confirmation from Writers of the Early and English Church. By the same. Fcp. 4s.

Readings for Every Day in Lent, compiled from the Writings of Bishop JEREMY TAYLOR. By the same Author. Fcp. 5s.

Preparation for the Holy Communion; the Devotions chiefly from the works of JEREMY TAYLOR. By the same. 32mo. 3s.

Principles of Education drawn from Nature and Revelation, and Applied to Female Education in the Upper Classes. By the same Author. 2 vols. fcp. 12s. 6d.

Bishop Jeremy Taylor's Entire Works: with 'Life by BISHOP HEBER. Revised and corrected by the Rev. C. P. EDEN. 10 vols. £5 5s.

England and Christendom. By ARCHBISHOP MANNING, D.D. Post 8vo. price 10s. 6d.

The Wife's Manual; or, Prayers, Thoughts, and Songs on Several Occasions of a Matron's Life. By the Rev. W. CALVERT, M.A. Crown 8vo. 10s. 6d.

Singers and Songs of the Church: being Biographical Sketches of the Hymn-Writers in all the principal Collections; with Notes on their Psalms and Hymns. By JOSIAH MILLER, M.A. Second Edition, enlarged. Post 8vo. 10s. 6d.

'Spiritual Songs' for the Sundays and Holidays throughout the Year. By J. S. B. MONSELL, LL.D. Vicar of Egham and Rural Dean. Fourth Edition, Sixth Thousand. Fcp. price 4s. 6d.

The Beatitudes. By the same Author. Third Edition, revised. Fcp. 3s. 6d.

His Presence not his Memory, 1855. By the same Author, in memory of his SON. Sixth Edition. 16mo. 1s.

Lyra Germanica, translated from the German by Miss C. WINKWORTH. FIRST SERIES, the Christian Year, Hymns for the Sundays and Chief Festivals of the Church; SECOND SERIES, the Christian Life. Fcp. 8vo. price 3s. 6d. each SERIES.

Lyra Eucharistica; Hymns and Verses on the Holy Communion, Ancient and Modern: with other Poems. Edited by the Rev. ORBY SHIPLEY, M.A. Second Edition. Fcp. 5s.

Shipley's Lyra Messianica. Fcp. 5s.

Shipley's Lyra Mystica. Fcp. 5s.

Endeavours after the Christian Life: Discourses. By JAMES MARTINEAU. Fourth Edition, carefully revised. Post 8vo. 7s. 6d.

Invocation of Saints and Angels; for the use of Members of the English Church. Edited by the Rev. ORBY SHIPLEY, M.A. 24mo. 3s. 6d.

Travels, Voyages, &c.

The Playground of Europe. By LESLIE STEPHEN, late President of the Alpine Club. Post 8vo. with Frontispiece. [*Just ready.*

Westward by Rail: the New Route to the East. By W. F. RAE. Post 8vo. with Map, price 10s. 6d.

Travels in the Central Caucasus and Bashan, including Visits to Ararat and Tabreez and Ascents of Kazbek and Elbruz. By DOUGLAS W. FRESHFIELD. Square crown 8vo. with Maps, &c., 18s.

Cadore or Titian's Country. By JOSIAH GILBERT, one of the Authors of the 'Dolomite Mountains.' With Map, Facsimile, and 40 Illustrations. Imp.8vo.31s.6d.

Zigzagging amongst Dolomites; with more than 300 Illustrations by the Author. By the Author of 'How we Spent the Summer.' Oblong 4to. price 15s.

The Dolomite Mountains. Excursions through Tyrol, Carinthia, Carniola, and Friuli. By J. GILBERT and G. C. CHURCHILL, F.R.G.S. With numerous Illustrations. Square crown 8vo. 21s.

Pilgrimages in the Pyrenees and Landes. By DENYS SHYNE LAWLOR. Crown 8vo. with Frontispiece and Vignette, price 15s.

How we Spent the Summer; or, a Voyage en Zigzag in Switzerland and Tyrol with some Members of the ALPINE CLUB. Third Edition, re-drawn. In oblong 4to. with about 300 Illustrations, 15s.

Pictures in Tyrol and Elsewhere. From a Family Sketch-Book. By the same Author. Second Edition. 4to. with many Illustrations, 21s.

Beaten Tracks; or, Pen and Pencil Sketches in Italy. By the same Author. With 42 Plates of Sketches. 8vo. 16s.

The Alpine Club Map of the Chain of Mont Blanc, from an actual Survey in 1863—1864. By A. ADAMS-REILLY, F.R.G.S. M.A.C. In Chromolithography on extra stout drawing paper 28in. × 17in. price 10s. or mounted on canvas in a folding case, 12s. 6d.

England to Delhi; a Narrative of Indian Travel. By JOHN MATHESON, Glasgow. With Map and 82 Woodcut Illustrations. 4to. 31s. 6d.

History of Discovery in our Australasian Colonies, Australia, Tasmania, and New Zealand, from the Earliest Date to the Present Day. By WILLIAM HOWITT. 2 vols. 8vo. with 3 Maps, 20s.

The Capital of the Tycoon; a Narrative of a 3 Years' Residence in Japan. By Sir RUTHERFORD ALCOCK, K.C.B. 2 vols. 8vo. with numerous Illustrations, 42s.

Guide to the Pyrenees, for the use of Mountaineers. By CHARLES PACKE. Second Edition, with Maps, &c. and Appendix. Crown 8vo. 7s. 6d.

The Alpine Guide. By JOHN BALL, M.R.I.A. late President of the Alpine Club. Post 8vo. with Maps and other Illustrations.
Guide to the Eastern Alps, price 10s. 6d.
Guide to the Western Alps, including Mont Blanc, Monte Rosa, Zermatt, &c. price 6s. 6d.
Guide to the Central Alps, including all the Oberland District, price 7s. 6d.
Introduction on Alpine Travelling in general, and on the Geology of the Alps, price 1s. Either of the Three Volumes or Parts of the *Alpine Guide* may be had with this INTRODUCTION prefixed, price 1s. extra.

Roma Sotterranea; or, an Account of the Roman Catacombs, especially of the Cemetery of San Callisto. Compiled from the Works of Commendatore G. B. DE ROSSI, by the Rev. J. S. NORTHCOTE, D.D. and the Rev. W. B. BROWNLOW. With Plans and numerous other Illustrations. 8vo. 31s. 6d.

Memorials of London and London Life in the 13th, 14th, and 15th Centuries; being a Series of Extracts, Local, Social, and Political, from the Archives of the City of London, A.D. 1276–1419. Selected, translated, and edited by H. T. RILEY, M.A. Royal 8vo. 21s.

Commentaries on the History, Constitution, and Chartered Franchises of the City of London. By GEORGE NORTON, formerly one of the Common Pleaders of the City of London. Third Edition. 8vo. 14s.

The Northern Heights of London; or, Historical Associations of Hampstead, Highgate, Muswell Hill, Hornsey, and Islington. By WILLIAM HOWITT. With about 40 Woodcuts. Square crown 8vo. 21s.

The Rural Life of England. By the same Author. With Woodcuts by Bewick and Williams. Medium 8vo. 12s. 6d.

Visits to Remarkable Places: Old Halls, Battle-Fields, and Scenes illustrative of striking Passages in English History and Poetry. By the same Author. 2 vols. square crown 8vo. with Wood Engravings, 25s.

Narrative of the Euphrates Expedition carried on by Order of the British Government during the years 1835, 1836, and 1837. By General F. R. CHESNEY, F.R.S. With 2 Maps, 45 Plates, and 16 Woodcuts. 8vo. 24s.

Works of Fiction.

Lothair. By the Right Hon. B. DISRAELI, Cabinet Edition (the Eighth), complete in One Volume, with a Portrait of the Author, and a new General Preface. Crown 8vo. price 6s.—By the same Author, Cabinet Editions, revised, uniform with the above:—

CONINGSBY, 6s.	ALROY; IXION; the
SYBIL, 6s.	INFERNAL MAR-
TANCRED, 6s.	RIAGE; and PO-
VENETIA, 6s.	PANILLA. Price 6s.
HENRIETTA TEMPLE,	YOUNG DUKE and
6s.	COUNT ALARCOS,
CONTARINI FLEMING	6s.
and RISE OF IS-	VIVIAN GREY, 6s.
KANDER, 6s.	

The Modern Novelist's Library. Each Work, in crown 8vo. complete in a Single Volume:—

MELVILLE'S GLADIATORS, 2s. boards; 2s. 6d. cloth.
——— GOOD FOR NOTHING, 2s. boards; 2s. 6d. cloth.
——— HOLMBY HOUSE, 2s. boards; 2s. 6d. cloth.
——— INTERPRETER, 2s. boards; 2s. 6d. cloth.
——— QUEEN'S MARIES, 2s. boards; 2s. 6d. cloth.
TROLLOPE'S WARDEN, 1s. 6d. boards; 2s. cloth.
——— BARCHESTER TOWERS, 2s. boards; 2s. 6d. cloth.
BRAMLEY-MOORE'S SIX SISTERS OF THE VALLEYS, 2s. boards; 2s. 6d. cloth.

Stories and Tales by the Author of 'Amy Herbert,' uniform Edition:—

AMY HERBERT, 2s. 6d.	KATHARINE ASHTON,
GERTRUDE, 2s. 6d.	3s. 6d.
EARL'S DAUGHTER,	MARGARET PERCI-
2s. 6d.	NAL, 5s.
EXPERIENCE OF LIFE,	LANETON PARSON-
2s. 6d.	AGE, 4s. 6d.
CLEVE HALL, 3s. 6d.	URSULA, 4s. 6d.
IVORS, 3s. 6d.	

A Glimpse of the World. Fcp. 7s. 6d.
Journal of a Home Life. Post 8vo. 9s. 6d.
After Life; a Sequel to the 'Journal of a Home Life.' Post 8vo. 10s. 6d.

A Visit to my Discontented Cousin. Reprinted, with some Additions, from *Fraser's Magazine.* Crown 8vo. price 7s. 6d.

Ierne; a Tale. By W. STEUART TRENCH, Author of 'Realities of Irish Life.' 2 vols post 8vo. [*Just ready.*]

Three Weddings. By the Author of 'Dorothy,' &c. Fcp. 8vo. 5s.

The Giant; a Witch's Story for English Boys. Edited by ELIZABETH M. SEWELL, Author of 'Amy Herbert,' &c. Fcp. 8vo. price 5s.

Uncle Peter's Fairy Tale for the XIXth Century. By the same Author and Editor. Fcp. 8vo. 7s. 6d.

Vikram and the Vampire; or, Tales of Hindu Devilry. Adapted by RICHARD F. BURTON, F.R.G.S. &c. With 33 Illustrations. Crown 8vo. 9s.

Becker's Gallus; or, Roman Scenes of the Time of Augustus. Post 8vo. 7s. 6d.

Becker's Charicles: Illustrative of Private Life of the Ancient Greeks. Post 8vo. 7s. 6d.

Tales of Ancient Greece. By GEORGE W. COX, M.A. late Scholar of Trin. Coll. Oxford. Being a collective Edition of the Author's Classical Series and Tales, complete in One Volume. Crown 8vo. 6s. 6d.

Cabinet Edition of Novels and Tales by G. J. WHYTE MELVILLE:—

THE GLADIATORS, 5s.	HOLMBY HOUSE, 5s.
DIGBY GRAND, 5s.	GOOD FOR NOTHING, 6s.
KATE COVENTRY, 5s.	QUEEN'S MARIES, 6s.
GENERAL BOUNCE, 5s.	THE INTERPRETER, 5s.

Our Children's Story. By One of their Gossips. By the Author of 'Voyage en Zigzag,' &c. Small 4to. with Sixty Illustrations by the Author, price 10s. 6d.

Wonderful Stories from Norway, Sweden, and Iceland. Adapted and arranged by JULIA GODDARD. With an Introductory Essay by the Rev. G. W. COX, M.A. and Six Illustrations. Square post 8vo. 6s.

C

Poetry and *The Drama.*

Thomas Moore's Poetical Works, the only Editions containing the Author's last Copyright Additions :—

Shamrock Edition, price 3s. 6d.
Ruby Edition, with Portrait, 6s.
Cabinet Edition, 10 vols. fcp. 8vo. 35s.
People's Edition, Portrait, &c. 10s. 6d.
Library Edition, Portrait & Vignette, 14s.

Moore's Lalla Rookh, Tenniel's Edition, with 68 Wood Engravings from Original Drawings and other Illustrations. Fcp. 4to. 21s.

Moore's Irish Melodies, Maclise's Edition, with 161 Steel Plates from Original Drawings. Super-royal 8vo. 31s. 6d.

Miniature Edition of Moore's Irish *Melodies,* with Maclise's Illustrations (as above), reduced in Lithography. Imp. 16mo. 10s. 6d.

Southey's Poetical Works, with the Author's last Corrections and copyright Additions. Library Edition. Medium 8vo. with Portrait and Vignette, 14s.

Lays of Ancient Rome ; with *Ivry* and the *Armada.* By the Right Hon. Lord Macaulay. 16mo. 4s. 6d.

Lord Macaulay's Lays of Ancient Rome. With 90 Illustrations on Wood, Original and from the Antique, from Drawings by G. Scharf. Fcp. 4to. 21s.

Miniature Edition of Lord Macaulay's Lays of Ancient Rome, with Scharf's Illustrations (as above) reduced in Lithography. Imp. 16mo. 10s. 6d.

Goldsmith's Poetical Works, Illustrated with Wood Engravings from Designs by Members of the Etching Club. Imp. 16mo. 7s. 6d.

Poems of Bygone Years. Edited by the Author of 'Amy Herbert. Fcp. 8vo. 5s.

Poems, Descriptive and Lyrical. By Thomas Cox. New Edition. Fcp. 8vo. price 5s.

' Shew moral propriety, mental culture, and no slight acquaintance with the technicalities of song.' ATHENÆUM.

Madrigals, Songs, and Sonnets. By John Arthur Blaikie and Edmund William Gosse. Fcp. 8vo. price 5s.

Poems. By Jean Ingelow. Fifteenth Edition. Fcp. 8vo. 5s.

Poems by Jean Ingelow. With nearly 100 Illustrations by Eminent Artists, engraved on Wood by Dalziel Brothers. Fcp. 4to. 21s.

Mopsa the Fairy. By Jean Ingelow. With Eight Illustrations engraved on Wood. Fcp. 8vo. 6s.

A Story of Doom, and other Poems. By Jean Ingelow. Third Edition. Fcp. price 5s.

Glaphyra, and other Poems. By Francis Reynolds, Author of 'Alice Rushton.' 16mo. 5s.

Bowdler's Family Shakspeare, cheaper Genuine Edition, complete in 1 vol. large type, with 36 Woodcut Illustrations, price 14s. or in 6 pocket vols. 3s. 6d. each.

Arundines Cami. Collegit atque edidit H. Drury, M.A. Editio Sexta, curavit H. J. Hodgson, M.A. Crown 8vo. price 7s. 6d.

Horatii Opera, Pocket Edition, with carefully corrected Text, Marginal References, and Introduction. Edited by the Rev. J. E. Yonge, M.A. Square 18mo. 4s. 6d.

Horatii Opera, Library Edition, with Copious English Notes, Marginal References and Various Readings. Edited by the Rev. J. E. Yonge, M.A. 8vo. 21s.

The Æneid of Virgil Translated into English Verse. By John Conington, M.A. Corpus Professor of Latin in the University of Oxford. New Edition. Crown 8vo. 9s.

The Story of Sir Richard Whit-tington, Thrice Lord Mayor of London, A.D. 1397, 1406-7, and 1419. Written in Verse and Illustrated by E. Carr. With Eleven Plates. Royal 4to. 21s.

Hunting Songs and Miscella-neous Verses. By R. E. Egerton Warburton. Second Edition. Fcp. 8vo. 5s.

Works by Edward Yardley :—
Fantastic Stories, fcp. 3s. 6d.
Melusine and other Poems, fcp. 5s.
Horace's Odes translated into English Verse, crown 8vo. 6s.
Supplementary Stories and Poems, fcp. 3s. 6d.

Rural Sports, &c.

Encyclopædia of Rural Sports; a Complete Account, Historical, Practical, and Descriptive, of Hunting, Shooting, Fishing, Racing, &c. By D. P. BLAINE. With above 600 Woodcuts (20 from Designs by JOHN LEECH). 8vo. 21s.

The Dead Shot, or Sportsman's Complete Guide; a Treatise on the Use of the Gun, Dog-breaking, Pigeon-shooting, &c. By MARKSMAN. Fcp. with Plates, 5s.

A Book on Angling: being a Complete Treatise on the Art of Angling in every branch, including full Illustrated Lists of Salmon Flies. By FRANCIS FRANCIS. Second Edition, with Portrait and 15 other Plates, plain and coloured. Post 8vo. 15s.

Wilcocks's Sea-Fisherman: comprising the Chief Methods of Hook and Line Fishing in the British and other Seas, a glance at Nets, and remarks on Boats and Boating. Second Edition, enlarged, with 80 Woodcuts. Post 8vo. 12s. 6d.

The Fly-Fisher's Entomology. By ALFRED RONALDS. With coloured Representations of the Natural and Artificial Insect. Sixth Edition, with 20 coloured Plates. 8vo. 14s.

The Book of the Roach. By GREVILLE FENNELL, of 'The Field.' Fcp. 8vo. price 2s. 6d.

Blaine's Veterinary Art: a Treatise on the Anatomy, Physiology, and Curative Treatment of the Diseases of the Horse, Neat Cattle, and Sheep. Seventh Edition, revised and enlarged by C. STEEL. 8vo. with Plates and Woodcuts, 18s.

Horses and Stables. By Colonel F. FITZWYGRAM, XV. the King's Hussars. Pp. 624; with 24 Plates of Illustrations, containing very numerous Figures engraved on Wood. 8vo. 15s.

Youatt on the Horse. Revised and enlarged by W. WATSON, M.R.C.V.S. 8vo. with numerous Woodcuts, 12s. 6d.

Youatt on the Dog. (By the same Author.) 8vo. with numerous Woodcuts, 6s.

The Horse's Foot, and how to keep it Sound. By W. MILES, Esq. Ninth Edition, with Illustrations. Imp. 8vo. 12s. 6d.

A Plain Treatise on Horse-shoeing. By the same Author. Sixth Edition, post 8vo. with Illustrations, 2s. 6d.

Stables and Stable Fittings. By the same. Imp. 8vo. with 13 Plates, 15s.

Remarks on Horses' Teeth, addressed to Purchasers. By the same. Post 8vo. 1s. 6d.

Robbins's Cavalry Catechism; or, Instructions on Cavalry Exercise and Field Movements, Brigade Movements, Out-post Duty, Cavalry supporting Artillery, Artillery attached to Cavalry. 12mo. 5s.

The Dog in Health and Disease. By STONEHENGE. With 70 Wood Engravings. New Edition. Square crown 8vo. 10s. 6d.

The Greyhound. By the same Author. Revised Edition, with 24 Portraits of Greyhounds. Square crown 8vo. 10s. 6d.

The Ox, his Diseases and their Treatment; with an Essay on Parturition in the Cow. By J. R. DOBSON, M.R.C.V.S. Crown 8vo. with Illustrations, 7s. 6d.

Commerce, Navigation, and Mercantile Affairs.

The Elements of Banking. By HENRY DUNNING MACLEOD, M.A. of Trinity College, Cambridge, and of the Inner Temple, Barrister-at-Law. Post 8vo. [Nearly ready.

The Law of Nations Considered as Independent Political Communities. By Sir TRAVERS TWISS, D.C.L. 2 vols. 8vo. 80s. or separately, PART I. Peace, 12s. PART II. War, 18s.

The Theory and Practice of Banking. By HENRY DUNNING MACLEOD, M.A. Barrister-at-Law. Second Edition. entirely remodelled. 2 vols. 8vo. 30s.

M'Culloch's Dictionary, Practical, Theoretical, and Historical, of Commerce and Commercial Navigation. New Edition, revised throughout and corrected to the Present Time; with a Biographical Notice of the Author. Edited by H. G. REID, Secretary to Mr. M'Culloch for many years. 8vo. price 63s. cloth.

Works of Utility and General Information.

Modern Cookery for Private
Families, reduced to a System of Easy
Practice in a Series of carefully-tested Re-
ceipts. By ELIZA ACTON. Newly revised
and enlarged; with 8 Plates, Figures, and
150 Woodcuts. Fcp. 6s.

A Practical Treatise on Brewing;
with Formulæ for Public Brewers, and Instructions for Private Families. By W.
BLACK. Fifth Edition. 8vo. 10s. 6d.

Chess Openings. By F. W. LONGMAN,
Balliol College, Oxford. Fcp. 8vo. 2s. 6d.

The Cabinet Lawyer; a Popular
Digest of the Laws of England, Civil,
Criminal, and Constitutional. 25th Edition;
with Supplements of the Acts of the Parliamentary Session of 1870. Fcp. 10s. 6d.

The Philosophy of Health; or, an
Exposition of the Physiological and Sanitary
Conditions conducive to Human Longevity
and Happiness. By SOUTHWOOD SMITH,
M.D. Eleventh Edition, revised and en-
larged; with 113 Woodcuts. 8vo. 7s. 6d.

Maunder's Treasury of Know-
ledge and Library of Reference: comprising
an English Dictionary and Grammar, Universal Gazetteer, Classical Dictionary,
Chronology, Law Dictionary, Synopsis of
the Peerage, Useful Tables, &c. Fcp. 6s.

Hints to Mothers on the Manage-
ment of their Health during the Period of
Pregnancy and in the Lying-in Room. By
T. BULL, M.D. Fcp. 5s.

The Maternal Management of
Children in Health and Disease. By THOMAS
BULL, M.D. Fcp. 5s.

How to Nurse Sick Children;
containing Directions which may be found
of service to all who have charge of the
Young. By CHARLES WEST, M.D. Second
Edition. Fcp. 8vo. 1s. 6d.

Notes on Hospitals. By FLORENCE
NIGHTINGALE. Third Edition, enlarged;
with 13 Plans. Post 4to. 18s.

Pewtner's Comprehensive Speci-
fier; a Guide to the Practical Specification
of every kind of Building-Artificer's Work:
with Forms of Building Conditions and
Agreements, an Appendix, Foot-Notes, and
Index. Edited by W. YOUNG. Architect.
Crown 8vo. 6s.

Tidd Pratt's Law relating to
Benefit Building Societies; with Practical
Observations on the Act and all the Cases
decided thereon, also a Form of Rules and
Forms of Mortgages. Fcp. 3s. 6d.

Collieries and Colliers: a Handbook
of the Law and Leading Cases relating
thereto. By J. C. FOWLER, of the Inner
Temple, Barrister, Stipendiary Magistrate.
Second Edition. Fcp. 8vo. 7s. 6d.

Willich's Popular Tables for Ascertaining the Value of Lifehold, Leasehold,
and Church Property, Renewal Fines, &c.;
the Public Funds; Annual Average Price
and Interest on Consols from 1731 to 1867;
Chemical, Geographical, Astronomical,
Trigonometrical Tables, &c. Post 8vo. 10s.

Coulthart's Decimal Interest
Tables at Twenty-four Different Rates not
exceeding Five per Cent. Calculated for the
use of Bankers. To which are added Com-
mission Tables at One-eighth and One-
fourth per Cent. 8vo. 15s.

Periodical Publications.

The Edinburgh Review, or Cri-
tical Journal, published Quarterly in January, April, July, and October. 8vo. price
6s. each Number.

Notes on Books: An Analysis of the
Works published during each Quarter by
Messrs. LONGMANS & Co. The object is to
enable Bookbuyers to obtain such information regarding the various works as is
usually afforded by tables of contents and
explanatory prefaces. 4to. Quarterly.
Gratis.

Fraser's Magazine. Edited by JAMES
ANTHONY FROUDE, M.A. New Series,
published on the 1st of each Month. 8vo.
price 2s. 6d. each Number.

The Alpine Journal: A Record of
Mountain Adventure and Scientific Obser-
vation. By Members of the Alpine Club.
Edited by LESLIE STEPHEN. Published
Quarterly, May 31, Aug. 31, Nov. 30, Feb.
28. 8vo. price 1s. 6d. each No.

INDEX.

SPOTTISWOODE AND CO., PRINTERS, NEW-STREET SQUARE, LONDON.

CPSIA information can be obtained
at www.ICGtesting.com
Printed in the USA
BVHW061240160819
556068BV00020B/1991/P